PROTEIN SCIENCE AND ENGINEERING

RIBOSOMAL PROTEINS AND PROTEIN ENGINEERING: DESIGN, SELECTION AND APPLICATIONS

PROTEIN SCIENCE AND ENGINEERING

Additional books in this series can be found on Nova's website at:

https://www.novapublishers.com/catalog/index.php?cPath=23_29&seriesp=
Protein+Science+and+Engineering

Additional e-books in this series can be found on Nova's website at:

https://www.novapublishers.com/catalog/index.php?cPath=23_29&seriespe=
Protein+Science+and+Engineering

PROTEIN SCIENCE AND ENGINEERING

RIBOSOMAL PROTEINS AND PROTEIN ENGINEERING: DESIGN, SELECTION AND APPLICATIONS

VALTER ORTENDHAL
AND
HJALMAR SALCHOW
EDITORS

Nova Biomedical
Nova Science Publishers, Inc.
New York

For permission to use material from this book please contact us:
Telephone 631-231-7269; Fax 631-231-8175
Web Site: http://www.novapublishers.com

LIBRARY OF CONGRESS CATALOGING-IN-PUBLICATION DATA

Ribosomal proteins and protein engineering : design, selection, and applications / editors, Valter Ortendhal and Hjalmar Salchow.
p. ; cm.
Includes bibliographical references and index.
ISBN 978-1-60741-005-8 (hardcover)
1. Ribosomes. 2. Protein engineering. I. Ortendhal, Valter. II. Salchow, Hjalmar.
[DNLM: 1. Ribosomal Proteins. 2. Protein Engineering--methods. QU 55 R486 2009]
QH603.R5R48 2009
572'.6--dc22

2009042006

Published by Nova Science Publishers, Inc. ✦ New York

CONTENTS

PREFACE

Ribosome is an important cellular organelle that is necessary for the basic cellular process, protein synthesis. Furthermore, ribosomal protein is a group of proteins that is important in biomedical research. The authors of this book present and review important data on ribosomal proteins, important in tropical diseases such as malaria and tuberculosis. In addition, some functions of ribosomal proteins, other than translation, are called extraribosomal function or extraribosomal activity. In this book, the extraribosomal functions of ribosomal protein S19 (RPS19) is discussed in detail. Other chapters in this book examine the structures of the free and bound forms of proteins that have experimentally been found to be essential for the first steps of ribosome assembly. A summary of ribosomal protein mutations generated in eukaryotes is also provided and their proposed roles in the control of cell growth and proliferation, as well as their impact in human diseases. Important challenges we face in explaining life from the genome and proteome viewpoint is also discussed, including how to measure a living protein/RNA/DNA, how to restore the missing relation cross generations and how to correlate structure with function. Finally, the authors suggest possible solutions for such challenges.

Chapter 1 - Recombinant proteins are widely used in industry (pharmaceutical, chemical, food, cosmetics…), research and diagnostics; therefore, interest in their production is increasing. Due to low cost, high productivity, simple cultivation and rapid use, bacterial systems remain very interesting for recombinant protein production. One of the most commonly used bacterial hosts is still *Escherichia coli*. However, upon overexpression in *E. coli*, recombinant proteins tend to aggregate into inclusion bodies (IBs). For a long time, IBs were considered to be deposits of biologically inactive target proteins. The denaturation and renaturation step of protein isolation from IBs have to be optimized. However, the isolation process often results in poor recovery of biologically active proteins and accounts for a major cost in protein production, thus IBs were regarded as a bottleneck in protein production.

The temperature and induction regime were recognized as very important factors affecting the growth of the bacteria *E. coli*, as well as the recombinant protein yield and protein folding and thus protein quality inside IBs. After optimization of the bioprocess, the authors were able to identify and explore a new sub-type of inclusion bodies. These new IBs contain correctly folded protein. They show several significantly different properties from IBs previously described (solubility, fragility, irreversible contraction at low pH); and the authors have therefore proposed that they be given the designation "non classical inclusion bodies" (ncIBs).

Non-classical IBs are easily soluble in mild detergents, thus proteins can be extracted under non-denaturing conditions in biologically active form without applying any renaturation steps. Thus the isolation process is faster and economically advantageous and can be exploited for environment-friendly biotechnological production.

Such IBs composed of active proteins could also be used as pure nanoparticles in diagnostics, as biocatalysts in enzymatic processes, or even as biopharmaceuticals. Furthermore, studying IBs' formation inside prokaryotic cells could provide useful knowledge of aggregation mechanisms, applicable also to aggregates formed inside eukaryotic cells, responsible for numerous degenerative diseases.

Chapter 2 - Some ribosomal proteins play roles in activities other than translation. Such functions of ribosomal proteins are called extraribosomal function or extraribosomal activity in common. Ribosomal protein S19 (RPS19) is a component of the small subunit of eukaryotic and archaeobacterial ribosomes having 145 amino acid residues. RPS19 exhibits various extraribosomal functions, which basically correlate to the apoptosis or programmed cell death. In the apoptosis-initiated cells, RPS19 molecules are cross-linked intermolecularly with an isopeptide bond between Lys122 and Gln137 by a transglutaminase-catalyzed reaction. This results in homo-dimerization gaining a new molecular conformation capable of binding the C5a receptor, a G protein-coupled heptahelical cell surface receptor. The dimer is then released from the apoptotic cells. Meanwhile, the apoptosis-initiated cells *de novo* synthesize the C5a receptor. The C5a receptor was initially identified as a leukocyte chemotactic receptor for the complement-derived C5a. Due to a molecular mimicry between the RPS19 dimer and C5a, both molecules can ligate the C5a receptor exhibiting some difference between them. The RPS19 dimer has a potency to cause at least three cellular events through the C5a receptor ligation. One is apoptosis promotion by an autocrine/paracrine effect. On the other hand, the RPS19 dimer recruits monocytes/macrophages by ligating the C5a receptor on these leukocytes. The monocytes/macrophages thus infiltrated engulf the apoptotic cells and digest them. The macrophages then move to regional lymph nodes *via* lymphatics and present apoptotic cell-derived antigenic peptides to T lymphocytes. The acquired immune response may then start on demand. The apoptotic cell clearance is basically an event outside the acute inflammatory response in which neutrophils play major roles. The RPS19 dimer inhibits neutrophil infiltration by ligating the neutrophil C5a receptor apparently as an antagonist. Other extraribosomal functions of the RPS19 dimer have been observed. There is a hypothesis that some elements of the apoptotic program would be used in terminal differentiation of erythroblasts when they lose their nuclei, and macrophages would be responsible for clearing the erythroblast-derived nuclei. The pro-apoptotic effect and macrophage recruiting effect of the RPS19 dimer seem to also function in the erythropoiesis. Interestingly enough, there is a molecule indistinguishable from RPS19 in plasma. The RPS19-like molecule is cross-linked by coagulation factor XIIIa, the plasma transglutaminase, in the blood coagulation process. The RPS19 dimer-like molecule seems play a role in resolution of coagula due to recruiting monocytes/macrophages.

Chapter 3 - The recent crystallographic structures of the ribosomal subunits have revealed the picture of the final product of a complex assembly process that condenses the rRNA and the ribosomal proteins into active ribosomes. The folding of rRNA is the rate-limiting step of this complicate assembly pathway. Some ribosomal proteins are required for facilitating the structural rearrangements of rRNA and avoiding the kinetic traps that frequently impede RNA folding. It is thought that the long basic r-protein extensions that penetrate deeply into the subunit cores play a key role in this process through disorder-order transitions and/or co-folding mechanisms. A current view is that such structural transitions may facilitate the proper rRNA folding. In this paper, the structures of the free and bound forms of proteins that have been experimentally found to be essential for the first steps of ribosome assembly have been compared. It is shown that the extensions of L3, L4, L20 and L22, have different structural and dynamics properties that probably correlate with different functions. This study also suggests that the specific coil-helix transition that occurs in a phylogenetically conserved cluster of basic residues of the L20 extension is strictly required for the large subunit assembly in eubacteria. In contrast, in showing that the ordering of the inner loop of L4 upon rRNA binding is not required for the assembly function of L4, this study indicates that different categories of disorder-order transitions are associated with different biological functions.

Chapter 4 - There is much concentration in recent years in understanding the coordinated role of ribosomal protein expression in the regulation of ribosome biogenesis, translational control, and cellular functions. Although ribosomal proteins are dispensable for catalyzing the formation of peptidic bonds, they provide a conformational environment essential for the process of protein biosynthesis. Imbalance in the production of ribosomal proteins results in non-equimolar amounts of protein for each subunit and induces defects in ribosome assembly. For these reasons, delineation of the precise function of specific ribosomal protein gene products is rendered difficult as their gene targeting is accompanied by severe growth defects resulting in cellular death. The objective of this commentary is to provide a summary of ribosomal protein mutations generated in eukaryotes and discuss their proposed roles in the control of cell growth and proliferation and impact in human diseases.

Chapter 5 - The extent of current research in ancient biology is not restricted by the presence of fossil records, owing to the developments in computational analysis and engineering techniques of proteins. Ancestral sequences of extant genes/proteins can be inferred with the molecular phylogeny reconstruction methods, which are mainly based on maximum-likelihood estimation. Once an ancestral sequence is determined, a reconstruction of the ancestral protein is not difficult by using the current protein engineering techniques, and reproduced protein can be experimentally tested either biochemically or biophysically. To date, such 'experimental molecular archeology' has been examined on several enzymes, elongation factor, photo-reactive proteins, nuclear receptor, and lectin, and largely helped to elucidate evolutionary processes of stability, specificity, and structure of the proteins. Although this method would promise a new field in protein engineering, several problems still remain to be solved, the largest among which should be accuracy of inferred ancestral sequence as a whole. In this chapter, the reported studies of ancestral protein reconstruction are reviewed with a focus on the authors' recent results on the ancestral congerin, which is the β-galactoside specific lectin of fish involved in the biological defense system. The reconstruction of ancestral congerin is important because two extant isoforms of this protein, congerins I and II, have been rapidly evolving, and their molecular properties and structures have been differentiating due to the natural selection pressures operating on the genes.

Chapter 6 - The ribosome is an important cellular organelle that is necessary for the basic cellular process, protein synthesis. Ribosomal protein is a group of proteins that is presented with a focus on biomedical research. In this article, the author will review and present important data on the ribosomal proteins in important tropical diseases including malaria and tuberculosis.

Chapter 7 - In this chapter, the authors discuss three challenges we face in explaining the life from genome and proteome viewpoint, that is, how to measure a living protein/RNA/DNA, how to restore the missing relation cross generations and how to correlate the structure with its function. Finally, the authors suggest the possible solutions for these three challenges.

Chapter 8 – Self-similar properties of the proteins in the ribosome, the ribosomal proteins, are investigated in terms of various types of fractal dimensions. The authors especially estimate both the chain fractal and capacity dimensions of backbones of the ribosomal proteins, together with the mass fractal dimension of the ribosomal proteins at the atomic level. The chain fractal and capacity dimensions of the ribosomal proteins are compared with that of the self-avoiding walk, which is a typical model of a polymer without interaction between monomers. These fractal dimensions of the ribosomal protein support the existence of the long and extended domain, which is hardly seen in the globular protein. The authors demonstrate that the mass fractal dimension of the ribosomal proteins is pertinent to explain their structural characteristics. The self-similarity also upholds the fact that the ribosomal proteins function primarily to stabilize the structure of the ribosome by both the long-extended domain of the protein penetrating into the inside of the RNA, and the globular domain interacting with the RNA on the exterior of it. These results partially, if not whole, unravel the structural characteristics of the proteins in the ribosome.

Chapter 9 - For tissue growth, organisation and repair, interaction between cells and their extracellular matrix (ECM) is critical. Fibronectin ECM renders a polyvalent ligand display for cell adhesion through binding to cell membrane integrin receptors. Mimicking this polyvalent display and designing ligands which bind specific integrin receptor subtypes will be discussed in this chapter. The central peptide motif for integrin binding is RGD (arginine-glycine-aspartic acid) but the integrin $\alpha5\beta1$ subtype requires the synergy site on the 9th type III fibronectin domain (FIII) in addition to RGD on the 10th FIII domain. Polyvalent integrin $\alpha5\beta1$ ligands have been synthesised using a protein engineering approach, designing chimeras consisting of a N-terminal 9th-10th FIII domain pair, IgG-derived hinge and leucine zipper-derived helix; the latter mutated to yield di-, tri- and tetrameric coiled coils and thus self-assembling, multimeric integrin $\alpha5\beta1$ ligands. A unique C-terminal cysteine was appended to the helix to facilitate 'anchoring' of the chimeras with a defined orientation on a surface. Careful characterization of such self-assembling chimeras is required and can be achieved through the use of complementary techniques including size exclusion chromatography, analytical ultracentrifugation and circular dichroism. Surface adsorption and specific immobilization can then be unambiguously characterized by neutron reflectivity, relating the determined models of the adsorbed protein layers to their biological function. Application of these engineered proteins and cell adhesive surfaces as scaffolds or other biomedical devices has the potential to promote tissue repair and regeneration, for a wide variety of tissues including bone and skin. These applications will be discussed in the context of the future outlook for these ECM protein engineering approaches.

Chapter 10 - Halophilic organisms thrive in a wide variety of environments, ranging from mildly saline to saturated brine. Extremely saline conditions are very harsh to proteins and can result in problems such as loss of flexibility, which leads to loss of catalytic activity, or aggregation and precipitation due to salting-out effects. Halophilic proteins have therefore acquired specific adaptations to prevent such problems, and that is clearly reflected in their amino acid composition. As a consequence, halophilic proteins do not function properly in low salt conditions, and many even unfold and denature without sufficient amounts of salt. One of the most obvious adaptations found in halophilic proteins is an excess of acidic residues, while other adaptations include a low lysine content and an increase in small hydrophobic amino acids, the latter of which is at the expense of large hydrophobic residues. Different functions have been suggested for the role of the acidic residues. They are, most likely, important for binding of essential water molecules, but may play also a critical role in binding of salt or prevention of aggregation. The other changes in amino acid composition in halophilic proteins are probably important for the reduction of hydrophobicity, in order to prevent aggregation and maintain flexibility in high salt concentrations. Structural features that may be important for halophilic adaptation are also found, such as an increased number of salt-bridges. That is, however, not used by all halophilic proteins, indicating that different strategies have evolved to cope with very saline conditions.

Chapter 11 - Alkaliphilic microorganisms (alkaliphiles) produce a variety of proteins, mostly enzymes, that are stable and optimally active at alkaline pH. The alkaline enzymes represent most of the currently known proteins adapted to high pH. Alkaline enzymes are widely used for industrial or commercial purposes, for example, as additives to laundry detergents. They have been improved in the course of evolution to be resistant to low proton density conditions that would cause protein denaturation or dissociation. In order to elucidate the mechanism of high-pH adaptation, the alkaline enzymes have been characterized and their structures have been compared with homologous non-alkaliphilic proteins. These comparative studies, together with several experimental and theoretical studies, have suggested three major strategies for alkaline adaptation: p*K*a modulation, Asp+Glu gain, and Asp+Lys loss–Glu+Arg gain. The history of representative alkaline enzymes, proposed strategies for high-pH adaptation, and protein engineering studies performed in order to adapt proteins to alkaline conditions will be reviewed in this chapter.

In: Ribosomal Proteins and Protein Engineering… ISBN 978-1-60741-005-8
Editors: V. Ortendhal and H. Salchow, pp. 1-20

Chapter 1

DESIGNING NON-CLASSICAL INCLUSION BODIES

Špela Peternel[1]
Department of Biosynthesis and Biotransformation, National Institute of Chemistry, Ljubljana, Slovenia

ABSTRACT

Recombinant proteins are widely used in industry (pharmaceutical, chemical, food, cosmetics…), research and diagnostics; therefore, interest in their production is increasing. Due to low cost, high productivity, simple cultivation and rapid use, bacterial systems remain very interesting for recombinant protein production. One of the most commonly used bacterial hosts is still *Escherichia coli*. However, upon overexpression in *E. coli*, recombinant proteins tend to aggregate into inclusion bodies (IBs). For a long time, IBs were considered to be deposits of biologically inactive target proteins. The denaturation and renaturation step of protein isolation from IBs have to be optimized. However, the isolation process often results in poor recovery of biologically active proteins and accounts for a major cost in protein production, thus IBs were regarded as a bottleneck in protein production.

The temperature and induction regime were recognized as very important factors affecting the growth of the bacteria *E. coli*, as well as the recombinant protein yield and protein folding and thus protein quality inside IBs. After optimization of the bioprocess, we were able to identify and explore a new sub-type of inclusion bodies. These new IBs contain correctly folded protein. They show several significantly different properties from IBs previously described (solubility, fragility, irreversible contraction at low pH); and we have therefore proposed that they be given the designation "non classical inclusion bodies" (ncIBs).

Non-classical IBs are easily soluble in mild detergents, thus proteins can be extracted under non-denaturing conditions in biologically active form without applying any renaturation steps. Thus the isolation process is faster and economically advantageous and can be exploited for environment-friendly biotechnological production.

[1] Špela Peternel, Laboratory of Biosynthesis and Biotransformation, National Institute of Chemistry, Hajdrihova 19, 1000 Ljubljana, Slovenia, Telephone num.: +386 1 4760262, Fax: +386 1 4760300, e-mail: spela.peternel@ki.si.

Such IBs composed of active proteins could also be used as pure nanoparticles in diagnostics, as biocatalysts in enzymatic processes, or even as biopharmaceuticals. Furthermore, studying IBs' formation inside prokaryotic cells could provide useful knowledge of aggregation mechanisms, applicable also to aggregates formed inside eukaryotic cells, responsible for numerous degenerative diseases.

INTRODUCTION

Since the requirement for proteins in various areas of modern society increases, there is a need for improved and more efficient recombinant protein production in host organisms.

In order to achieve high protein production yields and the quality of the final product, the whole bioprocess should be optimized. Recombinant protein production represents stress for the host cell [1,2]. The whole cell machinery has to adapt for over-expression of foreign protein. Therefore, in process planning, an understanding of the host organism and its response to changes in growth conditions is necessary. Only in such a way is it possible to get as much and as active a target protein as possible already in the first production phase, during biosynthesis. However, all the following steps (cell disruption, protein isolation) should also be optimised. Therefore, understanding the variability of the bioprocess and its critical points enables us to predict the consequence of changes in the process. Process planning with quality designed or built into the process/product [3] has become known as Quality by Design [4].

Escherichia coli, as one of the most thoroughly studied microorganisms, has been used for protein production since the late 1970s when the first recombinant peptide, somatostatin, was produced [5], and it is still commonly used for recombinant protein production [6-8]. However, one of the main limitations in foreign protein production is the inability of recombinant protein to fold into its native conformation inside *E. coli*, while it has simple protein folding machinery and lacks posttranslational modifications. Therefore, foreign proteins often deposit into aggregates called inclusion bodies (IBs) [9,10]. IBs were for a long time regarded as inactive deposits of misfolded proteins, and protein isolation from them represented the main obstacle in protein production. They were considered as useless for biotechnology. Thus many aggregate prone proteins have been disregarded for commercialization [11]. Since recent studies have seriously challenged this thesis, by demonstrating that IBs can contain native like structures, protein aggregation has become an important subject in many fields, including biology, medicine and biotechnology [12]. Here strategies for designing a protein production process with increased protein productivity and quality is described.

INCLUSION BODIES STRUCTURE

IBs are formed in the cytoplasm as well as in the periplasm of *E. coli* [13,14]. They can be observed under the microscope as dense refractive particles [15]. When observing cross-sections of bacterial cells under the electron microscope, IBs appears as electro-dense, amorphous inclusions [16,17]. Isolated IBs appear as spheroid to cylindrical particles and their surface observed under a scanning electron microscope reveals a highly porous structure (Figure 1) [17-19]. On the other hand, when the IBs are shortly treated with a low concentration of the detergent or exposed to proteolytic enzymes, their inner granular architecture becomes visible [16,17]. It seem as they are composed of small granules (proto-aggregates) imbedded in a cotton-like matrix (Figure 2) [17].

They have a diameter of 0.5–1,3 μm inside the bacterial cytoplasm [13,17,20] and can fill the whole bacterial cell diameter when fully grown. In the *E. coli* cell, usually one IB can be observed [17], sometimes two, usually immediately after induction [20]. Two IBs could also be observed in the cells that simultaneously co-express two different proteins. Under such conditions, two types of IBs are formed, each enriched in one type of recombinant protein [21].

IBs are composed predominantly of the target protein (from 50 up to 90% of all protein inside IBs) [22] although other cytoplasm proteins, outer and inner membrane proteins, proteolytic fragments of recombinant protein, phospholipids and even nucleic acids are also trapped inside [23-26].

For a long time IBs were regarded as intracellular deposits of misfolded, biologically inactive proteins [27], thus they were considered to be a bottleneck in protein production. Several approaches as to how to avoid time consuming and often unfavorable renaturation procedures had been described. Higher growth temperatures, higher salt concentration and lower airflow in the medium results in higher ratio of aggregated vs. soluble target proteins. Choosing a suitable *E. coli* strain [28] and altering the production process can increase protein solubility in the cytoplasm, as well as IBs formation and composition [29-32]. Increased solubility of proteins produced in bacteria can also be achieved either by co-expression of chaperones [33,34] or by altering the target protein (point mutations, fusion proteins) [35-37]. Nevertheless, protein aggregation into IBs is often unavoidable [38].

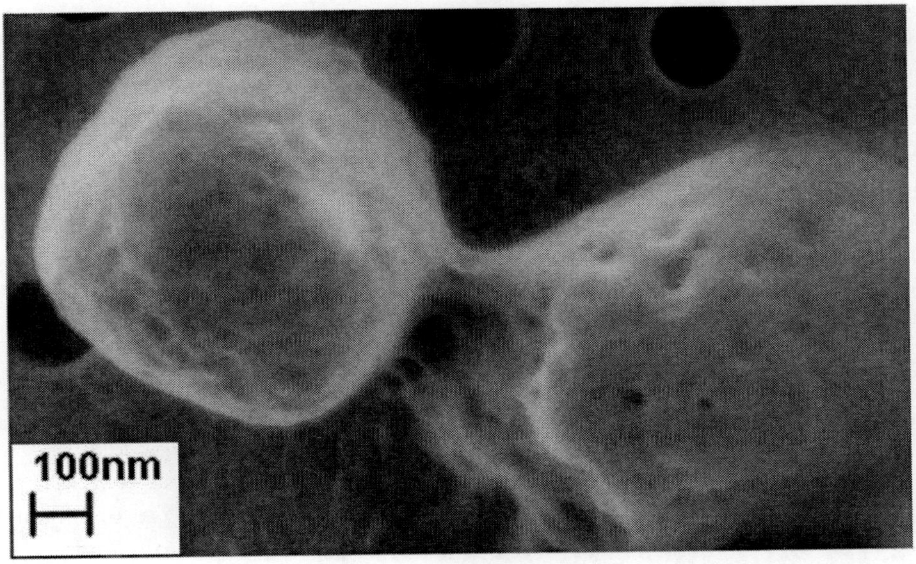

Figure 1. Porosity of IBs. Isolated IBs thoroughly washed in pure water observed under scanning electron microscope. Highly porous structure of IBs can be observed.

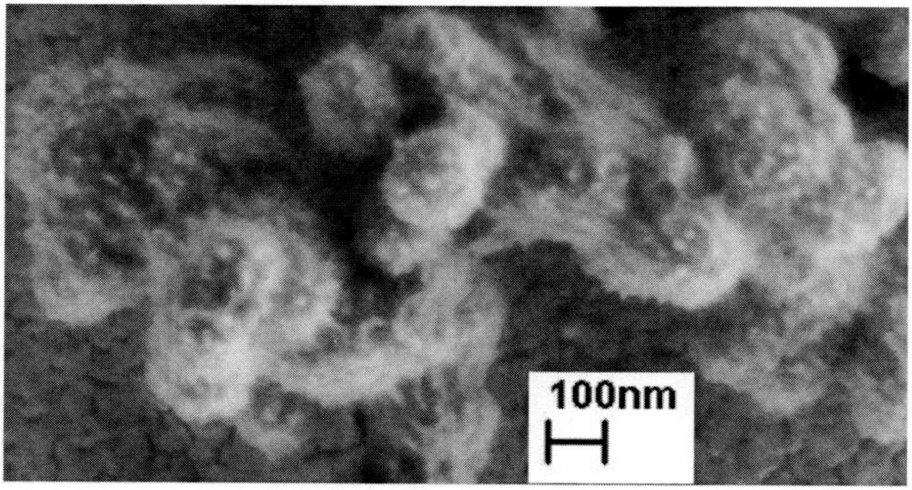

Figure 2. Granular structure of IBs. Isolated IBs thoroughly washed in pure water and additionally rapidly washed with mild detergent. Smaller granules (proto-aggregates) are clearly visible imbedded in a cotton-like matrix.

In the last few years, with the increasing number of reports about the presence of correctly folded protein precursors inside IBs [39], the view on IBs is slowly changing. The first observations of enzymatic activity in proteins inside IBs [29,40] were reported more than 15 years ago, but the phenomenon was not further explored and the usefulness and biotechnological importance were not recognized. Recently, several groups have reported the presence of biologically active proteins inside IBs [17,41-45]. The presence of residual native-like structures detected by Fourier transform infrared (FT-IR) spectroscopy has been reported by several authors [43,45-50].

The first thorough studies about protein activity inside IBs were reported in the last few years, when it was discovered that during production of GFP in *E. coli* at lower growth temperatures, fluorescent IBs are formed [42,51,52].

These new IBs contain a correctly folded protein precursor which can be extracted under non-denaturing conditions [43]. As they show several significantly different properties from IBs previously described they were given the designation "non-classical inclusion bodies" (ncIBs) [17].

Bacterial Cell Cultivation and Recombinant Protein Production

E. coli is one of the most widely used hosts for the production of recombinant proteins due to its well characterized genetics and availability of large numbers of cloning vectors and mutant host strains [53] as well as fast growth rate, high cell density production, inexpensive substrates and simple cultivation.

Choosing optimal expression system is vital for an efficient protein production process. Foreign genes are expressed from plasmids under the control of inducible promoters. Recently *E. coli* B strains, such as BL21, combined with pET vectors, became popular due to their efficiency in recombinant protein production [6,53]. Here, target genes are positioned downstream of the bacteriophage T7 promoter. Usually production host contain also a prophage (DE3) encoding the enzyme under control of the isopropyl β-D-1-thiogalactopyranoside (IPTG) inducible lacUV5 promoter. Thus large amounts of mRNA are synthesized and target proteins are accumulated in the host organism at very high concentrations (40-50% of total cell proteins). At the beginning such strains had some drawbacks (e.g. leaky expression, plasmid toxicity, plasmid instability) [53] that were later improved with the selection of cAMP-deficient BL12(DE3) mutants. Such strains exhibit enhanced expression (can exceed 50% of total cell proteins) along with plasmid stability and are thus especially recommended for the production of toxic proteins [54]. This *E. coli* strain was chosen for production of recombinant proteins inside ncIBs [17,43,45].

After the selection of the suitable bacterial strain, production media should also be optimized. In order to ensure highly reproducible protein production, quality inoculum should be prepared. The quality of the inoculum is very important for the reproducibility of the production process, since the growth rate and the final biomass achieved by a batch culture can be manipulated by altering its density during the early exponential phase [55]. Complex media such as LB (Luria-Bertani) or TB (Terrific broth) are usually used for protein production in *E. coli*, since they are cheap and easy to prepare. Nevertheless, Jevsevar and coworkers [56] report that LB media without glucose trigger unexpectedly high leaky expression, where without IPTG induction accumulation of target protein exceeds 15% of total cell proteins. However, it seems that glucose supplement efficiently represses expression of the foreign genes; therefore glucose addition to the medium is essential for high quality inoculum preparation. The addition of antibiotic as a selection pressure is also necessary.

Since the LB media lacks some essential microelements, GYSP media was tested for protein production [43,45]. It was reported, that GYSP medium containing glucose, enables higher accumulation of recombinant proteins, even without addition of antibiotic into the production medium. It enables higher cell density production, hence GYSP medium with addition of glucose was chosen as the optimal medium for recombinant protein production [56].

Optimal growth temperature of *E. coli* is 37°C; however, some laboratory strains can grow at much higher temperatures, up to 49°C [57]. On the other hand, growth temperatures can also be reduced and for recombinant protein production even as low as 16°C was reported to be successful [51,58]. Lowering cultivation temperature was proven to effectively limit the *in vivo* aggregation of recombinant proteins [59], thus higher amounts of soluble proteins are formed [60]. Besides affecting protein distribution between soluble and insoluble fraction low cultivation temperature influences also the quality of the proteins in both cell fractions, also of those trapped inside IBs [43,45,51,61,62]. At lower cultivation temperature, all the cell processes are slower, including transcription and translation [59] thus there is more time for proper protein folding. The extended time period between synthesis and deposition as well as rapidly exceeded solubility of target protein results in aggregation of better folded proteins [44,51]. For production of such non-classical IBs, entrapping properly folded proteins, temperatures between 20 and 25°C are optimal. At higher temperatures the amount of properly folded protein inside IBs is insignificant while, on the other hand, at temperatures below 20°C the IBs are hardly formed [43,45,51,61,62].

Optimization of all the parameters influencing protein production inside *E. coli* results in high expression yields of the target protein. Satisfactory yields could be achieved in the shake flask culture, although extremely high yields could only be obtained by high cell density fermentation, using finely tuned expression systems [63]. However, unusually high transcription rates and thus strong production of recombinant proteins could result in a stressful situation for the host cell [2]. This could have a profound negative effect on productivity and protein quality [64]. Therefore, during bacterial cell cultivation fragile balance between high culture density and high protein yield versus high protein quality should be maintained for optimal results.

IBs Isolation from Bacterial Cells

Since IBs are highly pure in composition and are enriched in the recombinant protein they are very suitable for protein isolation. Inside *E. coli* target protein expression can exceed 50% of total cell proteins. During IBs isolation the target protein is further enriched, as it represents majority of the proteins inside IBs (up to 90 %). Therefore IBs are a convenient source of easily extractable proteins and are largely exploited for biotechnology purposes [65].

After bacterial culture production, bacterial cells are separated from the medium by centrifugation and resuspended in a suitable buffer (PBS, Tris buffer). With the intention of IBs isolation the cells are than disrupted by non mechanical (e.g. chemical permeabilization, enzymatic lysis (Lysozyme), osmotic shock…) or mechanical methods (e.g. ultrasonic disruption, homogenization, French press, ball mill…).

Cell disruption step should be carefully optimized as well. Proteins are very sensitive and most chemicals used for the chemical permeabilization, such as organic solvents and chaotropic agents, can destroy protein structure and should thus be avoided [66]. In addition, non-mechanical methods used for cell disruption often require additional reagents or removal process to continue with further downstream applications.

Not only chemical but also mechanical methods, such as ultrasonic cell disruption (sonication), were reported to alter the proteins. During sonication aggregates, resembling properties of amyloid, can be formed [67]. When using ultrasonication, careful optimization of the process should be made regarding the acoustic power [68] as well as time and quantity of consecutive sonication treatments [69]. Alternatively, a less conflict method that does not alter the structure of ncIBs and/or the proteins could be used for cell disruption. Homogenization was successfully used for ncIBs isolation from bacterial cells [17,43,45].

After cell disruption IBs can be separated from soluble cytoplasm proteins and other cell debris by centrifugation while they sediment faster and the first purification step is thus achieved [27].

Higher purity of IBs could be achieved by washing IBs with low concentrations of chaotropic agents (low molar Urea or *Guanidinium hydrochloride)* or detergents (SDS, TritonX-100, Na-deoxycholate…) [70,71]. However non-classical IBs are soluble in chaotropic agents as well as in buffers containing even very low detergent concentrations [17,43]. In detergents usually used for washing IBs (0.5 % TritonX-100, 1% Na-deoxycholate) up to 20% of target protein can be lost from IBs in a single washing step. Even lower detergent concentrations, e.g. 0.1% deoxycholate, results in significant protein loss [44]. Therefore, only washing in pure water or mild buffers (e.g. TRIS, PBS) with pH 7-8 would be appropriate. However, optimization of centrifugation as well as repeated homogenization described by Wong and coworkers [72] could also be applied. After IBs are isolated from bacterial cells, their properties can be studied and target proteins can be isolated.

PROTEIN ISOLATION FROM IBS

Classical protocols for protein isolation from IBs includes solubilization of IBs in denaturing agents followed by renaturation of the protein [73-75]. Most protocols requires solubilization in strong denaturants (e.g. 8M Urea, 6M *Guanidinium hydrochloride*) [71,76,77], while some authors reports the treatment of IBs in somewhat milder denaturing conditions like use of various detergents, low molar denaturants, high pH or high hydrostatic pressure in combination with reducing agents [18,73,78-81]. Denaturation is followed by protein renaturation during or after chromatography [75,82]. Renaturation process should be optimized for each protein individually and recovery yields of active protein are often low, which represents the main bottleneck in protein isolation from IBs [83].

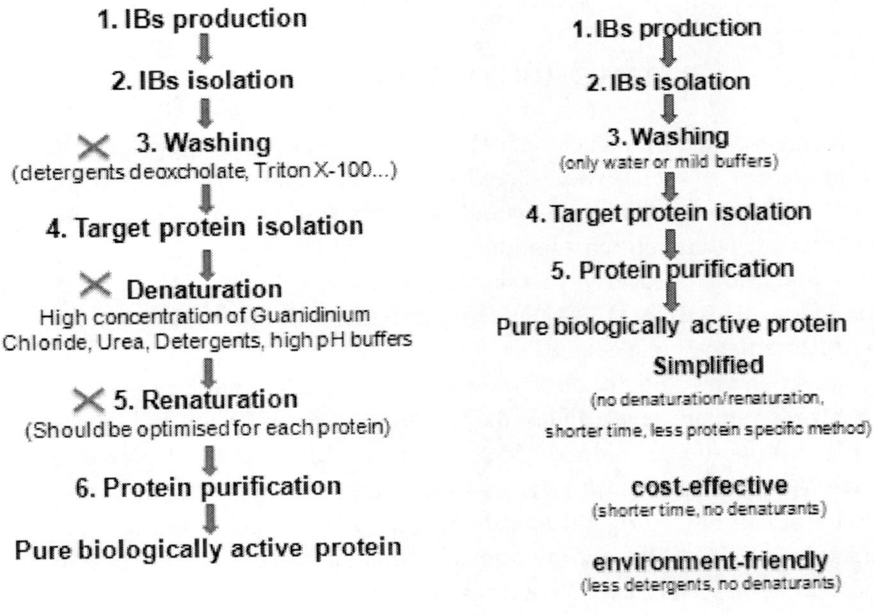

Figure 3. Protocol from protein isolation from IBs. On the left side protein isolation from classical inclusion bodies is described. After production, IBs are isolated from bacterial cells and washed with buffers containing detergents. Proteins from IBs are than solubilized in strongly denaturing conditions. Solubilization is followed by renaturation process that must be carefully optimized. With non-classical IBs, no detergents should be used for washing. Target proteins are isolated from ncIBs with mild detergents in non-denaturing conditions, therefore no refolding is needed. The protocol for downstream purification of target protein is thus simplified, more cost efficient and environment-friendlier.

On the other hand, ncIBs are composed of significant amount of properly folded protein [17,43,45]. As mentioned above, such IBs are very soluble in even very low detergent concentrations, so during washing, protein can be lost. However, this could also be used as an advantage. Protocol for target protein extraction from ncIBs in mild detergent (0.2% N-lauroyil sarcosine) was developed [84] and tested for application on various proteins [17,43-45]. As described by Jevsevar and coworkers [43] other non-denaturing agents, such as 0.05-0.2% N-lauroyil sarcosine, 5% dimethyl sulfoxide (DMSO) and 0.2% non-detergent sulphobetaines (NDSB) could also be used for efficient extraction of target proteins from ncIBs. Such extraction of target proteins from ncIBs without protein denaturation enables extraction of biologically active protein molecules from ncIBs with no need of renaturation. While denaturation/renaturation step is avoided, the protocol for downstream purification of target protein is simplified, more cost efficient and environment-friendlier (Figure 3).

Furthermore, during protein extraction in mild, non-denaturing conditions mainly properly folded proteins are extracted. Thus the target protein is even further enriched and therefore less downstream steps are necessary to purify the protein [45]. Downstream process for purification of target protein from ncIBs was developed by Gaberc-Porekar and Menart [85], where, in a single chromatography step, the production of a biologically active protein with a purity greater than 95% was achieved.

Contraction of Inclusion Bodies

After IBs are isolated from bacterial cells their properties can be studied. One of the most important properties of ncIBs was already mentioned above is their solubility in mild detergents. As already described, this could be used as an advantage for protein extraction from IBs without denaturation/renaturation.

Another interesting property that could be applied to all IBs is their irreversible contraction in low pH buffers [17,44]. When transferring IBs from pH 7 to pH 4 a significant reduction of IBs pellet volume occurs.

The contraction was initially observed upon centrifugation on macro-level and was later confirmed also on micro-level under the electron microscope [17]. Interestingly, ncIBs containing high proportion of properly folded protein are less sensitive to changes in pH than classical IBs. The volume of ncIBs reduces by one third while the volume of the classical IBs shrinks to almost half of the original size at pH 7, when buffer pH is reduced [17].

Contraction of IBs affects downstream isolation process, therefore it should be considered during IBs isolation. As a consequence of contraction of IBs at low pH, the density of IBs increases. This results in the sedimentation velocity changes during centrifugation. After centrifugation under the same conditions the pellet sediments faster and becomes more compact when IBs are previously treated with low pH buffer [44].

The contraction of IBs at low pH affects not only IBs density but also other IBs properties. As mentioned above ncIBs are very soluble in common washing solutions containing detergents such as 1% deoxycholate and up to 20% of target protein can be lost from IBs in a single washing step. However, ncIBs previously treated with low pH buffer are less sensitive to solubilization and washing of contracted IBs with 1% deoxycholate, could wash out only up to 5 % of target proteins in a single washing step [44].

This affects also protein extractability from ncIBs. They are easily soluble in0.2 % N-lauroyl sarcosine and more than 95 % of target protein can be dissolved from IBs. Using no renaturation procedure, up to 50 % of the extracted target protein molecules are present in the biologically active form [17]. However, when previously treated in low pH buffer, only up to 70 % of target protein can be dissolved from ncIBs [44].

As previously described, IBs are highly porous structure [17-19]. Our observations showed that as IBs are transferred to low pH buffer, not only the whole IBs diameter, but also the diameter of the pores contracts. Garcia-Fruitos and coworkers reports that porosity of IBs allows efficient diffusion of the substrate to catalytically active sites inside the aggregates [42]. Such IBs could be used as active nanoparticles in biotechnology or medicine. However, we might speculate that after contraction of the pores, accessibility of the catalytically active sites might be impaired and thus the activity of the nanoparticles would be lower or even completely lost.

Therefore, during downstream protein isolation optimization of every single step is indispensable. Even IBs washing, the step that seems fairly insignificant can result in a significant protein loss or can disable further protein extraction. The use of inappropriate washing buffers could even lead to inactivation of the IBs as nanoparticles.

DYNAMICS OF IBS AND PROTEIN QUALITY

In the last decade, several paradigms about the protein aggregation as well as about the IBs structure and composition became compromised. For a long time, IBs were described as intracellular aggregates of misfolded, biologically inactive proteins [27]. Still more and more authors reports about the presence of native-like secondary structures or even active proteins trapped inside IBs [17,29,40-48,61]. On the other hand, proteins solubility was considered as equivalent to protein conformational quality. Yet several studies confirmed that soluble proteins are not always in a biologically active form [42,45,51,86-89].

Furthermore it was believed that once formed, IBs stay in the cell as an unreactive deposit and are resistant to proteolysis [27]. On the contrary, Carrio and coworkers have shown that in the bacterial cell there is a dynamical balance between the construction and deconstruction of IBs, and that they are sensitive to proteolytic attack [20,25]. Furthermore, in the absence of protein synthesis, IBs are almost completely disintegrated in a few hours [90].

Hence, IBs could be described as protein reservoirs that are greatly incorporated in the protein quality system of the cell, and from which the proteins can be extracted [22]. Chaperones, small heat-shock proteins and proteases work together under stress conditions (e.g. heat shock, over-expression…) in order to enable proper protein folding upon translation. When proteins fail to fold into their native conformation, these elements of cell quality control machinery enable to unfold and refold these proteins, or degrade them when the proteins fail to fold. All these activities affect also the structure and composition of IBs [38,39].

Studies about the location of the active proteins inside IBs show that IBs are not a homogenous structure but it seems that the core is enriched in active proteins. It has been suggested, that the removal of the misfolded protein from the IBs surface results in a progressive enrichment of the native like and active proteins in the IB core [61].

MECHANISMS OF IBS FORMATION

Although the knowledge about protein aggregation rapidly increases and many physiological factors that influence the formation of IBs in bacteria are already known, some questions remains unanswered [19,32,38,39,91]. There are obstacles in the transcription and translation process, e.g. chaperone or protease shortage, rare codon and incorrect codon usage, lack of post-translational modifications as well as inability to form disulphide bonds [8,28,92]. Aggregation can also occur due to interactions among the *de novo* formed folding intermediates, which expose hydrophobic residues at their surface [93] and high local concentrations of the produced polypeptide [39].

Since a lot of structurally different proteins are able to form IBs without any common pattern, their formation has been considered to be driven by nonspecific intermolecular contacts [38]. However, recent studies showed that IBs formation results in protein-specific construction [37,94,95]. Specificity has also been observed upon co-expression of two different proteins in the same bacterial cell, where two types of cytoplasm IBs were formed [21].

Such specific construction of IBs formation could also be observed in amyloid aggregates. The detailed mechanism for the formation of fibrillar amyloid-like aggregates is not entirely clear. Nonetheless, the initiating event is protein misfolding that result in the formation of aggregation-prone structures. Folding intermediates assemble in a specific manner to form discrete structured oligomeres, which acts as seeds that further propagate protein misfolding [96]. They are than expanded to prefibrillar structures and then further matured to highly ordered fibrils of a β-sheet polypeptide chain arrangement [97,98]. This is the basis for the currently accepted nucleation-dependent polymerization model of amyloid formation [38,99-101]. Diverse proteins, associated with human conformational diseases (including Aβ, huntingtin, and α-synuclein), have been shown to follow this crystallization-like process.

Similar amyloid-like structure was also reported for IBs formed in bacterial cells [38,46,95]. Here an emerging IB acts as a nucleus to incorporate nascent proteins [38]. However, IBs containing high proportion of correctly folded protein (granulocyte colony stimulating factor – G-CSF) do not show amyloid-like structure. Considering that the G-CSF is a typical four-helix bundle protein, it is not unusual that no Congo red binding was observed [17]. Spectroscopy studies confirmed that G-CSF ncIBs show predominantly α-helical structure, whereas in classical G-CSF IBs, prepared at higher temperatures (where the proportion of properly folded target protein is insignificant), distinctive band at 1627 cm^{-1}characteristic for β-structure [77,102], increases [43]. Also other α-helical proteins, like IL-1β, contain significant amount of α-helix secondary structure in addition to characteristic intermolecular β-structure [47]. Such β-sheet rich structures might explain the increased resistance to proteases [16,103,104] as well as lower solubility of classical versus non-classical IBs in mild non-denaturing detergents [45].

Based on all the knowledge about protein synthesis and IBs properties, several potential mechanisms of IBs formations in bacterial cells were suggested [38,39,44,64,89]. On Figures 4, 5 and 6 these possible mechanisms are presented.

During over-expression of foreign genes, protein misfolding commonly occurs. Consequently, various transitional folding states of the target protein are simultaneously present in the cell. Some of the folding intermediates that fail to fold into a native conformation are immediately degraded, while others aggregate into small proto-aggregates. Aggregation occurs due to the hydrophobicity and rapidly exceeded solubility of these transitional forms (Figure 4) [38,44,64].

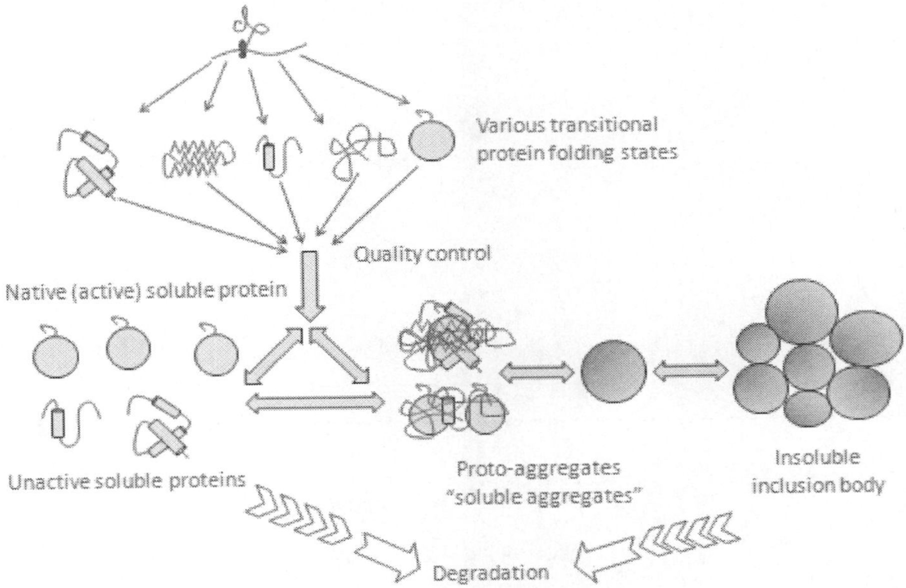

Figure 4. A potential mechanism of protein folding and aggregation in bacteria Escherichia coli. After protein synthesis many proteins fail to fold into their native conformation. Therefore various transitional folding intermediates as well as properly folded active proteins are simultaneously present in the cell. Protein folding is assisted with the cell quality control machinery (chaperones, small heat-shock proteins and proteases), which establishes a kinetic equilibrium between soluble and aggregated forms of the protein. Therefore soluble fraction is not composed solely of native protein but also of soluble aggregates. Inside these soluble aggregates native proteins are also trapped. Soluble aggregates are further aggregated into insoluble aggregates called inclusion bodies. The process is fully reversible. Protein release from inclusion bodies is controlled by the cell quality control machinery. Proteases can degrade both soluble and insoluble protein species that fail to fold to the native conformation.

The presence of such proto-aggregates (entitled "soluble aggregates") in the cytoplasm were confirmed by Schröder and de Marco [89]. These proto-aggregates are than fused together in an IB that continually grows further to form the final IB as depicted in Figure 5 [44]. It appears that proto-aggregates are imbedded into a cotton-like amorphous matrix, which probably belongs to a mass of misfolded proteins associated by hydrophobic interactions. This amorphous matrix fills the spaces both among and inside the aggregates. Inside the network, there may thus be amount of spaces large enough to allow properly folded protein precursors to be trapped [44].

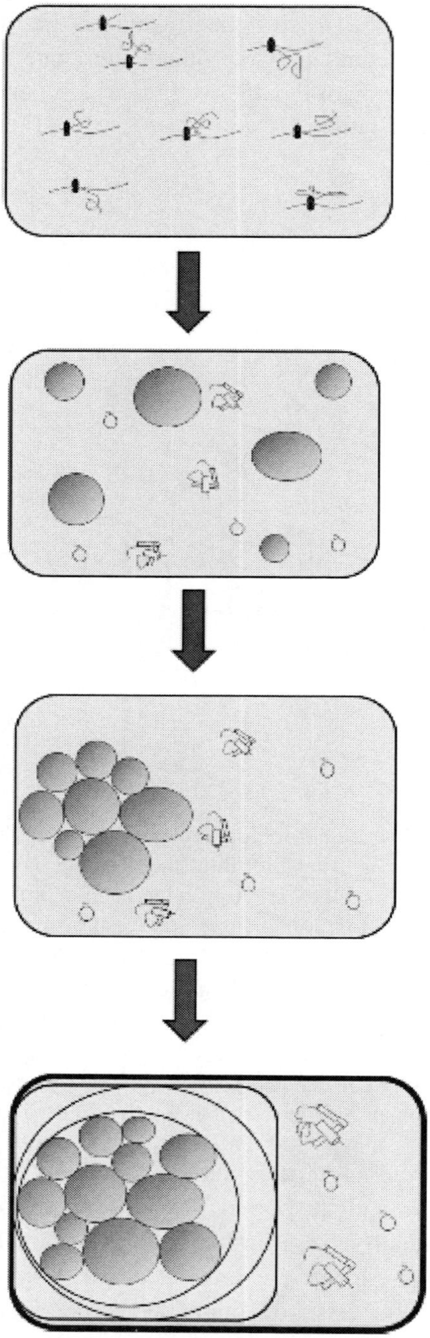

Figure 5. Suggested model for IBs formation. During over-expression of foreign genes, protein misfolding commonly occurs. Therefore various transitional folding states of the target protein are simultaneously present in the cell. Due to the hydrophobicity and rapidly exceeded solubility of these transitional forms, many small proto-aggregates are continuously formed. They are later glued together in an "attractor IB" which then grows further to form the final IB (colored in pale yellow on the figure). At the end the IB can fill a significant part of the whole bacterial cell.

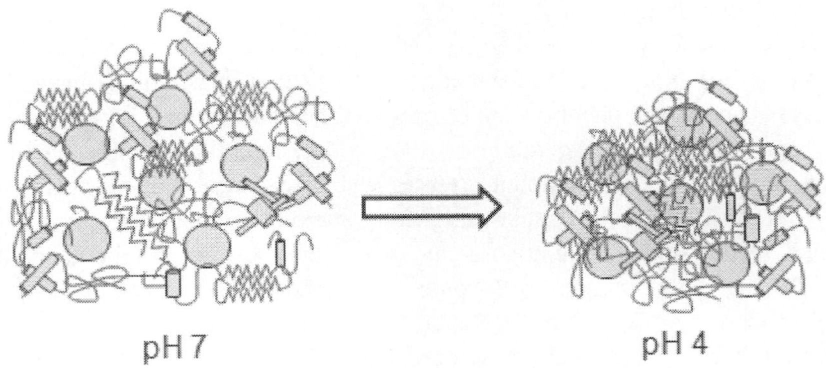

Figure 6. An assumed model for the internal structure of IBs. IBs are composed of the network of misfolded proteins inside which properly folded active proteins are trapped. At neutral pH, the whole network is loosely bound. By transferring IBs into acidic pH buffer (around 4), strong contraction of the network results in more compact IBs from which properly folded protein can no longer be efficiently extracted.

At neutral pH the whole network is loosely bound, allowing easy extraction of native-like precursor molecules as well as some still soluble proteins having biologically inactive conformations [43]. In contrast, by transferring IBs into buffer with acidic pH, the high proton concentration could induce a change in unfolded proteins, leading to strong contraction of the network and formation of more compact IBs where properly folded precursors are trapped and their extraction prevented (Figure 6) [44].

Conclusion: Impact in Biotechnology and Medicine

Understanding the structure and mechanisms of IBs formation as well as their properties could have a significant effect on biotechnology as well as on medicine.

The described properties of ncIBs, such as solubility, extractability, porosity and low pH contraction should be considered in biotechnology during development of the protein production process. Solubility of ncIBs in detergents, usually used for washing, might lead to target protein loss. On the other hand, it could be used as an advantage, as extraction of target proteins under non-denaturing conditions enables us to extract biologically active proteins from ncIBs in a single step [17,43].

In addition, low pH contraction should also be taken into account. In biotechnology, downstream processes, such as centrifugation, can be affected by density of IBs. For instance sedimentation velocity of compact IBs (prepared at pH 4.0) in comparison to less dense IBs (prepared at pH 7.0) is different and can be modulated by pH of the buffer. Even more importantly, the solubility/extractability of IBs depends very much on the buffer pH, where extraction of proteins from ncIBs is much less efficient in the pH region 3–5 than extraction undertaken in conditions of pH 7–8 [44].

The existing knowledge about ncIBs results in a better picture about their formation, giving rise to efficient technological solutions. Based on this knowledge, an economically advantageous biotechnological process for granulocyte colony stimulating factor production and purification was designed [85,105].

Such IBs composed of pure and active protein could also be used as active nanoparticles in diagnostics or as biocatalysts in enzymatic processes or even as biopharmaceuticals.

On the other hand, the phenomenon of irreversibly altered properties of IBs upon transfer to low pH as described above, might also have important implications in understanding mechanisms in human conformational diseases where IBs are usually formed and deposited [12,38,98,101,106-108]. In this latter case, cells attempt to control the level of denatured proteins and restrict their accumulation by use of two major protein degradation routes identified so far; the ubiqutine-proteasome and authopagy-lysosomal pathways [109]. It seems that protein inclusions stimulate endosomal-lysosomal activity in neurons [110]. We may hypothesize that in neuronal lysosomes, which show low pH, contraction of IBs might occur in a similar way to those observed in our experiments with G-CSF IBs. As a consequence, IBs destined to be degraded by autophagy, become resistant to proteolytic digestion and accumulate. This hypothetical explanation is still looking to be proved experimentally. However, it may represent a challenge to those investigating molecular mechanisms of human conformational diseases.

Acknowledgments

I would like to dedicate this work to Dr. Viktor Menart who played a crucial role in my initial studies of protein aggregation.

I would also like to apologize to all the authors whose exceptional work could not be cited.

References

[1] Grabherr R, Nilsson E, Striedner G, Bayer K: Stabilizing plasmid copy number to improve recombinant protein production. *Biotechnol. Bioeng.* 2002, 77: 142-147.

[2] Hoffmann F, Rinas U: Stress induced by recombinant protein production in Escherichia coli. *Adv. Biochem. Eng. Biotechnol* .2004, 89: 73-92.

[3] Garber CC, Witte DL: Quality for tomorrow: by design or by checking? *Clin. Chem.* 1997, 43: 864-865.

[4] Westgard JO: *Nothing but the Truth about Quality; Essays on Quality Management in the Healthcare Laboratory; Quality By Design*. Washington, DC: American Association for Clinical Chemistry *(AACC)*; 2004.

[5] Itakura K, Hirose T, Crea R, Riggs AD, Heyneker HL, Bolivar F *et al.*: Expression in Escherichia coli of a chemically synthesized gene for the hormone somatostatin. *Science* 1977, 198: 1056-1063.

[6] Graumann K., Premstaller A.: Manufacturing of recombinant therapeutic proteins in microbial systems. *Biotechnology Journal* 2006, 2006: 164-186.

[7] Schmidt FR: Recombinant expression systems in the pharmaceutical industry. *Appl. Microbiol. Biotechnol.* 2004, 65: 363-372.

[8] Terpe K: Overview of bacterial expression systems for heterologous protein production: from molecular and biochemical fundamentals to commercial systems. *Appl. Microbiol. Biotechnol.* 2006, 72: 211-222.

[9] Villaverde A, Carrio MM: Protein aggregation in recombinant bacteria: biological role of inclusion bodies. *Biotechnol. Lett.* 2003, 25: 1385-1395.

[10] Baneyx F, Mujacic M: Recombinant protein folding and misfolding in Escherichia coli. *Nat. Biotechnol.* 2004, 22: 1399-1408.

[11] de Groot NS, Espargaro A, Morell M, Ventura S: Studies on bacterial inclusion bodies. *Future Microbiol.* 2008, 3: 423-435.

[12] Smith A: Protein misfolding. *Nature* 2003, 426: 883.

[13] Bowden GA, Paredes AM, Georgiou G: Structure and morphology of protein inclusion bodies in Escherichia coli. *Biotechnology (N Y)* 1991, 9: 725-730.

[14] Arie JP, Miot M, Sassoon N, Betton JM: Formation of active inclusion bodies in the periplasm of Escherichia coli. *Mol. Microbiol.* 2006, 62: 427-437.

[15] Williams DC, Van Frank RM, Muth WL, Burnett JP: Cytoplasmic inclusion bodies in Escherichia coli producing biosynthetic human insulin proteins. *Science* 1982, 215: 687-689.

[16] Carrio MM, Cubarsi R, Villaverde A: Fine architecture of bacterial inclusion bodies. *FEBS Lett.* 2000, 471: 7-11.

[17] Peternel S, Jevsevar S, Bele M, Gaberc-Porekar V, Menart V: New properties of inclusion bodies with implications for biotechnology. *Biotechnol. Appl. Biochem* .2008, 49: 239-246.

[18] Singh SM, Panda AK: Solubilization and refolding of bacterial inclusion body proteins. *J. Biosci. Bioeng.* 2005, 99: 303-310.

[19] Carrio MM, Cubarsi R, Villaverde A: Fine architecture of bacterial inclusion bodies. *Febs. Lett.* 2000, 471: 7-11.

[20] Carrio MM, Corchero JL, Villaverde A: Dynamics of in vivo protein aggregation: building inclusion bodies in recombinant bacteria. *FEMS Microbiol. Lett.* 1998, 169: 9-15.

[21] Hart RA, Rinas U, Bailey JE: Protein composition of Vitreoscilla hemoglobin inclusion bodies produced in Escherichia coli. *J. Biol. Chem.* 1990, 265: 12728-12733.

[22] Carrio MM, Villaverde A: Construction and deconstruction of bacterial inclusion bodies. *J. Biotechnol.* 2002, 96: 3-12.

[23] Rinas U, Bailey JE: Protein compositional analysis of inclusion bodies produced in recombinant Escherichia coli. *Appl. Microbiol. Biotechnol.* 1992, 37: 609-614.

[24] Valax P, Georgiou G: Molecular characterization of beta-lactamase inclusion bodies produced in Escherichia coli. 1. Composition. *Biotechnol. Prog.* 1993, 9: 539-547.

[25] Carrio MM, Corchero JL, Villaverde A: Proteolytic digestion of bacterial inclusion body proteins during dynamic transition between soluble and insoluble forms. *Biochim. Biophys. Acta* 1999, 1434: 170-176.

[26] Rinas U, Hoffmann F, Betiku E, Estape D, Marten S: Inclusion body anatomy and functioning of chaperone-mediated in vivo inclusion body disassembly during high-level recombinant protein production in Escherichia coli. *J. Biotechnol.* 2007, 127: 244-257.

[27] Marston FA: The purification of eukaryotic polypeptides synthesized in Escherichia coli. *Biochem. J.* 1986, 240: 1-12.

[28] Sorensen HP, Mortensen KK: Advanced genetic strategies for recombinant protein expression in Escherichia coli. *J. Biotechnol.* 2005, 115: 113-128.

[29] Worrall DM, Goss NH: The formation of biologically active beta-galactosidase inclusion bodies in Escherichia coli. *Aust. J. Biotechnol.* 1989, 3: 28-32.

[30] Strandberg L, Enfors SO: Factors influencing inclusion body formation in the production of a fused protein in Escherichia coli. *Appl. Environ. Microbiol.* 1991, 57: 1669-1674.

[31] Sorensen HP, Mortensen KK: Soluble expression of recombinant proteins in the cytoplasm of Escherichia coli. *Microb. Cell Fact* .2005, 4: 1.

[32] Makrides SC: Strategies for achieving high-level expression of genes in Escherichia coli. *Microbiol. Rev.* 1996, 60: 512-538.

[33] De Marco A, Vigh L, Diamant S, Goloubinoff P: Native folding of aggregation-prone recombinant proteins in Escherichia coli by osmolytes, plasmid- or benzyl alcohol-overexpressed molecular chaperones. *Cell Stress Chaperones* .2005, 10: 329-339.

[34] Mogk A, Deuerling E, Vorderwulbecke S, Vierling E, Bukau B: Small heat shock proteins, ClpB and the DnaK system form a functional triade in reversing protein aggregation. *Mol. Microbiol* .2003, 50: 585-595.

[35] Rinas U, Tsai LB, Lyons D, Fox GM, Stearns G, Fieschko J *et al.*: Cysteine to serine substitutions in basic fibroblast growth factor: effect on inclusion body formation and proteolytic susceptibility during in vitro refolding. *Biotechnology (N Y)* 1992, 10: 435-440.

[36] De M, V, Stier G, Blandin S, De Marco A: The solubility and stability of recombinant proteins are increased by their fusion to NusA. *Biochem Biophys. Res. Commun.* 2004, 322: 766-771.

[37] Ventura S: Sequence determinants of protein aggregation: tools to increase protein solubility. *Microb. Cell Fact.* 2005, 4: 11.

[38] de Groot NS, Espargaro A, Morell M, Ventura S: Studies on bacterial inclusion bodies. *Future Microbiol* 2008, 3: 423-435.

[39] Ventura S, Villaverde A: Protein quality in bacterial inclusion bodies. *Trends Biotechnol.* 2006, 24: 179-185.

[40] Tokatlidis K, Dhurjati P, Millet J, Beguin P, Aubert JP: High activity of inclusion bodies formed in Escherichia coli overproducing Clostridium thermocellum endoglucanase D. *FEBS Lett.* 1991, 282: 205-208.

[41] Garcia-Fruitos E, Carrio MM, Aris A, Villaverde A: Folding of a misfolding-prone beta-galactosidase in absence of DnaK. *Biotechnol. Bioeng* .2005, 90: 869-875.

[42] Garcia-Fruitos E, Gonzalez-Montalban N, Morell M, Vera A, Ferraz RM, Aris A *et al.*: Aggregation as bacterial inclusion bodies does not imply inactivation of enzymes and fluorescent proteins. *Microb. Cell Fact* .2005, 4: 27.

[43] Jevsevar S, Gaberc-Porekar V, Fonda I, Podobnik B, Grdadolnik J, Menart V: Production of nonclassical inclusion bodies from which correctly folded protein can be extracted. *Biotechnol. Prog.* 2005, 21: 632-639.

[44] Peternel S, Bele M, Gaberc-Porekar V, Komel R: Contraction of inclusion bodies with implications in biotechnology . *Acta Chim. Slov.* 2008, 55: 608-612.

[45] Peternel S, Grdadolnik J, Gaberc-Porekar V, Komel R: Engineering inclusion bodies for non denaturing extraction of functional proteins. *Microb. Cell Fact* .2008, 7: 34.

[46] Ami D, Natalello A, Taylor G, Tonon G, Maria DS: Structural analysis of protein inclusion bodies by Fourier transform infrared microspectroscopy. *Biochim .Biophys. Acta* 2006, 1764: 793-799.

[47] Oberg K, Chrunyk BA, Wetzel R, Fink AL: Nativelike secondary structure in interleukin-1 beta inclusion bodies by attenuated total reflectance FTIR. *Biochemistry* 1994, 33: 2628-2634.

[48] Ami D, Natalello A, Gatti-Lafranconi P, Lotti M, Doglia SM: Kinetics of inclusion body formation studied in intact cells by FT-IR spectroscopy. *FEBS Lett* .2005, 579: 3433-3436.

[49] Doglia SM, Ami D, Natalello A, Gatti-Lafranconi P, Lotti M: Fourier transform infrared spectroscopy analysis of the conformational quality of recombinant proteins within inclusion bodies. *Biotechnol. J.* 2008, 3: 193-201.

[50] Gonzalez-Montalban N, Natalello A, Garcia-Fruitos E, Villaverde A, Doglia SM: In situ protein folding and activation in bacterial inclusion bodies. *Biotechnol. Bioeng.* 2008, 100: 797-802.

[51] Vera A, Gonzalez-Montalban N, Aris A, Villaverde A: The conformational quality of insoluble recombinant proteins is enhanced at low growth temperatures. *Biotechnol. Bioeng* .2007, 96: 1101-1106.

[52] de Groot NS, Ventura S: Protein activity in bacterial inclusion bodies correlates with predicted aggregation rates. *J. Biotechnol.* 2006, 125: 110-113.

[53] Baneyx F: Recombinant protein expression in Escherichia coli. *Curr. Opin. Biotechnol.* 1999, 10: 411-421.

[54] Grossman TH, Kawasaki ES, Punreddy SR, Osburne MS: Spontaneous cAMP-dependent derepression of gene expression in stationary phase plays a role in recombinant expression instability. *Gene* 1998, 209: 95-103.

[55] Carbonell X, Corchero JL, Cubarsi R, Vila P, Villaverde A: Control of Escherichia coli growth rate through cell density. *Microbiol. Res.* 2002, 157: 257-265.

[56] Jevsevar S, Palcic J, Jalen S, Pavko A: Influence of the Media Composition on Behavior of the pET Expression Systems. *Acta Chim. Slov.* 2007, 54: 360-365.

[57] Fotadar U, Zaveloff P, Terracio L: Growth of Escherichia coli at elevated temperatures. *J. Basic. Microbiol.* 2005, 45: 403-404.

[58] Mujacic M, Cooper KW, Baneyx F: Cold-inducible cloning vectors for low-temperature protein expression in Escherichia coli: application to the production of a toxic and proteolytically sensitive fusion protein. *Gene* 1999, 238: 325-332.

[59] Sorensen HP, Mortensen KK: Soluble expression of recombinant proteins in the cytoplasm of Escherichia coli. *Microb. Cell Fact.* 2005, 4: 1.

[60] Weickert MJ, Doherty DH, Best EA, Olins PO: Optimization of heterologous protein production in Escherichia coli. *Curr. Opin. Biotechnol.* 1996, 7: 494-499.

[61] Garcia-Fruitos E, Aris A, Villaverde A: Localization of functional polypeptides in bacterial inclusion bodies. *Appl. Environ. Microbiol.* 2007, 73: 289-294.

[62] de Groot NS, Ventura S: Effect of temperature on protein quality in bacterial inclusion bodies. *FEBS Lett.* 2006, 580: 6471-6476.

[63] Yin J, Li G, Ren X, Herrler G: Select what you need: a comparative evaluation of the advantages and limitations of frequently used expression systems for foreign genes. *J. Biotechnol.* 2007, 127: 335-347.

[64] Gasser B, Saloheimo M, Rinas U, Dragosits M, Rodriguez-Carmona E, Baumann K *et al.*: Protein folding and conformational stress in microbial cells producing recombinant proteins: a host comparative overview. *Microb. Cell Fact.* 2008, 7: 11.

[65] Vallejo LF, Rinas U: Strategies for the recovery of active proteins through refolding of bacterial inclusion body proteins. *Microb Cell Fact* 2004, 3: 11.

[66] Khmelnitsky YL, Mozhaev VV, Belova AB, Sergeeva MV, Martinek K: Denaturation capacity: a new quantitative criterion for selection of organic solvents as reaction media in biocatalysis. *Eur. J. Biochem.* 1991, 198: 31-41.

[67] Stathopulos PB, Scholz GA, Hwang YM, Rumfeldt JA, Lepock JR, Meiering EM: Sonication of proteins causes formation of aggregates that resemble amyloid. *Protein. Sci.* 2004, 13: 3017-3027.

[68] Feliu JX, Cubarsi R, Villaverde A: Optimized release of recombinant proteins by ultrasonication of E. coli cells. *Biotechnol. Bioeng.* 1998, 58: 536-540.

[69] Feliu JX, Villaverde A: An optimized ultrasonication protocol for bacterial cell disruption and recovery of beta-galactosidase fusion proteins . *Biotechn. Techn.* 1994, 8: 509-514.

[70] Bernardez Clark ED: Protein refolding for industrial processes. *Curr. Opin. Biotechnol* 2001, 12: 202-207.

[71] Middelberg AP: Preparative protein refolding. *Trends Biotechnol.* 2002, 20: 437-443.

[72] Wong HH, O'Neill BK, Middelberg AP: Centrifugal processing of cell debris and inclusion bodies from recombinant Escherichia coli. *Bioseparation* 1997, 6: 361-372.

[73] Patra AK, Mukhopadhyay R, Mukhija R, Krishnan A, Garg LC, Panda AK: Optimization of inclusion body solubilization and renaturation of recombinant human growth hormone from Escherichia coli. *Protein. Expr. Purif*.2000, 18: 182-192.

[74] Cabrita LD, Bottomley SP: Protein expression and refolding - A practical guide to getting the most out of inclusion bodies. *Biotechnol Annu. Rev.* 2004, 10: 31-50.

[75] Jungbauer A, Kaar W: Current status of technical protein refolding. *J Biotechnol* 2007, 128: 587-596.

[76] Rudolph R: Successful Protein Folding on an Industrial Scale. In *Protein Engineering: Principles and Practice*. Edited by Cleland JL, Craik CS. New York: Wiley-Liss, Inc.; 1996:283-298.

[77] Fabian H, Mäntele W: Infrared Spectroscopy of Proteins . In *Handbook of vibrational spectroscopy: Applications in Life, Pharmaceutical and Natural Sciences*. Edited by Chalmers JM, Griffiths PR. Wiley; 2001:3399-3425.

[78] Prakash S, Matouschek A: Protein unfolding in the cell. *Trends Biochem. Sci.* 2004, 29: 593-600.

[79] Tsumoto K, Umetsu M, Kumagai I, Ejima D, Arakawa T: Solubilization of active green fluorescent protein from insoluble particles by guanidine and arginine. *Biochem. Biophys. Res. Commun.* 2003, 312: 1383-1386.

[80] Umetsu M, Tsumoto K, Nitta S, Adschiri T, Ejima D, Arakawa T *et al.*: Nondenaturing solubilization of beta2 microglobulin from inclusion bodies by L-arginine. *Biochem. Biophys. Res. Commun.* 2005, 328: 189-197.

[81] Umetsu M, Tsumoto K, Ashish K, Nitta S, Tanaka Y, Adschiri T *et al.*: Structural characteristics and refolding of in vivo aggregated hyperthermophilic archaeon proteins. *FEBS Lett.* 2004, 557: 49-56.

[82] Swietnicki W: Folding aggregated proteins into functionally active forms. *Curr. Opin. Biotechnol.* 2006, 17: 367-372.

[83] Swartz JR: Advances in Escherichia coli production of therapeutic proteins. *Curr. Opin. Biotechnol.* 2001, 12: 195-201.

[84] Gaberc-Porekar V, Menart V: Perspectives of immobilized-metal affinity chromatography. *J Biochem Biophys Methods* 2001, 49: 335-360.

[85] Gaberc Porekar V, Menart V.: Process for the Purification And/Or Isolation Of Biologically Active Granulocyte Colony Stimulating Factor. *Patent: PCT/EP02/13810[WO 03/051922 A1]* 2003.

[86] Martinez-Alonso M, Gonzalez-Montalban N, Garcia-Fruitos E, Villaverde A: Learning about protein solubility from bacterial inclusion bodies. *Microb. Cell Fact* .2009, 8: 4.

[87] Martinez-Alonso M, Vera A, Villaverde A: Role of the chaperone DnaK in protein solubility and conformational quality in inclusion body-forming Escherichia coli cells. *FEMS Microbiol. Lett.* 2007, 273: 187-195.

[88] Martinez-Alonso M, Gonzalez-Montalban N, Garcia-Fruitos E, Villaverde A: The Functional quality of soluble recombinant polypeptides produced in Escherichia coli is defined by a wide conformational spectrum. *Appl. Environ. Microbiol.* 2008, 74: 7431-7433.

[89] Schrodel A, De Marco A: Characterization of the aggregates formed during recombinant protein expression in bacteria. *BMC Biochem.* 2005, 6: 10.

[90] Carrio MM, Villaverde A: Protein aggregation as bacterial inclusion bodies is reversible. *FEBS Lett.* 2001, 489: 29-33.

[91] Hoffmann F, Posten C, Rinas U: Kinetic model of in vivo folding and inclusion body formation in recombinant Escherichia coli. *Biotechnol. Bioeng.* 2001, 72: 315-322.

[92] Lilie H, Schwarz E, Rudolph R: Advances in refolding of proteins produced in E. coli. *Curr. Opin. Biotechnol.* 1998, 9: 497-501.

[93] Schlieker C, Bukau B, Mogk A: Prevention and reversion of protein aggregation by molecular chaperones in the E-coli cytosol: implications for their applicability in biotechnology. *J. Biotechnol.* 2002, 96: 13-21.

[94] Speed MA, Wang DIC, King J: Specific aggregation of partially folded polypeptide chains: the molecular basis of inclusion body composition. *Nat. Biotechnol.* 1996, 14: 1283-1287.

[95] Carrio M, Gonzalez-Montalban N, Vera A, Villaverde A, Ventura S: Amyloid-like properties of bacterial inclusion bodies. *J. Mol. Biol.* 2005, 347: 1025-1037.

[96] Soto C, Estrada L, Castilla J: Amyloids, prions and the inherent infectious nature of misfolded protein aggregates. *Trends Biochem. Sci.* 2006, 31: 150-155.

[97] Harper JD, Lansbury PT, Jr.: Models of amyloid seeding in Alzheimer's disease and scrapie: mechanistic truths and physiological consequences of the time-dependent solubility of amyloid proteins. *Annu. Rev. Biochem.* 1997, 66: 385-407.

[98] Dobson CM: Protein folding and misfolding. *Nature* 2003, 426: 884-890.

[99] Soto C, Estrada L, Castilla J: Amyloids, prions and the inherent infectious nature of misfolded protein aggregates. *Trends Biochem. Sci.* 2006, 31: 150-155.

[100] Soto C, Estrada LD: Protein misfolding and neurodegeneration. *Arch Neurol* 2008, 65: 184-189.

[101] Carrell RW, Lomas DA: Conformational disease. *Lancet* 1997, 350: 134-138.

[102] Bandekar J, KRIMM S: Normal mode of the spectrum of the parallel-chain beta-sheet. *Biopolymers* 1988, 27: 909-921.

[103] Cubarsi R, Carrio MM, Villaverde A: A mathematical approach to molecular organization and proteolytic disintegration of bacterial inclusion bodies. *Math. Med. Biol.* 2005, 22: 209-226.

[104] Cubarsi R, Carrio MM, Villaverde A: In situ proteolytic digestion of inclusion body polypeptides occurs as a cascade process. *Biochem. Biophys. Res. Commun.* 2001, 282: 436-441.

[105] Menart V., Gaberc-Porekar V., Jevsevar S.: Process For The Production of a Heterologous Protein. *Patent: PCT/EP2003/006134 (WO2004/015124)* 2004.

[106] Dobson CM: The structural basis of protein folding and its links with human disease. *Phil. Trans R. Soc. Lond B* 2001, 356: 133-145.

[107] Fornai F, Soldani P, Lazzeri G, di Poggio AB, Biagioni F, Fulceri F *et al.*: Neuronal inclusions in degenerative disorders Do they represent static features or a key to understand the dynamics of the disease? *Brain Res. Bull.* 2005, 65: 275-290.

[108] Soto C, Estrada LD: Protein misfolding and neurodegeneration. *Arch. Neurol.* 2008, 65: 184-189.

[109] Ravikumar B, Rubinsztein DC: Can autophagy protect against neurodegeneration caused by aggregate-prone proteins? *Neuroreport* 2004, 15: 2443-2445.

[110] Kegel KB, Kim M, Sapp E, McIntyre C, Castano JG, Aronin N *et al.*: Huntingtin expression stimulates endosomal-lysosomal activity, endosome tubulation, and autophagy. *J. Neurosci* .2000, 20: 7268-7278.

In: Ribosomal Proteins and Protein Engineering… ISBN 978-1-60741-005-8
Editors: V. Ortendhal and H. Salchow, pp. 21-39

Chapter 2

EXTRARIBOSOMAL FUNCTIONS OF RIBOSOMAL PROTEIN S19" IN MONOGRAPH RIBOSOMAL PROTEINS

Tetsuro Yamamoto[1] and Hiroshi Nishiura
Department of Molecular Pathology, Faculty of Medical and Pharmaceutical Sciences,
Kumamoto University Graduate School, Kumamoto, Japan

ABSTRACT

Some ribosomal proteins play roles in activities other than translation. Such functions of ribosomal proteins are called extraribosomal function or extraribosomal activity in common. Ribosomal protein S19 (RPS19) is a component of the small subunit of eukaryotic and archaeobacterial ribosomes having 145 amino acid residues. RPS19 exhibits various extraribosomal functions, which basically correlate to the apoptosis or programmed cell death. In the apoptosis-initiated cells, RPS19 molecules are cross-linked inter-molecularly with an isopeptide bond between Lys122 and Gln137 by a transglutaminase-catalyzed reaction. This results in homo-dimerization gaining a new molecular conformation capable of binding the C5a receptor, a G protein-coupled heptahelical cell surface receptor. The dimer is then released from the apoptotic cells. Meanwhile, the apoptosis-initiated cells *de novo* synthesize the C5a receptor. The C5a receptor was initially identified as a leukocyte chemotactic receptor for the complement-derived C5a. Due to a molecular mimicry between the RPS19 dimer and C5a, both molecules can ligate the C5a receptor exhibiting some difference between them. The RPS19 dimer has a potency to cause at least three cellular events through the C5a receptor ligation. One is apoptosis promotion by an autocrine/paracrine effect. On the other hand, the RPS19 dimer recruits monocytes/macrophages by ligating the C5a receptor on these leukocytes. The monocytes/macrophages thus infiltrated engulf the apoptotic cells and digest them. The macrophages then move to regional lymph nodes *via* lymphatics and present apoptotic cell-derived antigenic peptides to T lymphocytes. The acquired immune response may then start on demand. The apoptotic cell clearance is

[1] Correspondence: Tetsuro Yamamoto, MD, Department of Molecular Pathology, Faculty of Medical and Pharmaceutical Sciences, Kumamoto University Graduate School, 1-1-1 Honjo, Kumamoto 860-8556, Japan. Email: tetsu@gpo.kumamoto-u.ac.jp Fax: +81-96-373-5308 Phone: +81-96-373-5304.

basically an event outside the acute inflammatory response in which neutrophils play major roles. The RPS19 dimer inhibits neutrophil infiltration by ligating the neutrophil C5a receptor apparently as an antagonist. Other extraribosomal functions of the RPS19 dimer have been observed. There is a hypothesis that some elements of the apoptotic program would be used in terminal differentiation of erythroblasts when they lose their nuclei, and macrophages would be responsible for clearing the erythroblast-derived nuclei. The pro-apoptotic effect and macrophage recruiting effect of the RPS19 dimer seem to also function in the erythropoiesis. Interestingly enough, there is a molecule indistinguishable from RPS19 in plasma. The RPS19-like molecule is cross-linked by coagulation factor XIIIa, the plasma transglutaminase, in the blood coagulation process. The RPS19 dimer-like molecule seems play a role in resolution of coagula due to recruiting monocytes/macrophages.

INTRODUCTION

Ribosomal protein S19 (RPS19) is a component of the small (40S) subunit of the eukaryotic and archaebacterial ribosomes. It is absent in bacterial and mitochondrial ribosomes, and the prokaryotic RPS19 (S19p) is the homologue of eukaryotic RPS15 (S15e) [Lecompte et al., 2002]. RPS19 is a simple protein synthesized with 145 amino acid residues in which no cysteine residue is present. The NH_2-terminal Met is removed after translation by a methionine aminopeptidase, and the next Pro residue becomes the apparent first residue [Kondoh et al., 1992]. The primary structure of RPS19 is shown in Figure 1, and I number the amino acid residues of this protein from Met1 in this chapter. The crystal structure of RPS19 of *Pyrococcus abyssi*, an archaeobacterium, was recently solved.

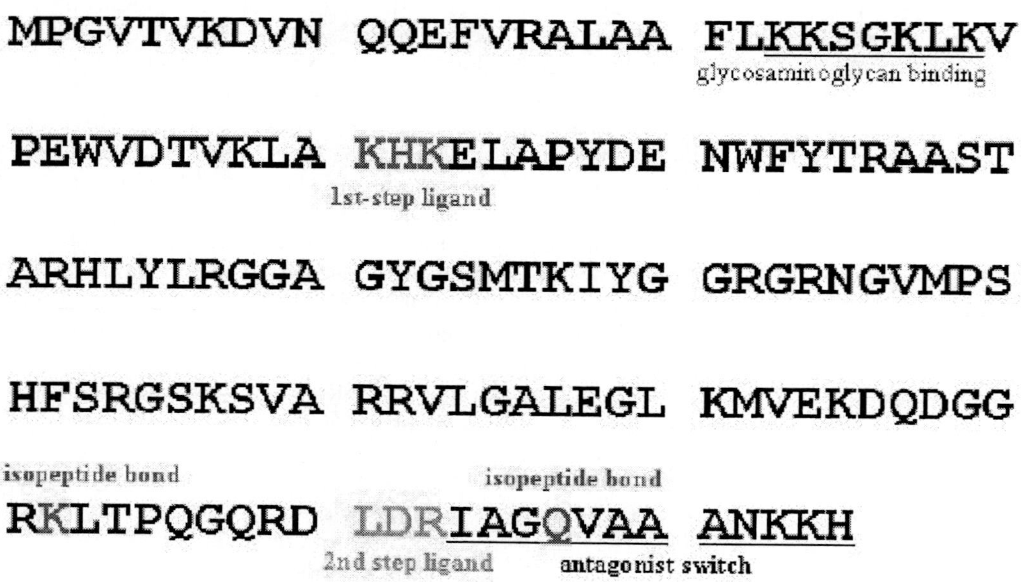

Figure 1. The primary structure of RPS 19.

The structure of RPS19 is almost entirely α-helical and folds a five α-helix bundle organized around a central amphipathic α-helix [Gregory et al., 2007]. *P. abyssi* RPS19 shares 36% identity and 57% similarity with human RPS19 in the amino acid sequence. One could assume that human RPS19 has a similar folding. RPS19 seems essential for the ribosome synthesis, because disruption of the RPS19 gene in human cells as well as yeast cells block maturation and production of the 40S ribosomal subunits [Flygare et al., 2007; Idol et al., 2007; Leger-Silvestre et al., 2005]. Role of RPS19 in the translation of mRNA by ribosomes seems unclear. A homozygous knockout experiment demonstrated that RPS19 was essential for early stage development of mice [Matsson et al., 2004]. The association of the RPS19 gene with Diamond-Blackfan anemia, which is an inherited disease characterized by erythroid aplasia, is known; this gene is found to be heterozygously mutated in 25% of the patients suffering from this anemia [Draptchinskaia et al., 1999; Willing et al., 1999]. The mechanical link between the RPS19 defect and the development of anemia is still unclear as discussed below.

Interestingly enough, RPS19 has functions and roles outside the translation. These are commonly called extraribosomal function or extraribosomal activity. I mainly describe extraribosomal functions of RPS19 in this chapter. Unfortunately, research on the extraribosomal function of RPS19 is apparently performed only in our laboratory. Therefore, the information on this subject is basically limited to our work, including working hypothesis and unpublished observation.

RPS19-LIKE MOLECULE IN PLASMA

Recently, we noticed the presence of a molecule indistinguishable from RPS19 in normal plasma. More than 20 years ago, we had reported a novel monocyte-selective chemoattracting factor, which was present in serum but not in plasma. The monocyte chemotactic factor was generated during clotting of blood or plasma depending upon the enzymatic activity of factor XIIIa, the plasma transglutaminase. At that time, we misunderstood that the origin of the chemotactic factor was complement component 5, C5, because the chemoattraction was caused through the C5a receptor and anti-C5 antibodies cross-reacted to the chemotactic factor. The chemotactic factor was distinguished from C5a, the authentic leukocyte chemotactic factor derived from C5 as described below, upon the monocyte-selectivity and the higher molecular size for several folds [Kukita et al., 1987; Okamoto et al., 1992]. Later we noticed that these characteristics were similar to those of another monocyte chemotactic factor separated from rheumatoid arthritis synovial lesion, which was identified to be a cross-linked homodimer of RPS19 as described below. Furthermore, the RPS19 dimer possesses antigen epitopes recognized by anti-C5a monoclonal antibodies [Nishiura et al., 1998]. Recently, we have revealed the precursor of the serum monocyte chemotactic factor was RPS19-like molecule present in plasma by means of immunochemistry and others. For example, when plasma was pretreated with anti-RPS19 antibody beads, the generation of the monocyte chemotactic factor usually occurred during the clotting together with platelets did not observed at all [Yamamoto et al., unpublished observation].

Although we do not know the accurate plasma concentration of the RPS19-like molecule, it seems to be around 100 ng/ml when judged from the chemotactic capacity of serum. Although we strongly assume that this molecule is RPS19 itself, we have no idea about the origin of RPS19 in plasma at this moment.

EXTRARIBOSOMAL FUNCTIONS

In 1990s, reports on the selective up- or down-regulations of the gene expression of several ribosomal proteins in association to cell growth and apoptosis or to tumor genesis were accumulated. At that opportunity, it was proposed that some ribosomal proteins might play roles in other than translation. Such expected functions were named extraribosomal function or extraribosomal activity in common [Naora and Naora, 1999]. However, many of the phenomena could not been clearly separated from the translation function of ribosomes. Whereas a few of them have been supported with evidence for their extraribosomal functions, they mainly relate to bacterial, fungal or plastid (chloroplast) ribosomal proteins [Wool, 1996, Trifa et al., 2000, Ma et al., 2004]. In mammalian ribosomal proteins, extraribosomal functions that have been proposed seem not to be substantial. Those of RPS19 are rare exception.

The extraribosomal functions of RPS19 are apparently various such as monocyte recruitment and apoptotic cell clearance, promotion of acquired immune response against apoptotic cells, acceleration of coagula resorption, inhibition of neutrophil infiltration, resolution of acute inflammation, acceleration of apoptosis, and promotion of erythrocyte maturation (Figure 2). However, all of the functions attribute to cross-linked homodimer of RPS19 that has gained binding capacity to the C5a receptor in association to its cross-linked dimerization.

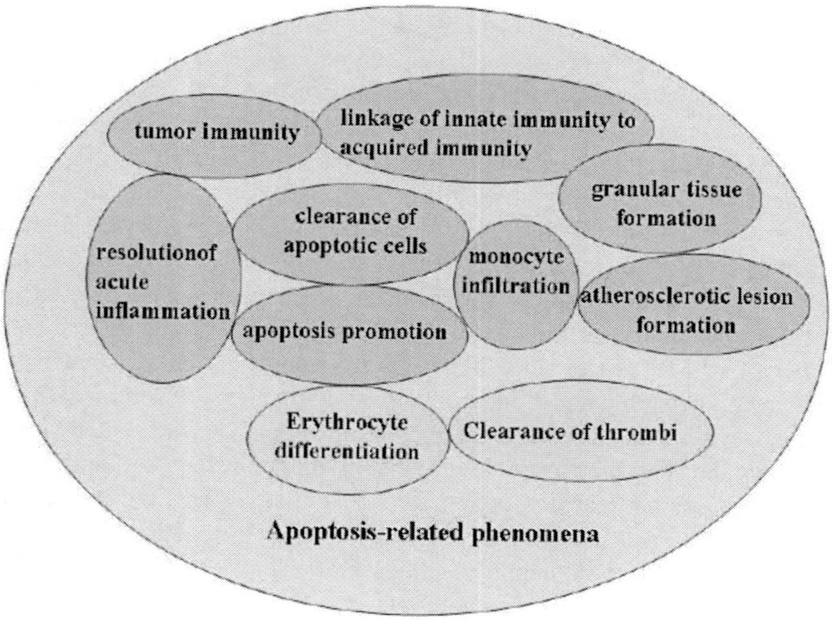

Figure 2. Bioogical phenomena involved by RP S19 dimer.

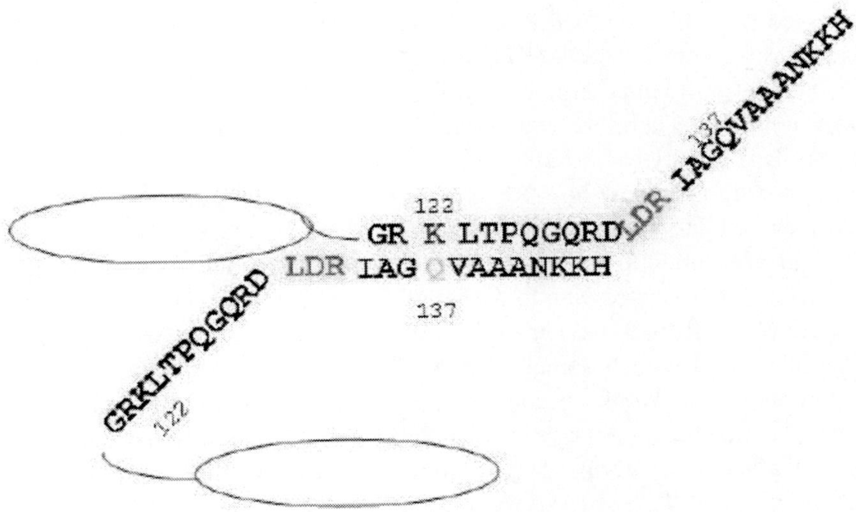

Figure 3. Schematic image of the cross-linked RP S19 homodimer.

Cross-Linked Dimerization of RPS19

The cross-linked RPS19 dimer was initially isolated from rheumatoid arthritis synovial lesion and later from atherosclerotic lesion of aorta as the major monocyte chemotactic factor [Nishiura et al., 1996, Shi et al., 2005]. As described above, RPS19 is composed of no Cys residue; therefore, the disulfide bridge is not involved in the dimerization. The homodimer is covalently maintained by an isopeptide-bond between Gln137 and Lys122. A schematic image of the cross-linked dimer is shown in Figure 3. The isopeptide-bond formation is catalyzed by mammalian transglutaminases such as type 2 transglutaminase and blood coagulation factor XIIIa. These transglutaminases are thiol enzymes with Ca^{2+} dependency. The enzyme forms a thioester-bond with Gln137 of a RPS19 molecule, liberating an NH_3 ion, at the first reaction step, then the enzyme returns free by nucleophilic attack of Lys122 in another RPS19 molecule resulting in the intermolecular isopeptide-bond formation between the Gln and Lys residues. This cross-linkage endows the homodimer with binding capacity to C5a receptor. More cross-linkages are sometimes formed among the RPS19 dimer resulting in homo-oligomer formation keeping the receptor-binding capacity. Instead, if the crosslinkage is formed intra-molecularly, it does not endow the cross-linked monomer with the receptor binding capacity. Even if we use a Gln137Asn mutant of RPS19 as the substrate, a RPS19 dimer is formed using another Gln residue but without the receptor-binding capacity.

1) Cross-Linked Dimerization of Rps19 in Apoptosis-Initiated Cells

During the apoptotic process the RPS19 dimer is intracellularly formed and extracellularly liberated around 24 hours after the apoptosis initiation regardless the initiation mechanism and cell-types [Horino et al., 1998, Nishimura et al., 2001, Nishiura et al., 2005,

Nishiura et al., 2008]. It is known that type 2 transglutaminase is a cytosolic enzyme ubiquitously expressed [Aeschlimann and Paulsson, 1994], and this enzyme is up-regulated during apoptosis to form apoptotic bodies [Knight et al., 1994, Cumings, 1996, Nishimura et al., 2001]. The functional up-regulation is caused by cytosolic Ca^{2+} influx, and quantitative one is by *de novo* synthesis. Therefore, we assume that the dimerization of RPS19 in the apoptosis-initiated cells is usually catalyzed by type 2 transglutaminase.

At this moment we do not know how RPS19 leaves the ribosome and how the RPS19 dimer is released extracellularly. Regarding the former, there is no evidence for *de novo* synthesis of RPS19 in association to apoptosis. It is known that the small subunit complex of ribosome is relatively labile in comparison to the large subunit complex. These let us assume that in addition to free form RPS19 molecules in equilibrium, those liberated from the small subunit due to degradation of the ribosomal RNA by RNase L and other RNases which are up-regulated during apoptosis would become the substrate of type 2 transglutaminase. Regarding the plasma membrane passage of the RPS19 dimer, we just speculate a co-transport with phosphatidylserine during the apoptotic plasma membrane flopping. Reasons for this speculation are that RPS19 possesses a sub-molecular high affinity region to negatively charged molecules and that in the dimerization of plasma RPS19-like molecule, phosphatidylserine on the activated platelets support the enzymatic reaction as scaffold, which is described subsequently.

2) Cross-Linked Dimerization of Rps19-Like Molecule in Plasma During Blood Coagulation

During the blood coagulation, the plasma RPS19-like molecule is dimerized by factor XIIIa, the plasma transglutaminase. This enzyme is present in plasma as a precursor form, factor XIII, which is a heterotetramer between the zymogen A subunits and the stabilizer B subunits. The A subunit is synthesized by megakaryocytes and by monocytes/macrophages; therefore, platelets also has this zymogen as the dimer of A subunit. Factor XIII (or factor XIIIA) is converted to the active form by thrombin due to a limited proteolysis. Before this activation, thrombin activates platelets due to a limited proteolysis of protease-activated receptor I. The platelet activation causes emergence of phosphatidylserine in the outer leaflet of plasma membrane. The protease cascade reaction in blood coagulation proceeds on the phosphatidylserine-rich membrane, and more thrombin molecules are generated from prothrombin. Thrombin thus generated converts fibrinogen to fibrin, liberating fibrinopeptides A and B; and the fibrin molecules polymerize together with fibronectin molecules. The factor XIIIa cross-links among the fibrin and fibronectin molecules in the polymers as the major substrates. Although we are not sure whether it occurs at meantime or subsequent to the fibrin cross-linkage, the RPS19-like molecule is also dimerized by factor XIIIa. However, we are almost sure that the dimerization reaction proceeds on phosphatidylserine molecules of the activated platelets, because the dimerization was blocked by the simultaneous presence of phosphatidylserine-binding molecules, and because the activated platelets could be substituted by phosphatidylserine- containing liposome but not by phosphatidylserine-free liposome [Yamamoto et al., unpublished observation].

In the artificial preparation process of the RPS19 dimer from a recombinant RPS19 using factor XIIIa, we noticed that the presence of heparin in the reaction mixture enhanced the generation of the dimer. We interpreted this as that the binding of RPS19 and factor XIIIa together to negatively charged heparin molecules made higher local concentrations both of the substrate and enzyme resulting in more reaction product, the RPS19 dimer in this case [Nishiura et al., 1999]. The basic cluster composed of Lys23, Lys24, Lys27 and Lys29 of RPS19 was revealed to be essential for the binding of this molecule to heparin. From this experience and the above experimental results, we speculate that the major role of the activated platelets in the generation of the RPS19 dimer-like molecule during the coagulation of platelet-rich plasma would be provision of scaffold for the enzymatic dimerization of RPS19 by factor XIIIa as in the case of heparin molecules (see Figure 6A).

Sub-Molecular Interaction Sites on RPS19 Dimer with C5a Receptor

The RPS19 dimer but not monomer possesses the capacity to ligate the C5a receptor [Nishiura et al., 1998]. The C5a receptor gene and the RPS19 gene were both mapped to the chromosome 19 q 13 [Gerard et al., 1993, Draptchinskaia et al., 1999]. The gene co-localization of them was also observed in the guinea pig [Umeda et al., 2004], although the biological significance of it is obscure.

The C5a receptor (CD88), which consists of 350 amino acid residues, is a heptahelical cell surface receptor coupled with heterotrimeric guanine nucleotide-binding proteins, or G proteins [Gerald and Gerald, 1991]. The C5a receptor bears an extracellular NH_2-terminal portion and an intracellular COOH-terminal site, and its transmembranous helical domains are connected by cytoplasmic loops I, II and III, and extracellular loops I, II and III. The C5a receptor would form a molecular cluster among them in a plasma membrane micro-domain, raft [Lee et al., 2003]. The G protein-coupled receptors are classified into two classes, A and B. Class A receptors bind ligands within their transmembrane core directly. Class B receptors have a two-step binding mechanism to their ligands: they initially bind ligands using their extracellular NH_2-terminal domain, and then bind them within their transmembrane core [Siciliano et al., 1994]. The second binding induces activation of the receptors. The activation of receptors in both classes A and B causes separation of the heterotrimer G protein complex into Gα and Gβγ molecules by facilitating the exchange of GDP in Gα to GTP. The separated G protein subunits raise intracellular signals to cause cellular responses including the chemotactic movement. Therefore, the active form of the receptors is a sort of guanine-nucleotide exchanger [Neubi and Siderovski, 2001].

As the naming is indicating, the C5a receptor was initially identified as the receptor of C5a, a chemoattractant/secretagogue to leukocytes. C5a, which is composed of 74 amino acid residues with 3 disulfide bonds, is liberated from the NH_2-terminal portion of α-subunit of C5 by C5 convertase. Therefore, C5a and RPS19 are totally different proteins with a calculated homology of 4%. However, in the tertiary structure there is a high similarity between them. C5a displays a pentahelical structure connected with loops [Zuiderweg et al., 1989, Zhang et al., 1997]. RPS19 expresses a similar hexahelical structure at least in the case of archaeobacterial RPS19 [Gregory et al., 2007]. Immunochemically, it was observed that two different monoclonal antibodies against C5a cross-reacted to the RPS19 dimer but not to the monomer form [Nishiura et al., 1998]. These suggest a molecular mimicry relationship between C5a and RPS19 dimer.

The C5a receptor is a class B receptor, having an acidic NH_2-terminal extracellular region composed of 37 amino acid residues including seven aspartate residues and two sulfated tyrosine residues. This portion binds a basic cluster with His15, Arg46 and Lys49 formed three dimensionally on the C5a molecule, or that containing a -Lys41-His42-Lys43- tandem sequence of the RPS19 dimer, at the first binding step. The second binding occurs between -Leu72-Gly73-Arg74-COOH of C5a, or the -Leu131-Asp132-Arg133- moiety of the RPS19 dimer [Shibuya et al., 2001], and transmembranous inter-helical regions of the receptor, probably Arg206 on helix V and Asp282 on helix VII [DeMartino et al.,1995, Cain et al., 2001]. The second binding site respectively locates at the COOH-terminal portion of C5a, and 12 residues upstream from the COOH-terminal of the RPS19 dimer (Figure 4, and see Figure 1). We assume the β-carboxyl group on Asp132 of RPS19 behaves as the α-carboxyl group on Arg74 of C5a [Shibuya et al., 2001]. These sites would represent a sort of molecular mimicry between C5a and the RPS19 dimer [Yamamoto, 2000].

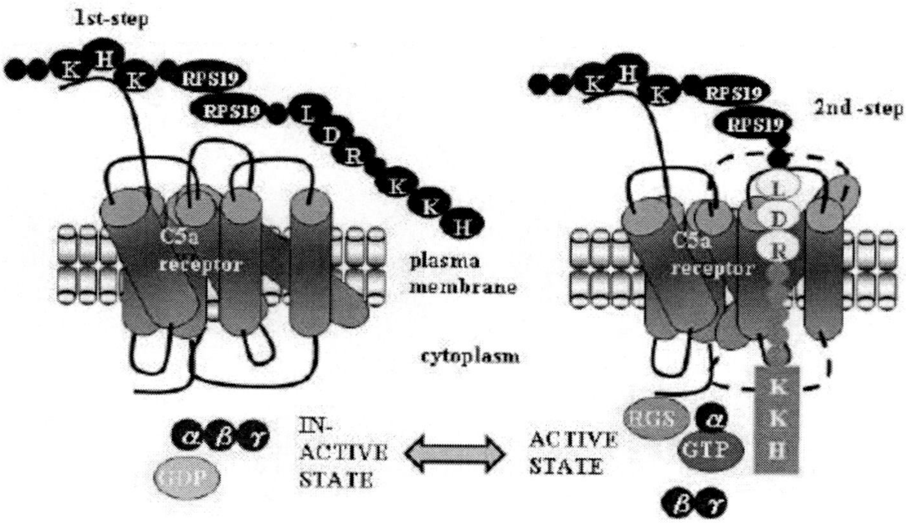

Figure 4. Two-step binding mechanism in C5a receptor activation by RPS 19 dimer.

Functional Difference between C5a and RPS19 Dimer, and C5a/Rps19 Chimera as Substitute Molecule of Rps19 Dimer

Whereas C5a and the RPS19 dimer equally ligate the C5a receptor, there is a significant difference in leukocyte preference; the former functions as a uniform C5a receptor agonist to both neutrophils and monocytes, however, the latter activates monocyte C5a receptor but apparently antagonizes the neutrophil receptor [Nishiura et al., 1998]. The COOH-terminal region of RPS19, $I_{134}AGQVAAANKKH_{145}$, functions as the molecular switch converting the agonist to the antagonist for the neutrophil C5a receptor [Shrestha et al., 2003] (see Figure 1).

We have currently prepared a recombinant C5a-chimeric protein bearing the C-terminus of RPS19 (C5a/RPS19 chimera) to be used as a substitute of the RPS19 dimer *in vitro* and *in vivo*. In this chimera, we changed Gly73 to Asp because of the reason described above. This chimera has perfectly reproduced the activities of the RPS19 dimer as far as we have examined [Oda et al., 2008].

The C5a receptor gene encodes C5a receptor in a two exon structure, with the 5' untranslated region and initiating codon in the first exon, and the remainder of the coding sequence and the 3' untranslated region in the second [Gerard et al., 1993]. This structure indicates the absence of alternative splice variants of the C5a receptor. On the other hand, the functional difference in response to the RPS19 dimer or C5a/RPS19 is present between the monocyte C5a receptor and the neutrophil one as described above. A similar difference is observed in the C5a receptor-related apoptosis regulation as described below. The mechanisms underlying the different functions and behaviors of the C5a receptor among cell kinds should be experimentally solved using different types of ligands. For this purpose, C5a/RPS19 must be useful, because it is not easy to prepare a large quantity of the RPS19 dimer; dimerization using transglutaminases *in vitro* results in more than 80 % of inactive RPS19 with an intra-molecular cross-linkage and only less than 20 % of the active RPS19 dimer with the inter-molecular cross-linkage.

Recruitment of Monocytes by RPS19 Dimer as Chemotactic Factor

As described above, the RPS19 dimer ligates the C5a receptor. The ligation of monocyte C5a receptor attracts the cell migration. The chemoattractant molecule with protein nature usually possesses a basic cluster region that binds sulfated glycosaminoglycans. The basic cluster is essential to exhibit the chemoattractant capacity *in vivo*; it provides a stable concentration gradient in the extravascular tissue space by binding to interstitial glycosaminoglycans, and an efficient presentation of the chemoattractant molecule to circulating leukocytes by binding endothelial surface glycosaminoglycans such as heparan sulfate. We assume that the basic cluster composed of Lys23, Lys24, Lys27 and Lys29 of the RPS19 dimer, which would have played the role to concentrate the monomer on phosphatidylserine to promote the intermolecular isopeptide bond formation, now owes the glycosaminoglycan-binding characteristics [Nishiura et al., 1998]. The RPS19 dimer rather inhibits the cell migration by ligating the neutrophil C5a receptor as an apparent receptor antagonist. The dual function on leukocyte chemoattraction of the RPS19 dimer, combining with the glycosaminoglycan-binding capacity, executes the monocyte-predominant infiltration *in vivo* [Nishiura et al., 1998].

1) Role in Clearance of Apoptotic Cells and Presentation of Apoptotic Cell-Derived Antigens to T Cell Repertoire

The RPS19 dimer and the C5a receptor of monocytes play roles in phagocytic clearance and immunologic monitoring of apoptotic cells without phlogistic stimulation (Yamamoto,

2000). As described above, the RPS19 dimer is commonly released from apoptotic cells. *In vivo* experiments, apoptotic HL-60 cells injected into the guinea pig skin or the rabbit one were rapidly cleared by infiltrating monocytes/macrophages in response to the RPS19 dimer liberated from the apoptotic cells [Horino et al., 1998]. During the apoptotic cell clearance, the inflammatory signs such as redness (hyperemia) and edema (plasma extravasation) were not obvious. It makes striking difference from the necrotic cell injection in which a typical acute inflammation is caused including the neutrophil infiltration [Tokita et al., 2004]. We have chased the macrophages that had engulfed the fluorescent-labeled apoptotic cells intradermally injected into footpads. The macrophages digesting engulfed apoptotic cells moved to the regional lymph nodes *via* lymph vessels and induced an acquired immune response by means of the presentation of the apoptotic cell-derived antigens to T cells [Shrethta et al., 1999].

We interpret this process from the self defense to be a coordinate work between the innate immune system and the acquired one against the viral infection and the neoplasm [Yamamoto, 2000]. Against the viral infection, we think that the apoptosis of infected cells is the first-line of host defense; the phagocytic digestion of the apoptotic cells including virus genome is the second-line; and the acquired immune response against the virus particles or the virus-infected cells is the third-line (Figure 5). In terms of the neoplasm, we have been studying the use of apoptotic cells as tumor vaccine. In mouse syngeneic tumor transplantation models using BALB/c mouse and Meth A fibroblastoma cell line and Colon-26 adenocarcinoma cell line, the apoptotic cell inoculations effectively induced cell line specific immunity [Yamamoto, 2007, Yamamoto et al., unpublished observation].

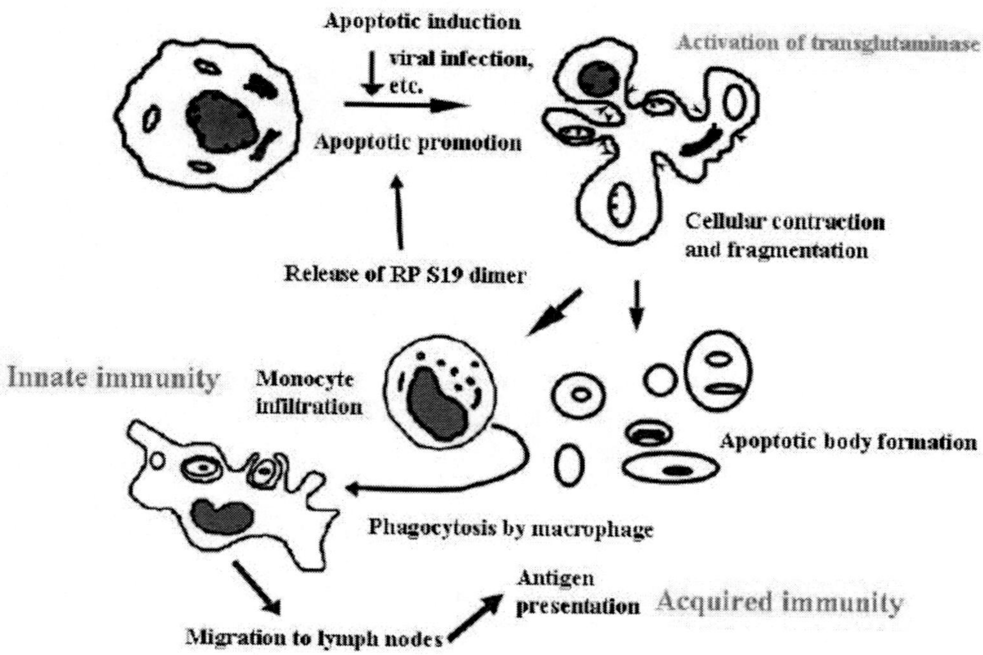

Figure 5. Association between innate and acquired immunities in apoptic cell clearance.

2) Role of Rps19 Dimer-like Molecule in Resolution of Blood Coagula

The first and second events in organization of intravascular blood coagula or thrombi are covering of the surface exposed to blood flow by monocytes and penetration of the coagula by monocytes/macrophages, respectively [Davies et al. 1975]. These indicate the generation of a chemoattractant that recruits monocytes in company with the blood coagulation. As described above, we have observed the generation of RPS19 dimer-like molecule during blood coagulation, therefore, we assumed that the RPS19 dimer would be the chemoattractant that recruits monocytes to thrombi (Figure 6A). We recently established an experimental model of blood coagula resorption in the guinea pig peritoneal cavity. In this model, the simultaneous presence of anti-RPS19 rabbit antibodies in the blood coagula significantly inhibited the surface covering by monocytes/macrophages and the resolution of the coagula intraperitoneally inserted (Figure 6B) [Yamamoto et al., unpublished data].

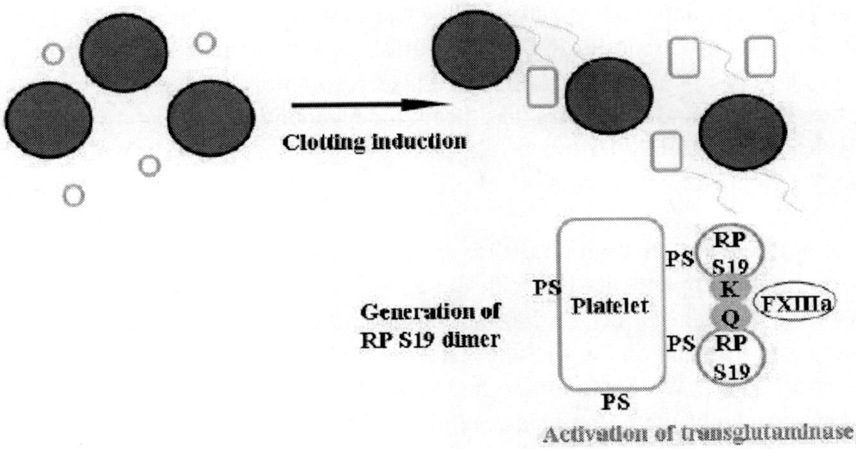

Figure 6A. Cross-linkage between PR S19 platelets.

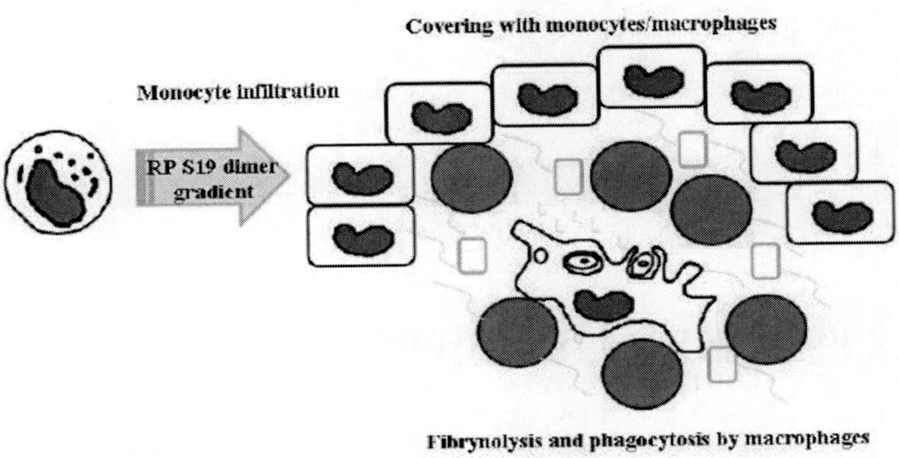

Figure 6B.Roles of Rp S19 dimer in coagular resorption.

PROMOTION OF APOPTOSIS TO APOPTOSIS-INITIATED CELLS

As described above, the RPS19 dimer is formed and extracellularly liberated during apoptosis. We initially thought that the RPS19 dimer only functioned to recruit monocytes/macrophages due to ligation of the monocyte C5a receptor. However, we recently found out another phenomenon mediated by the RPS19 dimer and C5a receptor in apoptosis. The RPS19 dimer accelerates apoptosis *via* the C5a receptor, even though the cells do not usually express the C5a receptor. Surprisingly, the C5a receptor is *de novo* synthesized by this cell type after the apoptosis initiation. It has been always reproduced as far as we have examined it in a mouse fibroblast cell line (NIH3T3), human leukemia cell lines (HL-60 and K562) and a human adenocarcinoma cell line (AsPC-1) under three kinds of apoptotic initiations such as hyperthermia, manganese (II)-loading, and thapsigargin-treatment [Nishiura et al., 2005, Nishiura et al., 2008]. Literature reports relevant to the C5a receptor gene provide support for our results. The promoter region of C5a receptor gene possesses a CCAAT sequence, which can be activated by the binding of nuclear factor Y [Hunt et al., 2005]. Nuclear factor-Y recruits RNA polymerase II and induces transcription of CCAAT box-containing genes [Kabe et al., 2005]. A close correlation between the nuclear factor-Y-CCAAT box system and apoptosis has been independently reported by several research groups [Bulter et al., 2002, Hirose et al., 2003, Hung et al., 2005]. We now assume the nuclear facto-Y-dependent expression of the C5a receptor gene in the apoptosis-initiated cells.

Interestingly enough, the length of life span of these apoptosis-initiated cells can change either shorter upon supplementation with the RPS19 dimer or longer upon supplementation with C5a *in vitro* [Nishiura et al., 2005]. The apoptotic process is sometimes divided into three phases such as the initiation, the effector and the execution [Eastman, 1999, and Gonzalez, et al, 2001]. We assume that the apoptosis-initiated cells collect information from their microenvironment using the C5a receptor and probably some other G protein-coupled heptahelical cell surface receptors in the effector phase to decide whether execution of apoptosis or survival. We recently found that regulator of G protein signaling (RGS) 3 was up-regulated by *de novo* synthesis and strongly promoted apoptosis upon the ligation of C5a receptor by the RPS19 dimer or C5a/RPS19 but not by C5a [Nishiura et al., 2008]. RGSs reduce heptahelical cell surface receptor-derived signals by accelerating the GTPase activity of Gα subunit resulted in reformation of Gαβγ complex loosing the signal transduction capacity. In a current view on the heptahelical cell surface receptor, the receptor molecules would swing between inactive conformation and active one, and therefore a proportion of the receptor molecules would raise intra-cellular signals, including the survival signal, even under the absence of ligation by agonist [Dragovich et al., 1998]. We assume that the up-regulated RGS3 would block such survival signals from heptahelical cell surface receptors.

RESOLUTION OF ACUTE INFLAMMATION DUE TO PROMOTION OF NEUTROPHIL APOPTOSIS BY RP S19 DIMER

It is well documented that neutrophils constitutively undergo apoptosis in senescence and that the neutrophil apoptosis and subsequent engulfment by monocytes/macrophages infiltrated later on is observed at the resolution stage of acute inflammation [Savill et al., 1989, Savill, 1997, Murai et al., 2003]. It is important to prevent undesired tissue destruction by infiltrated active neutrophils at the terminal stage of acute inflammation. However, the molecular mechanism that underlies the monocyte/macrophage infiltration to phagocytically clear the apoptotic neutrophils is poorly understood. Furthermore, it seems not clear whether any molecule accelerates neutrophil apoptosis *in situ* to promote the termination of inflammation, although several cytokines such as granulocyte-macrophage colony-stimulating factor and granulocyte colony-stimulating factor are known to extend lifespan of neutrophils and keep the inflammation [Simon, 2001].

To solve the above questions, we made a hypothesis that the RPS19 dimer is liberated from neutrophils undergoing apoptosis, then accelerates neutrophil apoptosis in a paracrine fashion, and recruits monocytes at meanwhile to clear the apoptotic neutrophils as in the cases of non-leukocytic apoptosis cell clearance as described above (see Figure 5). We have obtained supportive evidences so far that when HL-60 cells were differentiated into neutrophil-like cells, they became apoptosis; when HL-60 was transformed to produce Gln137Asn mutant RPS19, and then differentiated to neutrophil-like cells, in which generation of the functional RPS19 dimer was prevented, the spontaneous apoptosis of the cells was significantly retarded; in contrast to this, when HL-60 was transformed to over-produce wild type RPS19, and then differentiated to neutrophil-like cells, the spontaneous apoptosis was rather accelerated; when HL-60-derived neutrophil-like cells were injected into the guinea pig skin, the RPS19 over-producing cells were quickly cleared with a sufficient infiltration of phagocytic monocytes/macrophages; in contrast to this, the Gln137Asn-RPS19-producing cells retained and formed necrotic lesion with a few number of infiltrated monocytes/macrophages [Yamamoto, 2007, Nishiura et al., unpublished observation]. We are now examining the presence of the RPS19 dimer *in situ* at the resolution stage in an acute inflammation model.

Promotion of Erythrocyte Maturation by RPS19 Dimer

As described above, RPS19 was mutated in 25% of Diamond-Blackfan anemia patients. Recently, mutations of RPS24 and RPS17 were discovered in 1.4% and a single case, respectively, in Diamond-Blackfan anemia patients [Gazda et al., 2006, Cmejla et al., 2007]. The mutations in three different ribosomal proteins seem to support the hypothesis that the defects in ribosomal synthesis and subsequently impaired translation may be the main cause of pathogenesis of Diamond-Blackfan anemia. On the other hand, some other diseases accompanied by the defects in ribosomal synthesis such as Treacher Collins syndrome and anauxetic dysplasia do not show the bone marrow failure. Therefore, the opinion of Liu and Ellis [2006] sounds reasonable at this moment; it is not clear whether the defect in ribosome synthesis is the primary cause of the bone marrow failure or it is an enhancer of the disease severity by compounding the effects mediated through extraribosomal functions.

Apart from the relation to pathogenesis of Diamond-Blackfan anemia, we are studying a role of the RP S19 dimer in the erythrocyte differentiation. It is because of the presences of a hypothesis that some elements of the apoptotic program would be used in terminal differentiation of lens epithelial cells, erythroblasts and keratinocytes when they lose their nucleus and other organelles [Ishizaki et al., 1998], and of a report that DNase II in macrophages appears to be responsible for destroying the nuclear DNA expelled from erythroid precursor cells, and no live DNase II-null mice were born, owing to severe anemia [Kawane et al., 2001]. We have a working hypothesis that the RPS19 dimer would play two roles in the definitive erythropoiesis such as acceleration of erythroblast differentiation and recruitment of macrophages to form the hematopoietic islands and to clear the erythroblast-derived nuclei (Figure 7). We have observed the presence of a monocyte-selective chemotactic factor indistinguishable from the RPS19 dimer in erythropoietic bone marrow of guinea pigs [Yamamoto et al., unpublished observation], and an accelerating effect of C5a/RPS19 in erythrocyte differentiation of K562 cells under the presence of hemin *in vitro* [Nishiura et al., unpublished observation].

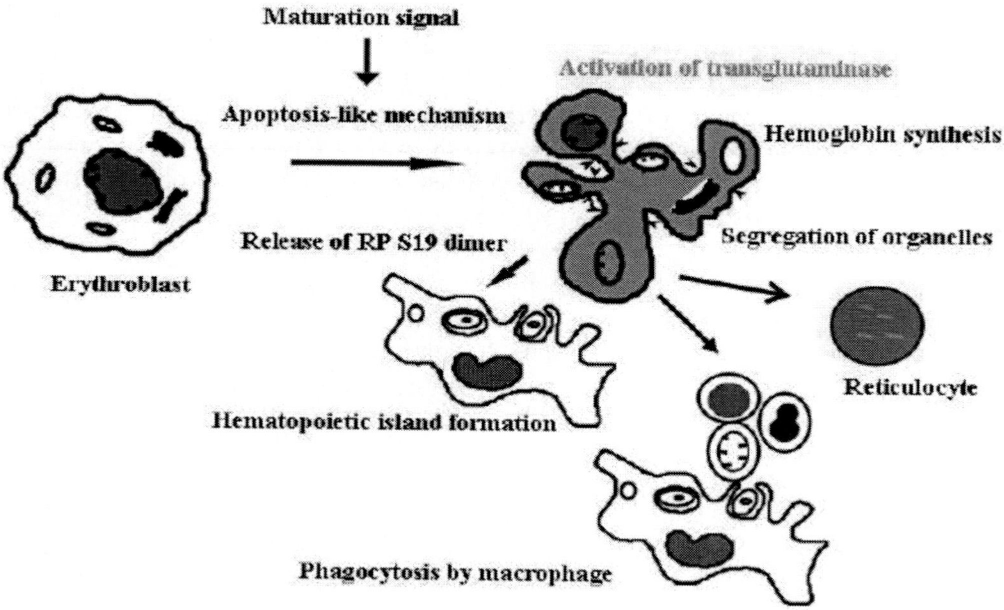

Figure 7. Apoptosis-like mechanism in erythropoiesis.

CONCLUSION

Ribosomal protein S19 (RPS19) has a potency to induce various functions outside the translation work of the ribosome. The extraribosomal functions are carried out by a cross-linked homodimer of RPS19 *via* ligation of the C5a receptor, a G protein-coupled heptahelical cell surface receptor. The extraribosomal functions relate to apoptosis or programmed cell death. Because one thousand billion (1,000,000,000,000) cells are in renewal every day and the apoptosis and apoptosis-related phenomena are deeply involved in the cell renewal, the extraribosomal functions of the RPS19 dimer must be critical to maintain our physiologic state including the host defense work.

REFERENCES

Aeschliman d; Paulsson M. Transglutaminases; protein cross-linking enzymes in tissue and body fluids. *Thrombos Haemostas* 1994 71, 402-415.

Butler LM; Zhou X; Xu WS; Scher HI; Rifkind RA; Marks PA; Richon VM. The histone deacetylase inhibitor SAHA arrests cancer cell growth, up-regulates thioredoxin-binding protein-2, and down-regulates thioredoxin. *Proc. Natl. Acad. Sci. USA* 2002 99, 11700-11705.

Cain SA; Coughlan T; Monk PN. Mapping the ligand-binding site on the C5a receptor: arginine[74] of C5a contacts aspartate[282] of the C5a receptor. *Biochemistry* 2001 40, 14047-52.

Cmejla R; Cmejlova J; Handrkova H; Petrak J; Pospisilova D. Ribosomal protein S17 gene (RPS17) is mutated in Diamond-Blackfan anemia. *Hum. Mutat.* 2007 28, 1178-1182.

Cumings M. Apoptosis of epithelial cells in vivo involves tissue transglutaminase upregulation. *J. Pathol.* 1996 179, 288-293.

Davies MJ; Ballantine SJ; Robertson WB; Woolf N. The ultrastructure of organizing experimental mural thrombi in the pig aorta. *J. Pathol.* 1975 117, 75–81.

DeMartino JA; Konteatis ZD; Siciliano SV; VanRiper G; Underwood DJ; Fisher PA; Springer MS. Arginine 206 of the C5a receptor is critical for ligand recognition and receptor activation by C-terminal hexapeptide analogs. *J. Biol. Chem.* 1995 270, 15966-15969.

Dragovich T; Rudin CM; Thompson CB. Signal transduction pathways that regulate cell survival and cell death. *Oncogene* 1998 17, 3207-3213.

Draptchinskaia N; Gustavsson P; Andersson B; Pettersson M; Willig T-N; Diazani I; Ball S; Tchernia G; Klar J; Matsson H; Tentler D; Mohandas N; Carlsson B; Dahl N. The gene encoding ribosomal protein S19 is mutated in Diamond-Blackfan anemia. *Nature genetics* 1999 21, 169-174.

Eastman A. The mechanism of action of cisplatin: From adducts to apoptosis. In Lippert B, editor. *Cisplatin. Chemistry and biochemistry of a leading anticancer drug.* Basel, Switzerland: Wiley-VCH; 1999; 111–134.

Flygare J; Aspesi A; Bailey JC; Miyake K; Caffrey JM; Karlsson S; Ellis SR. Human RPS19, the gene mutated in Diamond-Blackfan anemia, encodes a ribosomal protein required for the maturation of 40S ribosomal subunits. *Blood* 2007 109, 980–986.

Gazda HT; Grabowska A; Merida-Long LB; Latawiec E; Schneider HE; Lipton JM; Vlachos A; Atsidaftos E; Ball SE; Orfali KA; Niewiadomska E; Da Costa L; Tchernia G; Niemeyer C; Meerpohl JJ; Stahl J; Schratt G; Glader B; Backer K; Wong C; Nathan DG; Beggs AH; Sieff CA. Ribosomal protein S24 gene is mutated in Diamond-Blackfan anemia. *Am. J. Hum. Genet.* 2006 79, 1110–1118.

Gerard NP; Bao L; Xiao-Ping H; Eddy RL Jr; Shows TB; Gerard C. Human chemotaxis receptor genes cluster at 19q13.3-13.4. Characterization of the human C5a receptor gene. *Biochemistry* 1993 32, 1243-1250.

Gerard NP; Gerard C. The chemotactic receptor for human C5a anaphylatoxin. *Nature* 1991 349, 614-617.

Gonzalez VM; Fuertes MA; Alonso C; Perez JM. Is cisplatin-induced cell death always produced by apoptosis? *Mol. Pharmacol.* 2001 59, 657-663.

Gregory LA; Aguissa-Toure A-H; Pinaud N; Legrand P; Gleizes P-E; Fribourg S. Molecular basis of Diamond-Blackfan anemia: structure and function analysis of RPS19. *Nucl. Acid .Res.* 2007 35, 5913-5921.

Hirose T; Sowa Y; Takahashi S; Saito S; Yasuda C; Shindo N; Furuichi K; Sakai T. p53-independent induction of Gadd45 by histone deacetylase inhibitor: coordinate regulation by transcription factors Oct-1 and NF-Y. *Oncogene* 2003 22, 7762-7773.

Horino K; Nishiura H; Ohsako T; Shibuya Y; Hiraoka T; Kitamura N; Yamamoto T. A monocyte chemotactic factor, S19 ribosomal protein dimer, in phagocytic clearance of apoptotic cells. *Lab. Invest.* 1998 78, 603-617.

Huang W; Zhao S; Ammanamanchi S; Brattain M; Venkatasubbarao K; Freeman JW. Trichostatin A induces transforming growth factor beta type II receptor promoter activity and acetylation of Sp1 by recruitment of PCAF/p300 to a Sp1.NF-Y complex. *J. Biol. Chem.* 2005 280, 10047-10054.

Hunt JR; Martin CB; Martin BK. Transcriptional regulation of the murine C5a receptor gene: NF-Y is required for basal and LPS induced expression in macrophages and endothelial cells. *Mol. Immunol.* 2005 42, 1405-1415.

Idol RA; Robledo S; Du HY; Crimmins DL; Wilson DB; Ladenson JH; Bessler M; Mason PJ. Cells depleted for RPS19, a protein associated with Diamond Blackfan anemia, show defects in 18S ribosomal RNA synthesis and small ribosomal subunit production. *Blood Cells Mol. Dis.* 2007 39, 35–43.

Ishizaki Y; Jacobson MD; Raff MC. A role for caspases in lens fiber differentiation. *J. Cell Biol* .1998 140, 153-158.

Kabe Y; Yamada J; Uga H; Yamaguchi Y; Wada T; Handa H. NF-Y is essential for the recruitment of RNA polymerase II and inducible transcription of several CCAAT box-containing genes. *Mol. Cell Biol.* 2005 25, 512-522.

Kawane K; Fukuyama H; Kondoh G; Takeda J; Ohsawa Y; Uchiyama Y; Nagata S. Requirement of DNase II for definitive erythropoiesis in the mouse fetal liver. *Science* 2001 292, 1546-1549.

Knight CRL; Rees RC; Griffin M. Apoptosis: a potential role for cytosolic transglutaminase and its importance in tumor progression. *Biochim. Biophys. Acta* 1994 1096, 312-318.

Kondoh N; Schweinfest CW; Henderson KW; Papas TS. Differential expression of S19 ribosomal protein, laminin-binding protein, and human lymphocyte antigen class I messenger RNAs associated with colon carcinoma progression and differentiation. *Cancer Res.* 1992 52, 791-796.

Kukita I; Yamamoto T; Kawaguchi T; Kambara T. Fifth component of complement (C5)-derived high molecular weight macrophage chemotactic factor in normal guinea pig serum. *Inflammation* 1987 11, 459-479.

Lecompte O; Ripp R; Thierry JC; Moras D; Poch O. Comparative analysis of ribosomal proteins in complete genomes: an example of reductive evolution at the domain scale. *Nucl. Acid. Res.* 2002 30, 5382-5390.

Lee SP; O'Dowd BF; George SR. Homo- and hetero-oligomerization of G protein-coupled receptors. *Life Sci* .2003 74, 173-180.

Leger-Silvestre I; Caffrey JM; Dawaliby R; Alvarez-Arias DA; Gas N; Bertolone SJ; Gleizes PE; Ellis SR. Specific role for yeast homologs of the Diamond Blackfan anemia-associated Rps19 protein in ribosome synthesis. *J. Biol. Chem* .2005 280, 38177–38185.

Liu JM; Ellis SR; Ribosomes and marrow failure: coincidential association or molecular paradigm? *Blood* 2006 107, 4583-4588.

Ma Z; Dooner HK. A mutation in the nuclear-encoded plastid ribosomal protein S9 leads to early embryo lethality in maize. *Plant Journal.* 2004 37, 92-103.

Matsson H; Davey EJ; Draptchinskaia N; Hamaguchi I; Ooka A; Leveen P; Forsberg E; Karlsson S; Dahl N. Targeted disruption of the ribosomal protein S19 gene is lethal prior to implantation. *Mol. Cell Biol.* 2004 24, 4032–4037.

Murai N; Nagai K; Fujisawa H; Hatanaka K; Kawamura M; Harada Y. Concurrent evolution and resolution in an acute inflammatory model of rat carrageenin-induced pleurisy. *J. Leukoc. Biol* .2003 73, 456-463.

Naora H; Naora H. Involvement of ribosomal proteins in regulating cell growth and apoptosis: Translational modulation or recruitment for extraribosomal activity? *Immunol. Cell Biol.* 1999 77, 197-205,

Neubig RR; Siderovski DD. Regulators of G protein signaling as new central nervous system drug targets. *Natur. Rev.* 2002 1, 187-197.

Nishimura T; Horino K; Nishiura H; Shibuya Y; Hiraoka T; Tanase S; Yamamoto T. Apoptotic cells of an epithelial cell line, AsPC-1, release monocyte chemotactic S19 ribosomal protein dimer. *J. Biochem.* 2001 129, 445-454.

Nishiura H; Nonaka H; Revollo I; Semba U; Li Y; Irie A; Harada H; Kehrl JH; Yamamoto T. Ligand-dependent pro- and anti-apoptotic dual function of C5a receptor: Involvement of regulator of G protein signaling 3 and extracellular signal-regulated kinase. 2008 (submitted for publication)

Nishiura H; Shibuya Y; Matsubara S; Tanase S; Kambara T; Yamamoto T. Monocyte chemotactic factor in rheumatoid arthritis synovial tissue: Probably a cross-linked derivative of S19 ribosomal protein. *J. Biol. Chem.* 1996 271, 878-882.

Nishiura H; Shibuya Y; Yamamoto T. S19 ribosomal protein cross-linked dimer causes monocyte-predominant infiltration by means of molecular mimicry to complement C5a. *Lab. Invest.* 1998 78, 1615-1623.

Nishiura H; Tanase S; Shibuya Y; Futa N; Sakamoto T; Higginbottom A; Monk P; Zwirner J; Yamamoto T. S19 ribosomal protein dimer augments metal-induced apoptosis in a mouse fibroblastic cell line by ligation of the C5a receptor. *J .Cell Biochem.* 2005 94, 540-553.

Nishiura H; Tanase S; Sibuya Y; Nishimura T; Yamamoto T. Determination of the cross-linked residues in homo-dimerization of S19 ribosomal protein concomitant with exhibition of monocyte chemotactic activity. *Lab Invest* 1999 79, 915-923.

Oda Y; Tokita K; Ota Y; Li Y; Taniguchi K; Nishino N; Takagi K; Yamamoto T; Nishiura, H. Agonistic and antagonistic effects of C5a-chimera bearing S19 ribosomal protein Tail portion on the C5a receptor of monocytes and neutrophils, respectively. *J. Biochem.* 2008 (in press).

Okamoto M; Yamamoto T; Matsubara S; Kukita I; Takeya M; Miyauchi Y; Kambara T. Factor XIII-dependent generation of 5th complement component (C5)-derived monocyte chemotactic factor coinciding with plasma clotting. *Biochim. Biophys. Acta* 1992 1138, 53-61.

Savill J. Apoptosis in resolution of inflammation. *J. Leukoc. Biol.* 1997 61, 375-380.

Savill JS; Wyllie AH,; Henson JE; Walport MJ; Henson PM; Haslett C. Macrophage phagocytosis of aging neutrophils in inflammation; programmed cell death in the neutrophils leads to its recognission by macrophages. *J. Clin. Invest.* 1989 83, 865-875.

Shi L; Tsurusaki S; Futa N; Sakamoto T; Matsuda T; Nishino N; Kunitomo R; Kawasuji M; Tokita K; Yamamoto T. Monocyte chemotactic S19 ribosomal protein dimer in atherosclerotic vascular lesion. *Virchow Archiv.* 2005 447, 747-755.

Shibuya Y; Shiokawa M; Nishiura H; Nishimura T; Nishimura T; Nishino N; Okabe H; Takagi K; Yamamoto T. Identification of receptor binding sites of monocyte chemotactic S19 ribosomal protein dimer. *Am. J. Pathol.* 2001 159, 2293-2301.

Shrestha A; Horino K; Nishiura H; Yamamoto T. Acquired immune response as a consequence of the macrophage-dependent apoptotic cell clearance and role of the monocyte chemotactic S19 ribosomal protein dimer in this connection. *Lab. Invest.* 1999 79, 1629-1642.

Shrestha A; Shiokawa M; Nishimura T; Nishiura H; Tanaka Y; Nishino N; Shibuya Y; Yamamoto T. Switch moiety in agonist/antagonist dual effect of S19 ribosomal protein dimer on leukocyte chemotactic C5a receptor. *Am. J. Pathol.* 2003 162, 1381-1388.

Siciliano SJ; Rollins TE; DeMartino J; Konteatis ZD; Malkowitz L; Van Riper G; Bondy S; Rosen H; Springer MS. Two-site binding of C5a by its receptor: An alternative binding paradigm for G protein-coupled receptors. *Proc. Natl. Acad. Sci. USA* 1994 91, 1214-1218.

Simon H-U. Regulation of eosinophil and neutrophil apoptosis – similarities and differences. *Immunol. Rev.* 2001 179, 156-162.

Tokita K; Yamamoto T. Differential role of neutrophils and monocytes during subcutaneous plasma extravasation. *Lab Invest.* 2004 84, 1174-1184.

Trifa Y; Lerbs-Mache S. Extra-ribosomal function(s) of the plastid ribosomal protein L4 in the expression of ribosomal components in spinach. *Molecula General Genetics* 2000 263, 642-647.

Umeda Y; Shibuya Y; Semba U; Tokita K; Nishino N; Yamamoto T. Guinea pig S19 ribosomal protein as precursor of C5a receptor-directed monocyte-selective leukocyte chemotactic factor. *Inflamm. Res.* 2004 53, 623-630.

Willig TN; Draptchinskaia N; Dianzani I; Ball S; Niemeyer C; Ramenghi U; Orfali K; Gustavsson P; Garelli E; Brusco A; Tiemann C; Perignon JL; Bouchier C; Cicchiello L; Dahl N; Mohandas N; Tchernia G.. Mutations in ribosomal protein S19 gene and Diamond-Blackfan anemia: wide variations in phenotypic expression. *Blood* 1999 94, 4294–4306.

Wool IG. Extraribosomal functions of ribosomal proteins. *Trends Biochem. Sci.* 1996 21, 164-165.

Yamamoto T. Molecular mechanism of monocyte predominant infiltration in chronic inflammation: Mediation by a novel monocyte chemotactic factor, S19 ribosomal protein dimer. *Pathol. Internat.* 2000 50, 863-71.

Yamamoto T. Roles of the ribosomal protein S19 dimer and the C5a receptor in pathophysiological functions of phagocytic leukocytes. *Pathol. Internat.* 2007 57, 1–11.

Zhang X; Boyar W; Toth MW; Wennogle L; Gonnella N. Structural definition of the C5a C terminus by two-dimentional nuclear magnetic resonance spectroscopy. *Proteins* 1997 28, 261-267.

Zuiderweg ERP; Nettesheim DG; Mollison KW; Carter GW. Tertiary structure of human complement component C5a in solution from nuclear magnetic resonance data. *Biochemistry* 1989 28, 172-185.

In: Ribosomal Proteins and Protein Engineering...
Editors: V. Ortendhal and H. Salchow, pp. 41-54
ISBN 978-1-60741-005-8

Chapter 3

ORDER-DISORDER TRANSITIONS IN RIBOSOME ASSEMBLY

Youri Timsit*, Frédéric Allemand#, Claude Chiaruttini# and Mathias Springer#

*Information Génomique et Structurale CNRS - UPR2589
Parc Scientifique de Luminy
FR-13288 Marseille cedex 09 France
#Laboratoire de Biochimie, Institut de Biologie Physico-Chimique, CNRS[1]
13, rue Pierre et Marie Curie, 75005 Paris France

ABSTRACT

The recent crystallographic structures of the ribosomal subunits have revealed the picture of the final product of a complex assembly process that condenses the rRNA and the ribosomal proteins into active ribosomes. The folding of rRNA is the rate-limiting step of this complicate assembly pathway. Some ribosomal proteins are required for facilitating the structural rearrangements of rRNA and avoiding the kinetic traps that frequently impede RNA folding. It is thought that the long basic r-protein extensions that penetrate deeply into the subunit cores play a key role in this process through disorder-order transitions and/or co-folding mechanisms. A current view is that such structural transitions may facilitate the proper rRNA folding. In this paper, the structures of the free and bound forms of proteins that have been experimentally found to be essential for the first steps of ribosome assembly have been compared. It is shown that the extensions of L3, L4, L20 and L22, have different structural and dynamics properties that probably correlate with different functions. This study also suggests that the specific coil-helix transition that occurs in a phylogenetically conserved cluster of basic residues of the L20 extension is strictly required for the large subunit assembly in eubacteria. In contrast, in showing that the ordering of the inner loop of L4 upon rRNA binding is not required for the assembly function of L4, this study indicates that different categories of disorder-order transitions are associated with different biological functions.

---

[1] 13, rue Pierre et Marie Curie, 75005 Paris France.

INTRODUCTION

In addition to provide a considerable amount of structural data about the mechanisms of translation, the recent crystallographic structures of ribosomal particles (1-5) have brought to the scientific community a picture that could represent ”the origin of the world”. Similar to the Courbet’s painting, the spectacle of the most puzzling 3D labyrinth that nature has invented produces a kind of metaphysic vertigo that recurrently asks “how this so complex macromolecular structure assembles spontaneously in the cells?” Indeed, although it could seem a huge time relatively to the millisecond scale folding of a polypeptide chain into an active protein, the minute required for ribosome assembly *in vivo* is negligible relatively to the months or years that human brain needs for conceptualising its folding pathway. Perhaps the most fascinating issue of following the 23 S rRNA path within the 50 S large particle, is to realize that, at the end of a long travel along about 3000 nucleotides, and a visit of a large gallery of RNA motifs and folds, the exit is at close vicinity of the starting point. Adding the ribosomal proteins to this macromolecular daedal makes ribosomal assembly one of the most tricky problem of molecular biology. In contrast to other large molecular assemblies such as virus or flagels (6), ribosomal particles lack of an apparent symmetry. Symmetry greatly helps to rationalize how individual components are ordered into complex spatial arrangements. Also, while the shape of B-DNA double helix helps to organise its supramolecular assemblies into regular arrays (7), RNA double helices display a large conformational diversity that prevent a regular organisation. Fortunately, many biochemical, genetical and structural data may be put together for trying to understand how ribosomal proteins and rRNA co-fold into functional ribosomal particles. The purpose of the present paper is to try to elucidate how the extensions of the ribosomal proteins (L3, L4, L20 and L22) that are essential for the first steps of the eubacterial large subunit assembly, participate to the rRNA folding pathway.

RIBOSOME ASSEMBLY

The two bacterial subunits 30S and 50S assemble into a functional 70S particle that consists roughly two-thirds RNA and one third protein (8,9). The small subunit contains 16S RNA and 21 “S” ribosomal proteins. The large subunit 50 S is composed of two RNA, 23 S (2904 nt) and 5 S RNA (120 nt) and 33 proteins (10). All the information required for the *in vitro* assembly is contained within rRNA and ribosomal protein sequences. During the course of the assembly, a set of RNA conformational changes and protein binding leads to particles of increasing compactness. The process is highly interactive and the binding of many proteins depends on prior binding of other proteins. Similar basic principles have been found from the assembly of the 30S and 50S subunits. Primary binding proteins bind directly and independently to rRNA. Intermediate structures of increasing compactness have been identified *in vitro*. Since similar intermediates have been observed *in vivo*, it is thought that they reveal important aspects of RNA folding and protein binding during the assembly pathway. *In vivo*, the early assembly reactions already start with a small number of r-proteins shortly after the onset of rRNA synthesis. Thus, ribosomal assembly that is coupled with transcription only takes one minute. However, ribosome assembly *in vitro* takes several hours with the need of several steps of incubation at high temperature (11).

Assembly of 30 S that is simpler and faster than 50S is now well documented (8, 9, 12, 13). Incubation of 16 S RNA and a complete set of proteins at low temperature produces a 21 S reconstitution intermediate. It contains 16 S RNA and first and second binding proteins. Heating to 42°C induce a conformational change that result in a 26 S particle RI*. Addition the third binding proteins leads to the formation of the 30 S particle. The assembly landscape of the 30 S subunit proceeds through a global rate-limiting conformational change and traverses a landscape dotted with various local conformational transitions (12).

The assembly of the 50 S large particle is much more complex. An assembly map has been elaborated for the 50 S particle of *E. coli* ribosome (14,15). Three reconstitution intermediates have been found. $RI_{50}(1)$ 33 S, $RI_{50}*(1)$ 41 S, $RI_{50}(2)$ 48 S and 50 S. 22 proteins are incorporated into the first intermediate $RI_{50}(1)$. During the assembly gradient, five proteins essential for the early assembly reaction ($RI_{50}*(1)$) **bind exclusively near the 5'-end** of the 23S RNA. Among them, L4, L20, L22 and L24 that bind on the first rRNA domains **are essential. L24 that binds near the 5'-end and L3 that binds at the** 3'-end of the 23S RNA, are considered as initiator proteins since they bind independently to other r-proteins. The existence of two major protein assembly centres (L24 and L3) located at the ends of the 23S rRNA (I+II and V+VI) has been more recently confirmed by reconstitution experiments using separate transcripts of the six major structural domains of 23S RNA (16). This study indicates that the two centres assemble independently of each other around protein L24 and L3. Then five primary binding proteins L3, L4, L20, L22 and L24 play an essential role on the assembly of the first reconstitution intermediate. While post-transcriptional modifications of the 23 S RNA are not essential for the assembly (17), a recent work has pointed the importance of transient structures during post-transcriptional refolding of the pre-23S S RNA for the large subunit assembly (18).

The three dimensional structures of the ribosomal particles (1-5) are in good agreement with these biochemical data. A detailed analysis of the proteins of the large subunit of *H. marismortui* has provided interesting structural insights on 23S assembly. (i) Larger protein/RNA interface seems to correlate with proteins that bind early in the course of assembly. (ii) Component buried in the interior must bind the assembling ribosome earlier.

Do Ribosomal Protein Extensions Play a Role in Subunit Assembly?

One of the most surprising features of ribosomal proteins found in the crystal structures of large and small ribosome subunits is the finding that almost half of the proteins have globular bodies with extensions (1). Globular domains are found on the particle's exterior while the extensions penetrate deeply into the subunit core and are intertwined with rRNA helices (Figure 1). These extensions often lack obvious tertiary structure and in many regions are also devoid of significant secondary structure. As a consequence, most of the proteins that contain extensions do not crystallise at the free state. When their crystallisation is possible, the extensions are generally not visible in the electron density map since they are disordered. Although the biological role of the extensions is still unclear, it has been postulated that they

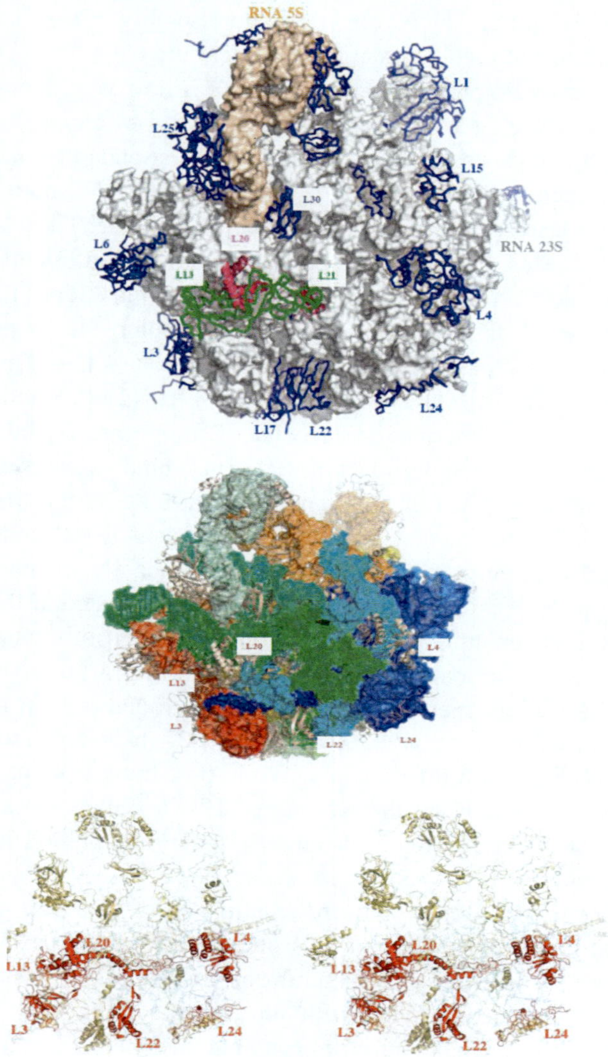

Figure 1.

a) The large subunit of *T. thermophilus* (4). The ribosomal proteins are depicted with blue ribbons, the 23 S RNA is represented in grey and the 16 S RNA is represented in pale pink. L20 protein is represented with pink ribbons; its neighbours L13 and L21 are depicted in green.

b) Same view of the large subunit with a colour scheme that puts in light the path and the domains of the 23 S RNA. **23 S RNA is coloured from blue to red from the 5'-end to the 3'-end.**

c) Stereo view of the large subunit showing the distribution of the six ribosomal proteins (L3, L4, L13, L20, L22 and L24) essential for the formation of the first reconstitution intermediate *in vitro*.

could participate to ribosome assembly on the basis of the crystal structures of the ribosome subunits. The detailed analysis of the ribosomal proteins of the large particle of *H. marismortui* has brought many structural insights that support this hypothesis (19). First, the extensions are basic and flexible, a property that make them candidate for assembling RNA segment during rRNA folding.

In both subunits, these extensions have a distinctive amino acid composition and they differ from the globular domains mainly in glycine (13.7 % vs. 8%), arginine (15,9 % vs. 7.5 %) and lysine (12,7 % vs. 5.1 %) (19,20). The basic nature of the extensions enables them to neutralize the highly negatively charged RNA backbone. The higher glycine contain is supposed to increase their flexibility and to avoid steric clashes in tightly packed RNA regions (19). Second, it has been noted that extensions that represent only 18 % of the proteins are responsible for 44 % of the total RNA surface buried by protein interaction. Because they make many contacts with rRNA and often interact with more than one domain of the RNA, it is thought that one role might be the stabilization of the proper RNA tertiary structure. Third, the finding of extensions in proteins essential (L3, L4, L22 and L20) for the formation of the first intermediate RI_{50} (1) in *in vitro* (14,15) reconstitution experiments has suggested that they may participate to ribosome assembly. Another possible role of the absence of secondary structure in the extensions is that it could allow interactions in major grooves of RNA double helices that are not wide enough to accommodate larger elements of protein secondary structures such as α-helices (19, 20).

This hypothesis fits well with current views on protein/RNA interactions in which induced fit or co-folding are required for the assembly (21,22). Many other examples of order-disorder transitions have been observed in protein/RNA interactions and growing evidences have shown that intrinsically unstructured proteins (IUPs) participate to many assembly and regulation functions (23-26). Following this view, co-folding or disorder/order transition in r-proteins extensions would help to avoid the kinetic traps that frequently impede the correct RNA folding during the course of ribosome assembly. The examination of subunit crystal structures also suggests that the globular domains of the assembly proteins bind first to rRNA (19, 20). Then, the extensions would bind additional segments in different domains, thus contributing to approach and seal distant rRNA regions. However, steric considerations require that proteins that contain extensions bind 23 S RNA at a stage prior to the formation of significant tertiary structure. Otherwise, the extensions would not have access to their binding sites. Therefore, the binding of extensions must occur before final assembly of the surrounding parts of the subunit.

However, a deeper analysis suggests that the picture may be more complex. Recent genetic, biochemical and structural data have indeed shown that r-protein extensions are not systematically involved in the subunit assembly. First, the observation that many proteins that possess extensions are not essential for ribosome assembly indicates that they are not strictly correlated with an assembly function. It is also important to note that in 30S subunit, none of the primary binding proteins has the extended basic tails. Rather, they appear to be typical globular proteins (20). Moreover, in the 50S particle, although the assembly initiator protein L24 is devoid of secondary and tertiary structure, it does not have an extension that penetrates in the ribosome core. L24 is bound at the ribosome surface similar to the other globular domains of other ribosomal proteins.

What about proteins essential for ribosome assembly? The effect of the deletion of the extensions of L4 and L22, two primary binding proteins that are essential for the 50 S subunit assembly has been tested *in vivo* (27). Both proteins bind initially to domain I and are essential for the formation of the first reconstitution intermediate RI_{50}*(1). Surprisingly, this study has shown that the extended loops of L4 and L22 are not only dispensable for assembly into 50 S ribosomal particle but also for the proper assembly of proteins that bind later in 50 S assembly pathway (27). These experiments provide a clear demonstration that the globular domains of these two proteins are sufficient to initiate the assembly of the large 50 S particles. In consequence, this finding does not support the general concept that extensions of ribosomal proteins play a role in ribosome assembly. Another study has also shown that C-terminal tails of S9 and S13 are not essential for ribosome functions (28). However, these two proteins are not essential for the early steps of the 30 S subunit assembly.

Among the proteins that are essential for the large subunit assembly, L20 is, to our knowledge, the sole example in which the extension is strictly required for correct folding of the 23 S RNA into an active ribosome subunit. L20 that is one of the most basic proteins of the eubacteria is a primary binding protein that belongs to the five proteins essential for first reconstitution steps *in vitro* (29,30). L20 can also replace the assembly initiator protein L24 for the initiation of assembly at permissive temperature (31). L20 has been also shown to be essential *in vivo*, as a deletion within its gene is lethal (32). Moreover, deletion experiments

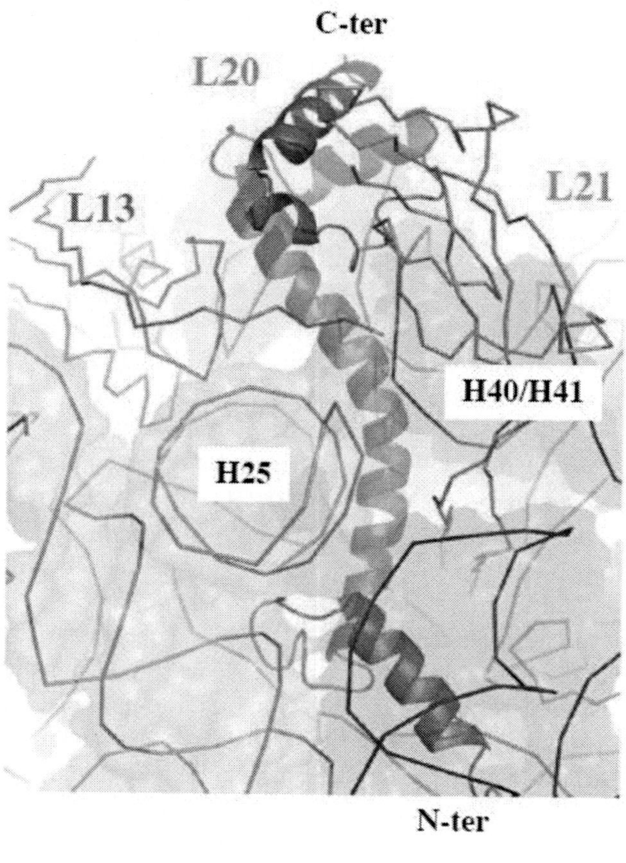

Figure 2. The path of L20 within the core of the large subunit of *T. thermophilus* (4) (as a representative structure for eubacteria).

have shown that the N-ter extension is strictly required for normal assembly (32). Within the eubacterial 50 S subunit, L20 is bound at the interface of two RNA domains and interacts with the helix H40/41 on one side and the helix H25 on the other side (Figure 2). Its globular C-ter domain interacts with L21 and L13 at the surface of the ribosome (Figure 1). Its spectacularly long α-helical extension penetrates deeply into the subunit core and seals the approach of two domains of the 23 S RNA (Figure 1c, 2). Biochemical data have shown that L20 extension is required for the assembly *in vivo* of the large ribosome subunit. What does distinguish the extension of L20 from the ones of L4 and L22 that could explain its specific function in ribosome assembly?

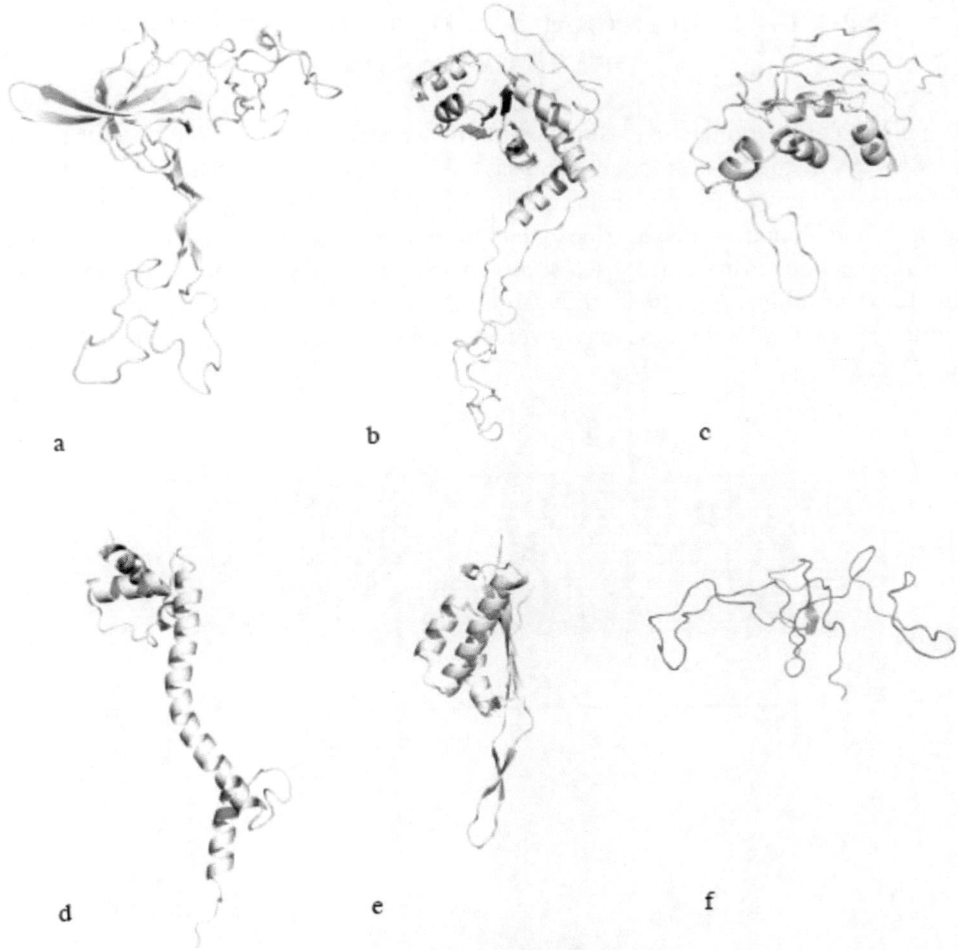

a b c

d e f

Figure 3. Comparison of the six proteins that are essential for the early steps of the assembly of the 50 S subunit of the eubacterial ribosome. a. L3. b. L4. c. L13. d. L20. e. L22. f. L24.

In contrast to the inner loops of L24 and L22, the extension of L20 (60 aas) is structured into a spectacularly long α-helix α2 and a smaller N-terminal α1 helix that penetrate deeply into the ribosome core (Figure 2). As shown in figure 3, the extensions of L4 and L22 adopt a different secondary structure within the ribosome. Although it is ordered through its interaction with the rRNA, the inner loop of L4 (68 aas, arg45-lys103) is totally devoid of secondary structure (Figure 1c, 3). The inner loop of L22 (23 aas, glu78-ser101) is smaller and is folded into a β-hairpin. Interestingly, the extension of L3, another protein essential for the formation of the first reconstitution intermediate, is similar to that of L4 and consists of a loop devoid of secondary structure. However, to our knowledge, deletion experiments have not been yet performed to test the role of L3 extension in ribosome assembly. This analysis suggests that the specific role of L20 extension may be related to its particular structural and dynamics properties. Another difference between these four proteins is the structure of the globular domain. L20 is unique in having a globular domain entirely made of α-helices.

Special Structural and Dynamics Properties of L20 Extensions

The picture of the protein extensions within the crystal structure of the ribosome particles provides a view of the final product of the assembly. These data are therefore insufficient to have a complete view of the molecular mechanisms of rRNA folding. Thus, the comparison of the free and bound forms of ribosomal proteins may provide useful insights on the molecular events occurring during rRNA-protein binding. They can help to understand how induced fit or co-folding with their RNA target may assist the subunit assembly. Indeed, key steps of the rRNA folding process may involve structural rearrangements upon rRNA/protein binding (21,22).

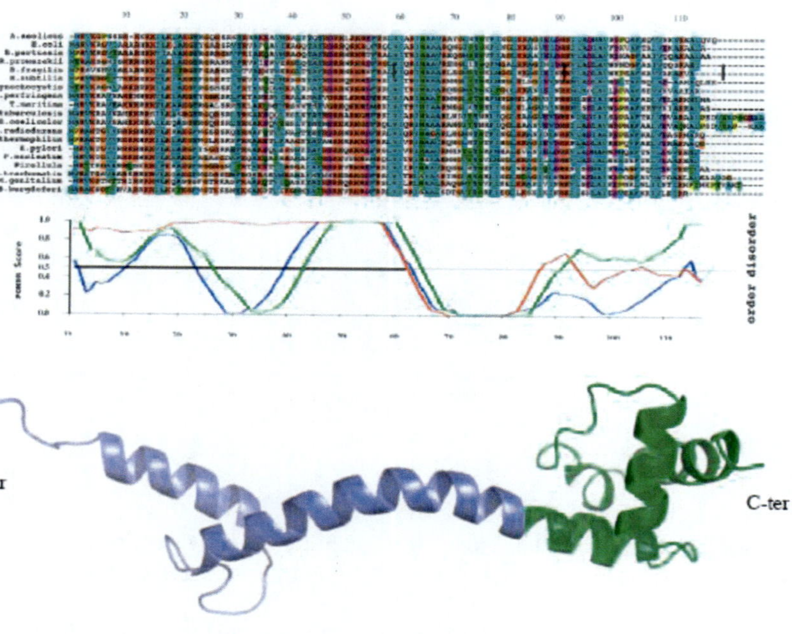

Figure 4. Sequence alignment of the ribosomal protein L20 of representative species of eubacteria and disorder prediction calculated with the program PondR for three species. Red: *A. aeolicus*. Blue; *E. coli*. Green: *D. radiodurans*.

Crystal structures of the free forms are available for four of the proteins (L4, L13, L22 and L20) essential for the early steps of assembly. The crystallographic structure of L4 shows that its inner loop is totally disordered in the crystal. Thus, the comparison of the free and bound forms suggests that a co-folding mechanism takes place during its binding on its rRNA target (33). However, deletion experiments have shown that the L4 extension is not required for the proper ribosome assembly (see above, 27). It can be therefore deduced that the structural transition that orders the L4 inner loop within the large particle does not play an essential role in the 23 S rRNA folding. On the other hand, the β-hairpin extension of L22 adopts a similar conformation in the free and bound states (34). However, here also, the extension is not required for the assembly. This provides another example in which an extension that is structured into a defined secondary structure in the free and the bound state is not required for assembly.

An NMR study has proposed that the extension of L20 is totally unstructured in the absence of rRNA (35). Disorder prediction programs such as PondR (36) also show that the extension displays many unstructured regions (Figure 4).

However, the crystal structure of the free L20 of *Aquifex aeolicus* has revealed a unique picture of the dynamic of the protein since two folding states of the protein have been trapped into the unit cell (37). This structure has shown for the first time the coexistence of a folded and a partially unfolded form of a protein in a crystal structure. Having them together in identical physico-chemical conditions allowed an unbiased structural comparison of two folding states of a protein. In the partially unfolded form, both C-ter globular domain and the N-ter extension display unstructured regions (Figure 5). Also, a region of the extension that corresponds to a cluster of phylogenetically basic residues (arg 48 – arg 57) has undergone a helix-coil transition. In the corresponding region of the folded form, the side chains of the basic amino acids point on the same side of the α-helix along three helix turns (Figure 5). This has indicated that the electrostatic repulsion between the positively charged side chains is responsible for the instability of this helical segment. Indeed, the spatial arrangement of the side chains along the same side of three α-helix turns generates a very high density of positive charges.

A similar organisation of negatively charged residues is found in the α-helical linker of calmodulin (38) (Figure 5). In calmodulin, this region is also characterized by a high instability that confers to the protein the ability to fit to many different binding partners (38). Thus, this study has shown L20 is specifically unfolded in an evolutionary conserved cluster of basic amino acids. It seems likely that the particular distribution of the charged residues along the α-helix is responsible for its higher flexibility that may have a functional role in ribosomal assembly. In addition, the comparison of the two forms has revealed that a switch mechanism allows a structural communication between the C-terminal domain and the N-terminal extension (Figure 6). Two transient salt bridges that involve two conserved basic amino acids (arg 90 and lys 91) at the interface of the two domains "stabilize" the partially unfolded form. Their formation seem to be related with the partial unfolding of the extension. The disruption of these salt bridges upon complete protein folding generates a dramatic reorganisation of the surface electrostatic potential (Figure 6 and 7).

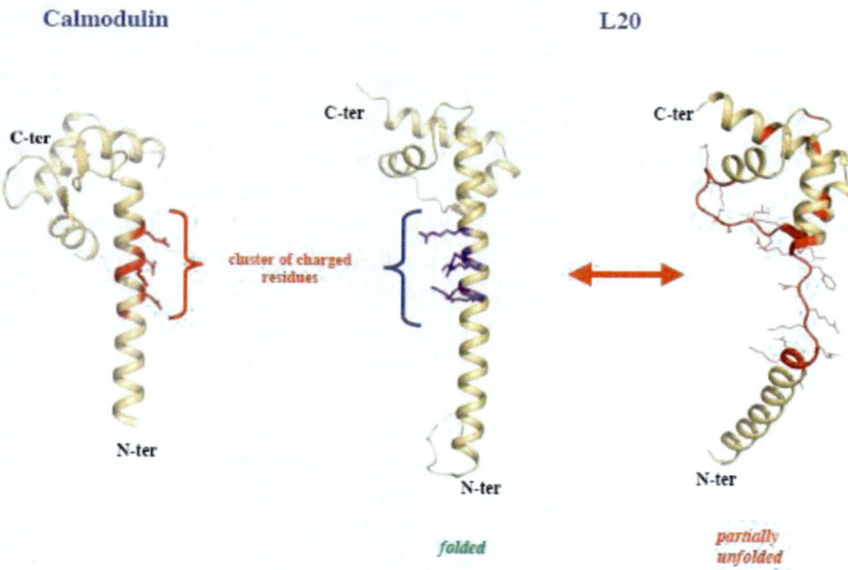

Figure 5. Comparison of the folded (middle) and the partially unfolded (right) forms of L20 in the crystal structure of the free L20 (37). The basic residues that point on the same side of the helix α2 of the folded form are depicted in purple. The calmodulin (left) display a similar distribution of negatively charged residues (red) along its α-helical linker.

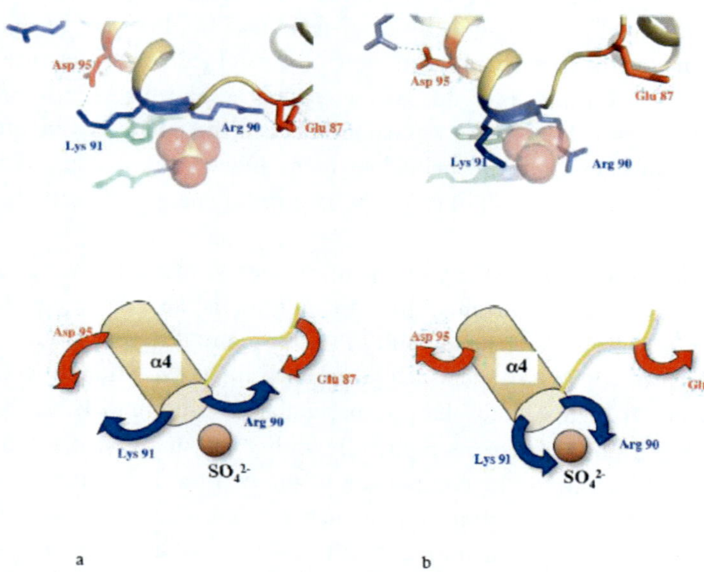

Figure 6. Structural rearrangements at the interface of the globular domain and the extension of L20 of *A. aeolicus*. a. The partially unfolded form is stabilised by the formation of two salt bridges that link two evolutionary conserved basic residues arg 90 and lys 91 to asp 95 and glu 87 located within the globular domain. b. In the folded form, the two salt bridges are disrupted and the side chains of arg 90 and lys 91 embrace a sulfate ion. The sulfate ion occupy an position identical to a phosphate group of the H40/41 junction within the 50 S subunit of the eubacterial ribosome.

Thus, the two forms have different binding properties and affinities for RNA. It has been speculated that the distinct structural and dynamics features of the two folding states play defined roles during the different steps of the folding pathway of the 23S RNA (37). Why is the coil helix transition that occurs in L20 extension essential for the early steps of 23 S RNA folding? Several hypotheses may be proposed.

i) The specific coil helix transition in L20 extension may help to avoid a kinetic trap during the folding of the helix H40/41 and H25. Alternatively, the co-folding process would stabilize a state of higher energy RNA binding site. This idea is supported by a recent NMR study that shows that the rRNA target of L20 adopts a different structure in the free and bound form (39). Thus, L20/RNA co-folding process would lower the energy required for the structural rearrangement of the RNA site required for the subsequent steps of 23 S RNA folding. Indeed, it is now admitted that the rate limiting steps for RNA folding pathways are RNA conformational changes
ii) Another possibility could be that the coil helix transition would be required for bringing distant rRNA segments into close proximity during the course of assembly (Figure 7). Indeed, at the beginning of the folding process rRNA is more flexible and less compact. Such a fishing mechanism has been proposed in endocytosis (37,40). Indeed, the size of an unstructured peptidic segment that is longer than an α-helical one can interact with more distant partners.
iii) The transient unfolding of the long α-helix may be required for fitting into the extremely narrow groove of the H40/41 helix junction? The resulting structure would shield the electrostatic repulsion between the phosphate groups that delineate the extremely narrow groove of the L20 binding site of the H40/41 junction

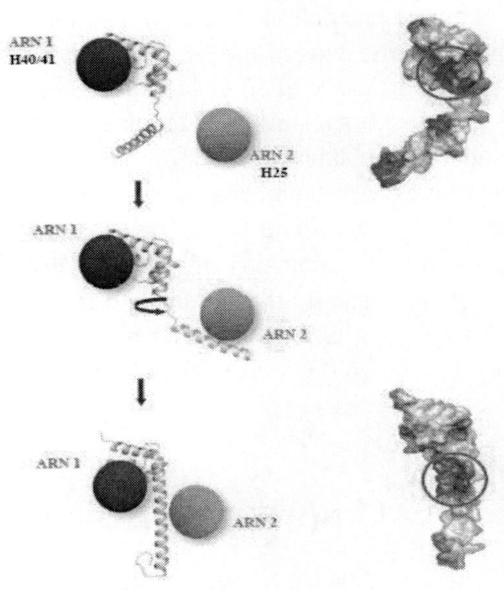

Figure 7. Fishing mechanism that can help to bring into proximity RNA segments coming from different rRNA domains. The two forms of L20 display very different electrostatic surface potential and affinity for RNA (top: partially unfolded, bottom: folded).

Conclusion

The ribosomal proteins L3, L4, L13, L20 and L22 that are essential for the early steps of the large subunit assembly have extensions with different structural and dynamics properties. Available biochemical and structural data suggest that the coil-helix transition observed in the long basic extension of L20 plays a specific and essential role in the 23 S RNA folding. This helix coil transition occurs in a discrete region whose flexibility is specifically tuned by phylogenetically conserved distribution of basic residues. This reveals that well-defined regions of structural "disorder" in protein may have specific functions. A similar helix-coil transition that occurs in the linker of calmodulin plays an essential role in the recognition of its multiple structurally distinct binding partners (38).

In contrast, the co-folding process that leads from a disordered loop to an ordered one upon L4/rRNA association seems to be not essential for ribosome assembly. It is interesting to suggest that such kind of transition may have a less specific role than coil-helix transitions during rRNA folding. One could imagine that the ordering of a disordered loop may be required for shielding the negatively charged RNA backbone. Thus, the deletion of this loop could be compensated by the binding of cations. It would be interesting to know what would be the phenotype associated with the deletion of the L3 extension. A prediction would be that the deletion of L3 extension that shares structural similarities with L4 would not affect ribosome assembly.

Open Questions

A question that seems until today unsolved is why L20 that is an essential protein in eubacteria is not present in archea and eukaryotes. How do archea and eukaryotes compensate the absence of L20 during the early steps of the large subunit assembly? A careful structural comparison of the large subunit structure of *H. marismortui* and the ones of eubacteria would provide some structural insights. Interestingly, H25, one of RNA binding site of L20, is longer in archea. It would be then interesting to introduce the missing RNA part within eubacterial ribosome that relieve the essential character of L20.

Another question about L20 is who bind first? The extension or the globular domain? A current view is that that the globular domains of r-proteins bind first to rRNA. However, knowing that the subunit assembly is co-transcriptionally dictated by the rRNA synthesis, it is possible that in the case of L20, the extension would bind first. Indeed, the binding site of L20 extension is the helix H25 which is well in 5' relatively to the H40/41 helix junction recognized by the globular domain.

Acknowledgments

This work was supported by the CNRS (UPR9080 and UPR9073), the University of Paris VII, and the ANR (ANR-05-BLAN-0159-01).

REFERENCES

[1] Ban, N., Nissen, P., Hansen, J., Moore, P., Steitz, T.A. (2000) *Science* 289, 905-920

[2] Wimberly, B.T., Brodersen, D.E., Clemons, W., Morgan-Warren, R.J., Carter, A.P., Vonrhein, C., Hartschk, T. and Ramakrishnan, V. (2000) *Nature* 407, 327-339

[3] Harms et al., (2001) *Cell* 107, 679-688

[4] Selmer, M., Dunham, F., Weixlbaumer, A., Petry, S., Kelley, A.C., Weir, J.R and Ramakrishnan, V. (2006) *Science* 313, 1935

[5] Schuwirth, B.S., Borovinskaya, M.A., Hau, C.A., Zhang, W., Vila-Sanjurjo, A., Holton, J.M., and Doudna Cate. J.H. (2005) *Science* 310, 827 – 834.

[6] Blamford, D., Gilbert, R.J.C., Grimes, J.M. and Stuart, D.I (2001) *Cur. Opin. Struct. Biol.* 11, 107-113

[7] Timsit, Y. and Moras, D. (1994) *EMBO J.* 13, 2737-2746

[8] Williamson, J.R. (2003) RNA 9, 165-167

[9] Kaczanowska, M., and Ryden-Aulin, M. (2007). Micr. Mol. Biol. Rev. 71, 477-494

[10] Noller, H.F. and Nomura, M. (1987) Ribosomes. In E. coli and S. thyphimurium. Cellular and molecular biology (ed. F.C. Neidhardt), pp 104-125 Am. Soc. For Microbiology Washington, DC

[11] Nomura, M. (1973) Science 179, 864-873

[12] Talkington, M.W.T, Siuzdak, G., and Williamson, J.R. (2005) Nature 438, 628-632

[13] Culver, G.M. and Kirthi, N. (2008) EcoSal Module 2.5.3 © 2006 ASM press

[14] Röhl, R. and Nierhaus, K.H. (1982) Proc. Natl. Acad. Sci. USA 79, 729-733

[15] Herold, M. and Nierhaus, K.H. (1987). *J. Biol. Chem.* 262, 8826-8833

[16] Ostergaard, P. et al. (1998) J. Mol. Biol. 284, 227-240

[17] Green, R. and Noller, H.F. (1999) Biochemistry 38, 17772-1779

[18] Liiv, A. and Remme, J. (2004) J. Mol. Biol 342, 725-741

[19] Klein, D.J., Moore, P.B. and Steitz, T.A (2004) J. Mol. Biol. 340, 141-177

[20] Brodersen, D., Clemons, W.M., Carter, A.P., Wimberly, B.T. and Ramakrishnan, V. (2002) J. Mol. Biol. 316, 725-768

[21] Leulliot, N. and Varani, G. (2001) Biochemistry 40, 7947-7956

[22] Williamson, J.R. (2000) Nat. Struct. Biol. 7, 834-837

[23] Dyson, H. J. and Wright, P. E. (2005). *Nat Rev Mol Cell Biol* 6, 197-208

[24] Wright P.E. and Dyson H.J. (1999) *J. Mol. Biol.* 269, 321-331

[25] Fink, A. (2005) *Curr Opin Struct Biol* 15, 35-41

[26] Uversky, V. N. (2002) *Protein Sci* 11, 739-56

[27] Zengel, J.M., Jerauld, A., Walker, A., Wahl, M., and Lindahl, L. (2003) RNA 9, 1188-1197

[28] Hoang, L., Frederick, K. and Noller, H.F. (2004) Proc. Natl. Acad. Sci. USA 101, 12439-12443

[29] Nowotny, V., and Nierhaus, K.H. (1980a) *Proc. Natl. Acad. Sci. USA* 79, 7238-7242

[30] Nowotny, V., and Nierhaus, K.H. (1980b) *J. Mol. Biol.* 137, 391-399

[31] Franceschi, F. and Nierhaus, K.H. (1982) *Biochemistry* 27, 7056-7059

[32] Guillier, M. et al.. *RNA* 11, 728-38 (2005).

[33] Worbs, M., Huber, R. and Wahl, M. (2000) *EMBO J.* 19, 807-818

[34] Unge, J. et al., (1998) *Structure* 6, 1577-1586

[35] Raibaut et al., (2002) *J. Mol. Biol.* 323, 143-151

[36] Romero, P.Obradovic, Z., Kissinger, C.R., Villafranca, J.E. and Dunker, A.K. *Proc. IEEE Int. Conf. Neural Netw.* 1, 90-95 (1997).

[37] Timsit, Y., Allemand, F., Chiarutini, C and Springer, M. (2006) *EMBO Reports* 7, 1013-1018

[38] Hoeflich, K.P. and Ikura, M. (2002) *Cell* **108**, 739-742.

[39] Shankar, N., Kennedy, S., Chen, G., Krugh, T. and Turner, D. (2006) *Biochemistry* 45, 11776-11789

[40] Dafforn, T.R. and Smith, C.J.I. (2004) EMBO Rep 5, 1046-1052

In: Ribosomal Proteins and Protein Engineering... ISBN 978-1-60741-005-8
Editors: V. Ortendhal and H. Salchow, pp. 55-63

Chapter 4

Disruption of Ribosomal Protein Gene Expression in Eukaryotes: A Class of Mutations Causing Similar Global Growth Defects with Cellular Variations

Catherine Kirn-Safran*
Department of Biological Sciences, University of Delaware, Newark, DE, USA

Abstract

There is much concentration in recent years in understanding the coordinated role of ribosomal protein expression in the regulation of ribosome biogenesis, translational control, and cellular functions. Although ribosomal proteins are dispensable for catalyzing the formation of peptidic bonds, they provide a conformational environment essential for the process of protein biosynthesis. Imbalance in the production of ribosomal proteins results in non-equimolar amounts of protein for each subunit and induces defects in ribosome assembly. For these reasons, delineation of the precise function of specific ribosomal protein gene products is rendered difficult as their gene targeting is accompanied by severe growth defects resulting in cellular death. The objective of this commentary is to provide a summary of ribosomal protein mutations generated in eukaryotes and discuss their proposed roles in the control of cell growth and proliferation and impact in human diseases.

Introduction

Although the basic mechanisms of protein synthesis are conserved throughout all kingdoms of life, the eukaryotic ribosome is considerably larger than its prokaryotic equivalent and consists of a small 40S subunit and a large 60S subunit composed of four

* Corresponding author: Catherine Kirn-Safran, Ph.D., University of Delaware, Dept. of Biological Sciences, 310 Wolf Hall, Newark, DE 19716, USA., email ckirn@udel.edu, phone: 302-831-3249, fax: 302-831-2281

distinct ribosomal RNAs (rRNA) and approximately 80 different ribosomal proteins (RPs). The approximate 50% increase in the amount of RPs in eukaryotes versus prokaryotes has not only been attributed to higher complexity in the regulation of rRNA folding and initiation of translation but also to organelle compartmentalization of major synthetic processes. This compartmentalization includes subunit biogenesis in the nucleolus followed by nuclear export and cytoplasmic assembly as well as the association of some ribosomal subunits to the endoplamic reticulum to engage in co-translational membrane processing.

Several integral proteins of the ribosome located at the periphery of the ribosome are loosely anchored and have extra-ribosomal functions outside their contribution to translation [1-3]. It is reasonable to assume that such RPs are important in modulating protein synthesis efficiency as well as interactions between the ribosome and its cellular environment. Due to the dynamic nature of these processes, a large amount of cellular energy is spent for the assembly/disassembly of these "accessory" RPs with core particles composed of peptide bond catalysts. For these reasons, tight regulation of individual *RP* gene expression is thought to be essential for the control of translation and cell growth.

I. Regulation of Ribosomal Protein Gene Expression

Recently, yeast RNA polymerase I (Pol I) was found to be a key determinant in the balanced production of ribosome components [4]. Although it is not clear how Pol I activity regulates the transcription of *RP* genes through Pol II, transcriptional co-regulation is thought to be the result of crosstalk between nuclear territories in the nucleolus. Unlike in bacteria, *RP* genes of eukaryotes are scattered throughout the genome along with numerous non-functional, intronless, processed pseudogenes that arose from the retrotransposition of mature mRNAs [5, 6]. We found that some of these RP processed pseudogenes can be transcribed but display premature stop codons, lack protein-coding ability, and are likely to be degraded by a non-sense mediated decay mechanism (unpublished observation). The expression of non-translated mRNA variants from processed pseudogenes could be responsible for regulating stability and expression of the paralogous functional *RP* genes [7, 8].

In mammals, RPs are generally encoded by single copy genes of conserved organization with promoters including common features supporting the concept of coordinated transcriptional activity [9]. Interestingly, *cis*-acting elements in the 3′-untranslated region (3′UTR) of some RP mRNAs have been proposed to be important for their targeting to specific subcellular localization and their post-transcriptional stability [10, 11]. In addition, RP mRNAs contain 5′ terminal oligopyrimidic (TOP) sequences displaying binding affinity for ribonucleoproteins that modulate the efficiency of RP translation [12, 13]. Furthermore, some RPs are capable of regulating their own expression by interfering with the splicing of their pre-mRNA through direct binding and production of splice variants that are substrates of non-sense mediated decay [14, 15].

Although the exact mechanisms allowing balanced production of RP proteins remain unknown, it is clear that expression of *RP* genes into proteins that assemble in a functional ribosome particle requires co-orchestration at multiple regulatory levels ranging from transcription, RNA maturation/processing, RNA stability and localization to translational control.

II. Genetic Analysis of Ribosomal Protein Function

Comparative sequence analysis of *RP* genes revealed that there are 32 RPs remarkably conserved between archaea, eukaryotes and bacteria, and the semi-atomic models of more than 40 eukaryotic RPs have been deduced from high resolution crystal structures of assembled or non-assembled bacterial ribosome subunits [16-20]. Although protein synthesis is primarily based on RNA-based mechanisms (tRNA, rRNA, mRNA, and more recently microRNA), the role of RPs is thought to be essential for proper functional conformation and to increase speed and fidelity of translation [21]. This concept is enforced in eukaryotes where an increased number of RPs are involved in intersubunit bridges, and important changes in the ribosome periphery were found to be due to RP expansion segments and the presence of additional RPs [2, 19]. Because abnormal expression of individual RPs is often deleterious and impedes normal cellular function, only a few spontaneous mutations have been reported to date. For example, in *Drosophila*, the *Minute* defect corresponds to a class of over 50 dominant mutations in genes scattered through the genome that exclusively encode for RPs [6]. The *Minute* phenotype is characterized by reduced body size and diminished fertility as a result of haploinsufficiency of a given RP. Thus, quantitative deficiencies of individual RPs can result in sub-lethal phenotypes due to inefficient kinetics of protein synthesis that can lead to ribosome stalling, decreased translation fidelity, and protein misfolding.

Recently, our group generated the first viable, homozygous, null mouse mutant lacking an individual RP [22]. These mutant mice lack the eukaryote-specific RPL29 and display significant global growth retardation, a phenotype reminiscent of *Minute* haploinsufficiencies in *Drosophila*. Interestingly, the RPL29 null defects are also similar to those observed in Rpl29 yeast mutants where the deletion is accompanied by delayed growth and reduced protein biosynthesis rates [23, 24]. Other consequences of this mutation include a high incidence of perinatal lethality, decreased fertility, and a short stature phenotype associated with increased bone fragility [23]. Current investigations indicate that RPL29 may modulate the rate of protein synthesis, possibly by influencing the assembly of the two ribosomal subunits. More recently, homozygous gene-trap disruption of the *Rpl22* gene encoding for another non-essential RP associating with the large subunit of the ribosome, has been found to result in a relatively mild phenotype that consists of selective developmental impairment of αβ T lymphocytes via specific induction of the tumor suppressor p53 in this T cell lineage [25, 26].

In contrast to accessory RPs, mutations of core *RP* genes typically result in severe defects, and natural mutations of RP deficiencies are rare in mammals. To date, homozygous null mutations of core *RP* genes engineered in mice all resulted in early embryonic death [27-29]. Thus, only mice heterozygous for the spontaneous *Rpl24* mutation called Belly spot (*Bst*) are viable, and reduced expression of the RPL24 protein results in pleiotropic effects including white spotting, eye abnormalities, and short stature defects [27].

Another of these core *RP* genes (*RPS19*) is widely mutated in patients with Diamond-Blackfan anemia (DBA). This bone marrow failure syndrome manifests in the first year of life and is thought to be due to defects in 18S rRNA processing, small ribosomal subunits production, and ribosome biogenesis [30-32]. Interestingly, mutant strains of RPS19 yeast orthologs also display defects in the processing of pre-rRNA and aberrant 40S subunit production indicating strong functional conservation for this RP [33]. Knock-down of RPS19 expression using a small interfering RNA (siRNA) experimental approach was accompanied by reduction in the steady-state levels of other RPs of the small ribosomal subunit confirming the idea that amounts of RPs are coordinately regulated by subunit assembly [34]. More recently, mutations in RPS24 and RPS17 have been associated with similar DBA pathologies and are proposed to cooperate with RPS19 in the regulation of ribosome biogenesis [35].

In another recent study, dark skin mouse phenotypes (*dsk*) were associated with missense mutations in highly conserved residues of RPS19 and RPS20. In these mutants, expression of melanocyte-specific genes was found to be associated with increased levels of p53 and its transcriptional targets [36]. Work carried out in zebrafish also correlated p53 transcriptional up-regulation with cell-specific effects of RPS19 deficiency on erythropoiesis and development [37]. Interestingly, keratinocytes-specific ablation of one allele of another core RP, *Rps6*, also caused a dark skin phenotype via stabilization of p53.

Gene targeting studies carried out in mice clearly demonstrate that normal levels of RPS6 expression are essential for life [29, 38]. To date, it is also the only known RP to undergo phosphorylation in response to external stimuli. Because both *Rps6* null and heterozygote mutations activate regulatory pathways that prevent embryonic development, gene targeting studies for this RP have usually been performed in specific tissues or cell types. Volarevic and colleagues have found that the decrease in cell proliferation observed after conditional knock-out of RPS6 in liver is a consequence of a cell cycle progression defect [38]. More recently, RPS6 haploinsufficiency in gastrula stage embryos was found to be due to the activation of a p53-dependent checkpoint that induces a specific block in the G2/M phase of the cell cycle and increases apoptosis [29, 39]. Notably, the tumor suppressor p53 is also found to be responsible for decreased cell survival of RPS6-heterozygous T lymphocytes in peripheral lymphoid organs [40].

III. Conclusions and Implications of RP Function in Human Health

Although mutations in genes encoding ubiquitously expressed RPs have been widely described to induce reduced cellular proliferation and growth deficiencies at the organismal level, it is also clear that individual *Rpl* or *Rps* mutations in vertebrates can trigger cell-specific effects with very distinct phenotypic outcomes. For example, hemizygosity for a panel of genes encoding different RPs lead to a rare type of malignant tumor of neuronal origin in fish, while dominant mutations in another set of *Rp* genes induce a dark skin phenotype restricted to ears, footpads, and tails in mice [36, 41]. The search for changes in the molecular signature of a number of RP mutants *versus* their wild type counterparts established in different systems a direct correlation between *Rp* gene dosage and regulation of p53 synthesis [37, 39, 42]. These observations suggest that p53 is a downstream target of a sensor which detects ribosome insufficiency (from defective ribosome biogenesis in the nucleolus or reduced protein synthesis in the cytoplasm).

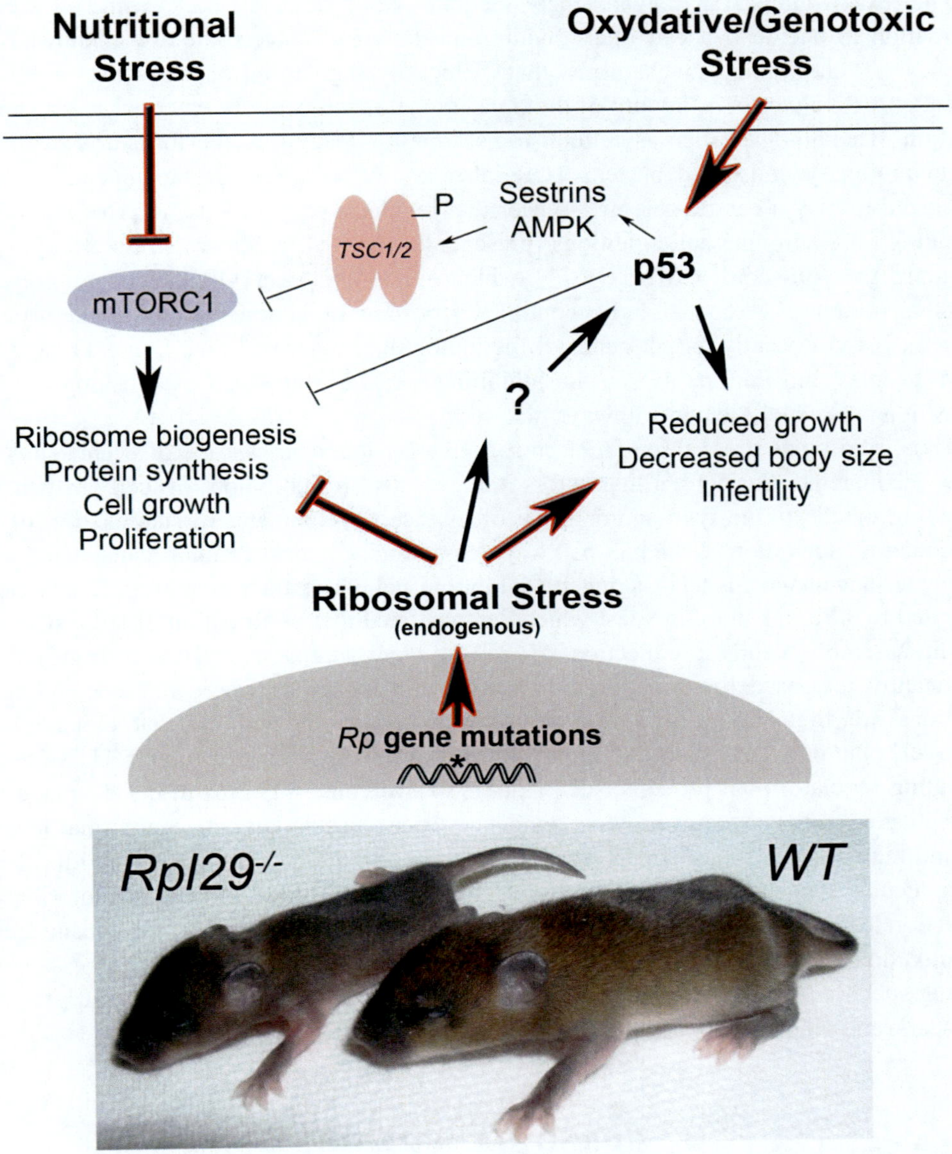

Figure 1: **Proposed model of effects induced by endogenous ribosomal stress**. Individual ribosomal protein (*Rp*) gene mutations are generally associated with global growth deficiencies. Like under genotoxic or oxidative stress, ribosomal insufficiency can trigger stabilization of p53. The molecular players involved in this activation are still unknown. Next, p53 is capable of inducing expression of two specific target genes (Sestrin1&2) which activate the AMP-responsive protein kinase (AMPK) known to phosphorylate TSC2 of the tuberous sclerosis complex which, in turn, inhibits the mTOR signaling pathway. Other potential links between the p53 and mTOR pathways have been described elsewhere [46]. This suggests that the mTORC1 complex is a sensor of both endogenous and environmental stresses and can be negatively regulated by either the lack of growth factors/nutrients and/or deficiencies in the translational machinery apparatus.

The exact molecular players that integrate environmental cues with intracellular information to decide if a cell must divide, quiesce, or ultimately die are unknown. The presence of checkpoint mechanisms that detect dysregulation of ribosome production associated with changes in fidelity of the translational machinery are essential to prevent the initiation of pathologies such as cellular transformation leading to the formation of tumors. The evolutionary conserved protein TOR, also known as target of rapamycin, has been recognized to play a central role of a sensor that regulates cell growth and protein synthesis depending on environmental conditions (presence of nutrient, oxidative stress etc.) [43]. In mammals, the TOR complex (mTORC1), which in addition to mTOR includes raptor and a few other proteins, is responsible for control of ribosome biogenesis and initiation of protein synthesis [44]. Recently, target genes of the tumor suppressor p53 (see figure 1) have been found to play an important role in inhibiting mTOR signaling, establishing a direct connection between these two pathways [45, 46].

Thus, ribosomal stress through RP mutations may induce a cascade of events similar to those encountered during chronic stress that results in inhibition of cell growth and, depending on the tissue type and/or stage of disease development, in stabilization of p53. Subsequently, activation of the p53 pathway helps protect against cellular damage and might present an advantage to delay aging by inhibiting cellular transformation. This notion is supported by a recent study in yeast which demonstrated that perturbation of large ribosomal subunit assembly using a collection of RPL mutant strains restricted cell growth but significantly increased lifespan [47]. In contrast, in situations where *RP* loss-of-function mutations are found associated with tumor formation, it is possible that clone of cells selectively shut down translation of tumor suppressor mRNA with structured 5′UTRs capable of binding regulatory RP proteins, such as p53, to overcome this growth disadvantage [48]. In conclusion, various studies provided evidence for an association between normal levels of RPs and regulation of tumor suppressor activity. Future studies will be necessary to identify the molecular sequence of events whereby alterations in dosage of RPs and of ribosome production lead to p53 modulation. The discovery of the exact molecular mechanisms behind the activation of different pathways downstream of ribosomal insufficiency is essential in defining specific targets for therapeutic intervention in Diamond-Blackfan anemia, some type of cancers, and other ribosomal stress-associated syndromes.

Acknowledgments

Dr. Jeffrey B. Safran is thanked for his careful reading of the commentary. Research in the Kirn-Safran laboratory is supported by the National Institute of Health (P20 RR016458-06)

References

[1] Wool IG. Extraribosomal functions of ribosomal proteins. *Trends Biochem Sci* 1996; 21: 164-165.

[2] Brodersen DE, Nissen P. The social life of ribosomal proteins. *Febs J* 2005; 272: 2098-2108.

[3] Mazumder B, Sampath P, Seshadri V, Maitra RK, DiCorleto PE, Fox PL. Regulated release of L13a from the 60S ribosomal subunit as a mechanism of transcript-specific translational control. *Cell* 2003; 115: 187-198.

[4] Laferte A, Favry E, Sentenac A, Riva M, Carles C, Chedin S. The transcriptional activity of RNA polymerase I is a key determinant for the level of all ribosome components. *Genes Dev* 2006; 20: 2030-2040.

[5] Kirn-Safran CB, Dayal S, Martin-DeLeon PA, Carson DD. Cloning, expression, and chromosome mapping of the murine Hip/Rpl29 gene. *Genomics* 2000; 68: 210-219.

[6] Uechi T, Tanaka T, Kenmochi N. A complete map of the human ribosomal protein genes: assignment of 80 genes to the cytogenetic map and implications for human disorders. *Genomics* 2001; 72: 223-230.

[7] Hirotsune S, Yoshida N, Chen A, Garrett L, Sugiyama F, Takahashi S, Yagami K, Wynshaw-Boris A, Yoshiki A. An expressed pseudogene regulates the messenger-RNA stability of its homologous coding gene. *Nature* 2003; 423: 91-96.

[8] Harrison PM, Zheng D, Zhang Z, Carriero N, Gerstein M. Transcribed processed pseudogenes in the human genome: an intermediate form of expressed retrosequence lacking protein-coding ability. *Nucleic Acids Res* 2005; 33: 2374-2383.

[9] Perry RP. The architecture of mammalian ribosomal protein promoters. *BMC Evol Biol* 2005; 5: 15.

[10] Russo A, Russo G, Cuccurese M, Garbi C, Pietropaolo C. The 3'-untranslated region directs ribosomal protein-encoding mRNAs to specific cytoplasmic regions. *Biochim Biophys Acta* 2006; 1763: 833-843.

[11] Ledda M, Di Croce M, Bedini B, Wannenes F, Corvaro M, Boyl PP, Caldarola S, Loreni F, Amaldi F. Effect of 3'UTR length on the translational regulation of 5'-terminal oligopyrimidine mRNAs. *Gene* 2005; 344: 213-220.

[12] Meyuhas O. Synthesis of the translational apparatus is regulated at the translational level. *Eur J Biochem* 2000; 267: 6321-6330.

[13] Schwartz EI, Intine RV, Maraia RJ. CK2 is responsible for phosphorylation of human La protein serine-366 and can modulate rpL37 5'-terminal oligopyrimidine mRNA metabolism. *Mol Cell Biol* 2004; 24: 9580-9591.

[14] Cuccurese M, Russo G, Russo A, Pietropaolo C. Alternative splicing and nonsense-mediated mRNA decay regulate mammalian ribosomal gene expression. Nucleic Acids Res 2005; 33: 5965-5977.

[15] Li B, Vilardell J, Warner JR. An RNA structure involved in feedback regulation of splicing and of translation is critical for biological fitness. *Proc Natl Acad Sci U S A* 1996; 93: 1596-1600.

[16] Ban N, Nissen P, Hansen J, Moore PB, Steitz TA. The complete atomic structure of the large ribosomal subunit at 2.4 A resolution. *Science* 2000; 289: 905-920.

[17] Yusupov MM, Yusupova GZ, Baucom A, Lieberman K, Earnest TN, Cate JH, Noller HF. Crystal structure of the ribosome at 5.5 A resolution. *Science* 2001; 292: 883-896.

[18] Wimberly BT, Brodersen DE, Clemons WM, Jr., Morgan-Warren RJ, Carter AP, Vonrhein C, Hartsch T, Ramakrishnan V. Structure of the 30S ribosomal subunit. *Nature* 2000; 407: 327-339.

[19] Spahn CM, Beckmann R, Eswar N, Penczek PA, Sali A, Blobel G, Frank J. Structure of the 80S ribosome from Saccharomyces cerevisiae--tRNA-ribosome and subunit-subunit interactions. *Cell* 2001; 107: 373-386.

[20] Ramakrishnan V. What we have learned from ribosome structures. *Biochem Soc Trans* 2008; 36: 567-574.

[21] Vasudevan S, Tong Y, Steitz JA. Switching from repression to activation: microRNAs can up-regulate translation. *Science* 2007; 318: 1931-1934.

[22] Kirn-Safran CB, Oristian DS, Focht RJ, Parker SG, Vivian JL, Carson DD. Global growth deficiencies in mice lacking the ribosomal protein HIP/RPL29. *Dev Dyn* 2007; 236: 447-460.

[23] Oristian DS, Sloofman LG, Zhou X, Wang L, Farach-Carson MC, Kirn-Safran CB. Ribosomal protein L29/HIP deficiency delays osteogenesis and increases fragility of adult bone in mice. *J Orthop Res* 2008.

[24] DeLabre ML, Kessl J, Karamanou S, Trumpower BL. RPL29 codes for a non-essential protein of the 60S ribosomal subunit in Saccharomyces cerevisiae and exhibits synthetic lethality with mutations in genes for proteins required for subunit coupling. *Biochim Biophys Acta* 2002; 1574: 255-261.

[25] Anderson SJ, Lauritsen JP, Hartman MG, Foushee AM, Lefebvre JM, Shinton SA, Gerhardt B, Hardy RR, Oravecz T, Wiest DL. Ablation of ribosomal protein L22 selectively impairs alphabeta T cell development by activation of a p53-dependent checkpoint. *Immunity* 2007; 26: 759-772.

[26] Shu-Nu C, Lin CH, Lin A. An acidic amino acid cluster regulates the nucleolar localization and ribosome assembly of human ribosomal protein L22. *FEBS Lett* 2000; 484: 22-28.

[27] Oliver ER, Saunders TL, Tarle SA, Glaser T. Ribosomal protein L24 defect in belly spot and tail (Bst), a mouse Minute. *Development* 2004; 131: 3907-3920.

[28] Matsson H, Davey EJ, Frojmark AS, Miyake K, Utsugisawa T, Flygare J, Zahou E, Byman I, Landin B, Ronquist G, Karlsson S, Dahl N. Erythropoiesis in the Rps19 disrupted mouse: Analysis of erythropoietin response and biochemical markers for Diamond-Blackfan anemia. *Blood Cells Mol Dis* 2006; 36: 259-264.

[29] Panic L, Tamarut S, Sticker-Jantscheff M, Barkic M, Solter D, Uzelac M, Grabusic K, Volarevic S. Ribosomal protein S6 gene haploinsufficiency is associated with activation of a p53-dependent checkpoint during gastrulation. *Mol Cell Biol* 2006; 26: 8880-8891.

[30] Choesmel V, Bacqueville D, Rouquette J, Noaillac-Depeyre J, Fribourg S, Cretien A, Leblanc T, Tchernia G, Da Costa L, Gleizes PE. Impaired ribosome biogenesis in Diamond-Blackfan anemia. *Blood* 2007; 109: 1275-1283.

[31] Draptchinskaia N, Gustavsson P, Andersson B, Pettersson M, Willig TN, Dianzani I, Ball S, Tchernia G, Klar J, Matsson H, Tentler D, Mohandas N, Carlsson B, Dahl N. The gene encoding ribosomal protein S19 is mutated in Diamond-Blackfan anaemia. *Nat Genet* 1999; 21: 169-175.

[32] Flygare J, Aspesi A, Bailey JC, Miyake K, Caffrey JM, Karlsson S, Ellis SR. Human RPS19, the gene mutated in Diamond-Blackfan anemia, encodes a ribosomal protein required for the maturation of 40S ribosomal subunits. *Blood* 2007; 109: 980-986.

[33] Leger-Silvestre I, Caffrey JM, Dawaliby R, Alvarez-Arias DA, Gas N, Bertolone SJ, Gleizes PE, Ellis SR. Specific Role for Yeast Homologs of the Diamond Blackfan

Anemia-associated Rps19 Protein in Ribosome Synthesis. *J Biol Chem* 2005; 280: 38177-38185.

[34] Idol RA, Robledo S, Du HY, Crimmins DL, Wilson DB, Ladenson JH, Bessler M, Mason PJ. Cells depleted for RPS19, a protein associated with Diamond Blackfan Anemia, show defects in 18S ribosomal RNA synthesis and small ribosomal subunit production. *Blood Cells Mol Dis* 2007; 39: 35-43.

[35] Choesmel V, Fribourg S, Aguissa-Toure AH, Pinaud N, Legrand P, Gazda HT, Gleizes PE. Mutation of ribosomal protein RPS24 in Diamond-Blackfan anemia results in a ribosome biogenesis disorder. *Hum Mol Genet* 2008; 17: 1253-1263.

[36] McGowan KA, Li JZ, Park CY, Beaudry V, Tabor HK, Sabnis AJ, Zhang W, Fuchs H, de Angelis MH, Myers RM, Attardi LD, Barsh GS. Ribosomal mutations cause p53-mediated dark skin and pleiotropic effects. *Nat Genet* 2008; 40: 963-970.

[37] Danilova N, Sakamoto KM, Lin S. Ribosomal protein S19 deficiency in zebrafish leads to developmental abnormalities and defective erythropoiesis through activation of p53 protein family. *Blood* 2008.

[38] Volarevic S, Stewart MJ, Ledermann B, Zilberman F, Terracciano L, Montini E, Grompe M, Kozma SC, Thomas G. Proliferation, but not growth, blocked by conditional deletion of 40S ribosomal protein S6. *Science* 2000; 288: 2045-2047.

[39] Panic L, Montagne J, Cokaric M, Volarevic S. S6-haploinsufficiency activates the p53 tumor suppressor. *Cell Cycle* 2007; 6: 20-24.

[40] Sulic S, Panic L, Barkic M, Mercep M, Uzelac M, Volarevic S. Inactivation of S6 ribosomal protein gene in T lymphocytes activates a p53-dependent checkpoint response. *Genes Dev* 2005; 19: 3070-3082.

[41] Amsterdam A, Sadler KC, Lai K, Farrington S, Bronson RT, Lees JA, Hopkins N. Many ribosomal protein genes are cancer genes in zebrafish. *PLoS Biol* 2004; 2: E139.

[42] MacInnes AW, Amsterdam A, Whittaker CA, Hopkins N, Lees JA. Loss of p53 synthesis in zebrafish tumors with ribosomal protein gene mutations. *Proc Natl Acad Sci U S A* 2008; 105: 10408-10413.

[43] Wullschleger S, Loewith R, Hall MN. TOR signaling in growth and metabolism. *Cell* 2006; 124: 471-484.

[44] Guertin DA, Sabatini DM. Defining the role of mTOR in cancer. Cancer *Cell* 2007; 12: 9-22.

[45] Budanov AV, Karin M. p53 target genes sestrin1 and sestrin2 connect genotoxic stress and mTOR signaling. *Cell* 2008; 134: 451-460.

[46] Feng Z, Hu W, de Stanchina E, Teresky AK, Jin S, Lowe S, Levine AJ. The regulation of AMPK beta1, TSC2, and PTEN expression by p53: stress, cell and tissue specificity, and the role of these gene products in modulating the IGF-1-AKT-mTOR pathways. *Cancer Res* 2007; 67: 3043-3053.

[47] Steffen KK, MacKay VL, Kerr EO, Tsuchiya M, Hu D, Fox LA, Dang N, Johnston ED, Oakes JA, Tchao BN, Pak DN, Fields S, Kennedy BK, Kaeberlein M. Yeast life span extension by depletion of 60s ribosomal subunits is mediated by Gcn4. *Cell* 2008; 133: 292-302.

[48] Takagi M, Absalon MJ, McLure KG, Kastan MB. Regulation of p53 translation and induction after DNA damage by ribosomal protein L26 and nucleolin. *Cell* 2005; 123: 49-63.

In: Ribosomal Proteins and Protein Engineering… ISBN 978-1-60741-005-8
Editors: V. Ortendhal and H. Salchow, pp. 65-90

Chapter 5

Ancestral Protein Reconstruction as a New Field in Protein Engineering

***Ayumu Konno*[1], *Tomohisa Ogawa*[1], *and Tsuyoshi Shirai*[2]**
[1]Tohoku University, Sendai, Japan
[2] Nagahama Institute of Bio-Science and Technology, and
Japan Science and Technology Agency, Nagahama, Shiga, Japan

Abstract

The extent of current research in ancient biology is not restricted by the presence of fossil records, owing to the developments in computational analysis and engineering techniques of proteins. Ancestral sequences of extant genes/proteins can be inferred with the molecular phylogeny reconstruction methods, which are mainly based on maximum-likelihood estimation. Once an ancestral sequence is determined, a reconstruction of the ancestral protein is not difficult by using the current protein engineering techniques, and reproduced protein can be experimentally tested either biochemically or biophysically. To date, such 'experimental molecular archeology' has been examined on several enzymes, elongation factor, photo-reactive proteins, nuclear receptor, and lectin, and largely helped to elucidate evolutionary processes of stability, specificity, and structure of the proteins. Although this method would promise a new field in protein engineering, several problems still remain to be solved, the largest among which should be accuracy of inferred ancestral sequence as a whole. In this chapter, the reported studies of ancestral protein reconstruction are reviewed with a focus on our recent results on the ancestral congerin, which is the β-galactoside specific lectin of fish involved in the biological defense system. The reconstruction of ancestral congerin is important because two extant isoforms of this protein, congerins I and II, have been rapidly evolving, and their molecular properties and structures have been differentiating due to the natural selection pressures operating on the genes.

INTRODUCTION

So far, the fossil records have been the only trace to elucidate the ancient forms of life, and they told us that a lot of organisms, which no longer exist, were once living on earth. Likewise, they have told that all of the extant organisms cannot be tracked down beyond a certain era, although similar species are found among those in the fossil records. These observations are the most intuitive clues of the evolution of organisms, and of the fact that the present species are offspring of ancestral organisms.

The same is also true for their genes; the genes of similar functional roles of different organisms are often similar to each other, though not exactly identical. These genes are thought to be derived from an ancestral gene, which does not currently exist. Although knowing the forms of ancient organisms is one of the major purposes of paleontology, it is also valuable to understand the present life, because it should be largely affected by the evolutional history. Though reconstructing a living organism out of fossils would be the ultimate methodology for paleontology, it is far beyond the technologies at hand. For the genes or their encoding proteins, however, reconstruction of an ancestor is a feasible technique already.

Production of ancestral or ancient proteins is highly attainable, though not always easy, if one knows its nucleotide/amino acid sequence to be synthesized, largely owing to the developments in molecular biology and protein engineering techniques. Once an ancestral protein is obtained, it can be tested in a laboratory with the biochemical or biophysical methods for its activity, stability, specificity, or even three-dimensional structure, which can never be applicable for fossil samples. To date, such 'experimental molecular archeology' has been examined on several enzymes [1-6], elongation factor [7], photo-reactive proteins [8, 9], nuclear receptor [10], and lectin [11].

Knowledge of the ancestral sequences of protein has a certain advantage in studying their molecular evolution. For example, if one has the sequence of protein X of a certain function, and also there is another sequence of homologous protein Y with a different function, the straight way to detect the origin of their functional difference is to compare their sequences directly. However, the apparent differences between two proteins are the result of multiple mutations accumulated all the way between X and Y. The mutations in the lineage of protein X might not be significantly relevant to the function of protein Y. If the sequence of the last common ancestor A of the proteins is available, a contrast of A to X, and A to Y would reveal the mutations directly important for each of them. We can also identify the patterns of amino acid substitutions in the course of evolution, and guess which function of two proteins has been the original one.

The problem still remained is how to know the sequence of ancestral protein. Since, in most cases, the ancestral genes do not presently exist anywhere, the ancestral sequences need to be reconstructed mainly *in silico* (by using a computer). Interestingly, no computational technique was ever developed specifically for this purpose, because the sophisticated mathematical tools devised for molecular phylogeny construction were available [12, 13]. In the early stage, ancestral sequence reconstructions with the molecular phylogeny technique were used in order to investigate the functional evolution of proteins, such as convergent evolution of lysozyme in ruminant stomach environments, or the alkaline adaptation of enzymes [14-16]. However, if an ancestral sequence is known, then the most straightforward way to conclude such research is reconstructing and experimentally testing the ancestral protein in a laboratory.

Ancestral protein reconstruction experiments are technically a plain site-directed mutagenesis to the present proteins. However, the non-ignorable difference between ancestor reconstruction and authentic site-directed mutagenesis experiments is that the former requires a series of mutations at the sites that might be far more than a few, and the target sites might be spread over the entire sequence. This has been the substantial impediment in practical studies, because researchers almost certainly have to synthesize entire genes. To date, this problem has been largely solved by the popularization and much lowered cost of solid-phase synthesis of long DNA molecules.

Ancestral protein reconstruction is the research tool that a combination of the developments in biochemistry, molecular biology, and bioinformatics has enabled. A new field in molecular evolutionary biology, and also in protein engineering, will be developed with this method. In this chapter, we will first introduce the methods in ancestral sequence reconstruction briefly. Then the important studies reported in this field will be reviewed. Finally, the case of ancestral fish galectin (congerin) will be discussed in detail in order to outline the benefits and problems in ancestral protein studies.

METHODS FOR ANCESTRAL SEQUENCE CALCULATION

Most of the ancestral sequences studied to date were calculated by using the existing computational methods developed for molecular phylogeny construction. Some of these methods, such as the maximum parsimony (MP), and the maximum likelihood (ML), have an integral procedure of ancestral sequence inference at each node of phylogenetic tree under construction [12, 13]. The MP method assumes that a phylogenetic tree with minimum substitutions should be most likely. Therefore, this method assigns possible nucleotide/amino acid for each site at every node of phylogeny to evaluate the minimum number of substitutions. Because of this assumption of parsimony, the MP method tends to underestimate the number of substitutions, if a branch is relatively long. This method is also fragile if the evolutionary rate is variable among branches.

The ML method, which does not require this assumption, is currently most used. The ML method evaluates posterior probability of nucleotide/amino acid at each node of a phylogenetic tree based on the empirical Bayesian statistics by using the present sequences and substitution probability matrix as inputs (observations). Therefore, the result would be significantly affected by the choice of input sequences and substitution probability matrix, and the probability of reconstructed sequence at a node might be low, when the node is connected to the present sequences through longer and/or more intervening branches. The ML method is popular in this field, largely owing to the presence of the excellent software package PAML [17]. Several other software applications have been also developed for this purpose, such as FastML [18], ANSESCON [19], and GASP [20]. These applications are mostly based on the ML method, except for the last one, which partly employs the MP method to enable ancestral state prediction at gapped sites in sequence alignment.

The accuracy of reconstructed ancestral sequence is most critical for studies. However, it is often difficult to obtain a highly accurate sequence as a whole. Because a molecular evolution is thought to be highly stochastic processe, there is no guarantee to identify the ancestral sequences without errors. Even if each residue site of a protein, which consists of one hundred residues, is identified with posterior probability of 0.99 (expectedly 99% correct), the sequence as a whole is expected to be accurate only with the probability of ~0.37 (i.e., 0.99^{100}). The site probabilities might be much lower in actual cases. This is the major problem in the ancestor reconstruction studies, and considerable efforts have been made to avoid incorrect conclusions due to imperfect reconstruction. In many of the reported studies, equally probable amino acid residus have been tested for such unreliable sites by introducing site-directed mutations to confirm that the molecular properties are not largely affected by those mutations. Despite these efforts, however, it is difficult to examine all of the possible combinations of alternative residues over suspicious sites.

THE EARLY STUDIES: RECONSTRUCTION OF PARTIAL ANCESTORS

The early studies exploiting the idea of ancestral protein reconstruction were the site-direct mutagenesis research, in which only a few amino acids were substituted into the anticipated ancestral status. For example, Benner and colleagues reconstructed the ribonuclease (RNase) of an extinct bovid ruminant [1]. They predicted four sequences of ancestral RNases using the MP method from five closely related bovids (ox, swamp buffalo, river buffalo, nilgai [21], and the primitive artiodactyl that lived in Pliocene [22]). The ancestor closest to the extant ox protein was selected for the target of experiment among the four probable ancestors, because it contained the mutation of residue 35 that was located close to Lys41, which was known to be important for catalysis [23]. Three ancestral mutants of the ox RNase, namely, Ala19Ser, Leu35Met, and Ala19Ser/Leu35Met were prepared (Figure 1). The purified mutants and ox RNases were examined for kinetic properties and temperature-dependency of the stabilities against tryptic digestions. However, no significant difference was found between the ancestral mutants and the modern ox RNase. The results suggested that these amino acid substitutions were the neutral evolution, although this conclusion should be limited to the extent of the examined properties.

Malcolm *et al.* have succeeded in identifying a non-neutral evolutional pathway of the game-bird lysozymes using ancestral lysozyme reconstructions predicted by the MP method [2]. They paid attention to three residue sites, namely, Thr40, Ile55, and Ser91, of game-bird lysozyme (Figure 2), because these residues were located in vicinity of the active site [24, 25]. Because the ancestral amino acid of these sites has been anticipated to be Ser40, Val55 and Thr91 [26], seven mutant proteins, in which these sites were combinatorically mutated into the ancestral status, were synthesized as possible intermediates in the evolutionary pathway of bird lysozyme. A comparison of the thermostabilities and the crystal structures of these mutants revealed a linear correlation between the side chain volume and the thermostability of the proteins. The Thr40Ser mutation increased the protein thermostability by more than 3°C, therefore this substitution was concluded to be non-neutral against natural selection.

a

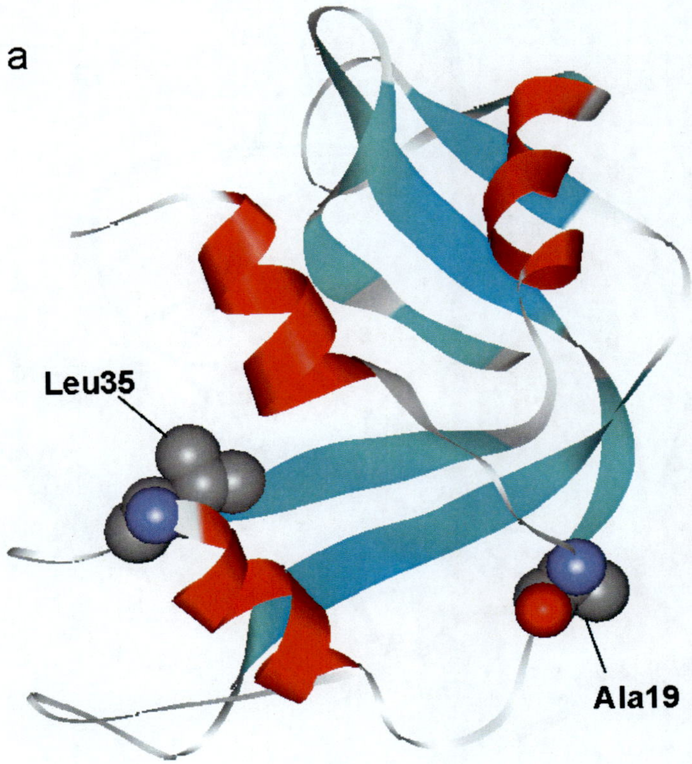

b

```
                 10         20         30         40         50
         ....|....| ....|....| ....|....| ....|....| ....|....|
Ox_RNase KETAAAKFER QHMDSSTSAA SSSNYCNQMM KSRNLTKDRC KPVNTFVHES  50
A19S     .......... ........S. .......... .......... ..........  50
L35M     .......... .......... .......... ....M..... ..........  50
A19S/L35M .......... ........S. .......... ....M..... ..........  50

                 60         70         80         90        100
         ....|....| ....|....| ....|....| ....|....| ....|....|
Ox_RNase LADVQAVCSQ KNVACKNGQT NCYQSYSTMS ITDCRETGSS KYPNCAYKTT 100
A19S     .......... .......... .......... .......... .......... 100
L35M     .......... .......... .......... .......... .......... 100
A19S/L35M .......... .......... .......... .......... .......... 100

                110        120
         ....|....| ....|....| ....
Ox_RNase QANKHIIVAC EGNPYVPVHF DASV 124
A19S     .......... .......... .... 124
L35M     .......... .......... .... 124
A19S/L35M .......... .......... .... 124
```

Figure 1. a. Three dimensional structure of bovine RNase A (PDB code 2e3w). The amino acid residues, which were mutated into the ancestral sequence, are shown in space-filling models. b. Amino acid sequence alignment of ox RNase and its ancestral mutants. The amino acid residues of the mutants are shown only for the sites different from the ox RNase sequence.

a

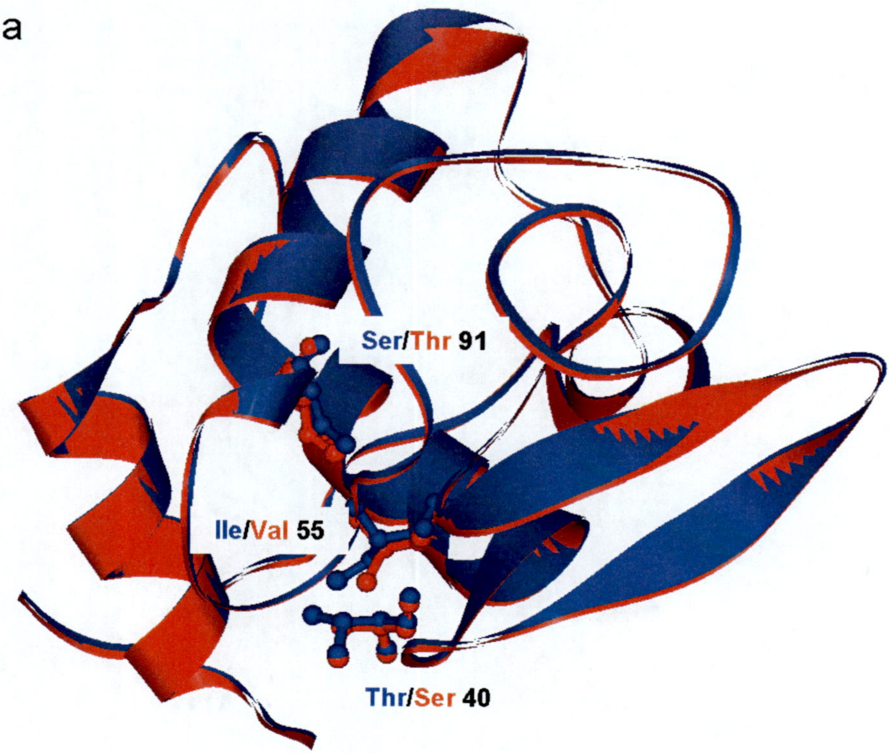

b

```
                   10         20         30         40         50
           ....|....| ....|....| ....|....| ....|....| ....|....|
Lysozyme   KVFGRCELAA AMKRHGLDNY RGYSLGNWVC AAKFESNFNT QATNRNTDGS  50
SVT        .......... .......... .......... .........S ..........  50

                   60         70         80         90        100
           ....|....| ....|....| ....|....| ....|....| ....|....|
Lysozyme   TDYGILQINS RWWCNDGRTP GSRNLCNIPC SALLSSDITA SVNCAKKIVS  100
SVT        ....V..... .......... .......... .......... T.........  100

                  110        120
           ....|....| ....|....| ....|....
Lysozyme   DGNGMNAWVA WRNRCKGTDV QAWIRGCRL 129
SVT        .......... .......... ......... 129
```

Figure 2. a. Structure of bird lysozymes in ribbon structure. Wild-type game-bird lysozyme (1hel) colored in blue and SVT mutants (1hep) colored in red. The amino acid residues, which were mutated, are shown in boll and stick models. b. Aligned amino acid sequences of game-bird lysozyme and its mutant SVT. SVT amino acid residues are shown only for sites different from the wild-type lysozyme sequence.

Yamagishi and colleagues used ancestral protein reconstructions [6, 27-30] to obtain direct evidence for the hypothesis that the common ancestor of all organisms was hyper-thermophilic [31]. Because the catalytic activities of 3-isopropylmalate dehydrogenase (IPMDH) and isocitrate dehydrogenase (ICDH) were similar to each other, and their three-dimensional structures were conserved, these proteins were thought to share a very ancient common ancestor (Figure 3) [32]. The sequence of their ancestor was inferred from a phylogenetic tree of IPMDH and ICDH of various species including thermophile (*Thermus thermophilus*) and extreme thermophile (*Sulfolobus* sp. strain 7) using the ML method. From the ancestral sequence, they selected seven sites located close to the substrate and cofactor-binding sites, and made seven single-mutants bearing the ancestral amino acid. The five out of the seven mutants showed higher thermostability than the wild type IPMDH of *Sulfolobus* sp. strain 7. Other two ancestral residues were also found to increase the thermostability [6, 29]. The successful reconstruction of the partially ancestral IPMDH, which had higher thermostability than that of the extreme thermophile, was thought to support the hypothesis of hyperthermophile common ancestor. Moreover, his group also reported the successful thermostabilization of ICDH [28] and Glycyl-tRNA synthetase [30] by using a similar method. Thus, they concluded that incorporation of ancestral residues into a modern protein could be used not only to test the evolutionary hypothesis, but also as a powerful protein engineering technique for protein thermostabilization.

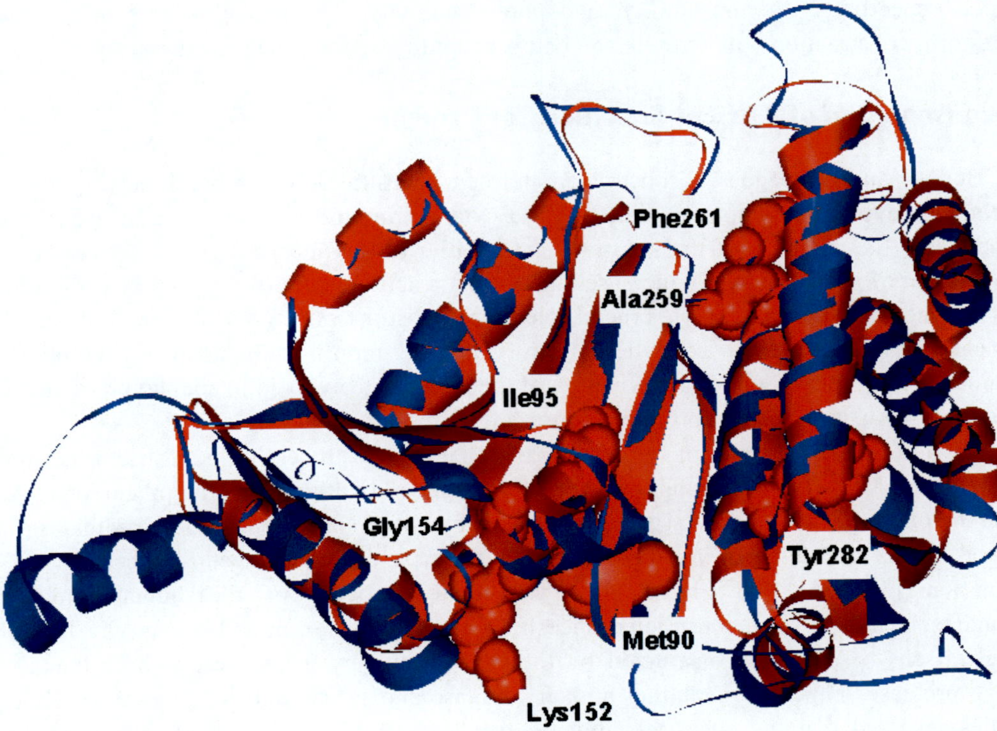

Figure 3. Three dimensional structures of IPMDH (PDB code 1wpw) and hypothetical ICDH (PDB code 2dht) of *Sulfolobus* sp. are shown with ribbon model, and colored in red and blue, respectively. Mutated residues in the study of Miyazaki *et al.* [27] are shown in space filling models.

These early studies were done by introducing a limited number of mutations into the present proteins, or by carefully selecting the ancestors that were separated from the present protein by only few substitutions. This might be partly because introducing mutations to many sites spread over gene was difficult and expensive. However, such ancestral reconstruction was incomplete, and the possibility, that the sites remained in non-ancestral status significantly affected the molecular property of interest, could not be ruled out.

Therefore, most of the recent studies were executed by substituting all sites in a target protein or domain into inferred ancestral amino acids.

Complete Ancestor Reconstruction

In most cases, the purpose of ancestral protein reconstruction is to detect the properties that might be advantageous at ancient conditions or disadvantageous (or primitive) in the light of present conditions. However, it is usually difficult to judge whether an amino acid substitution of a protein is neutral or not against natural selection, because it is unfeasible to examine each and every property of a protein, which might be responsible for selection. Therefore, most of the current ancestral reconstruction studies are designed to observe a **single definitive property of protein molecule, and there are several 'well-liked' properties,** namely, specificity, thermostability, and photo-reactivity. The recent studies of complete ancestor reconstruction will be reviewed below according to this categorization.

Reconstruction of Ancient Specificity of Protein

Benner and colleagues reconstructed thirteen ancestral RNases predicted by the MP method as evolutionary intermediates of the extant proteins from thirty-eight artiodactyls species [3]. Examination of the substrate specificities of the ancestral RNases revealed that substitution of Asp38 with Gly increased the catalytic activity toward duplex RNA, therefore, this residue was though to play a crucial role in the substrate selectivity. Although it was still unclear that this change in catalytic activity was how important for adaptive evolution of ruminants, they succeeded in identification of the mutations involved in adaptation of enzyme specificity by using ancestral protein reconstruction.

Chymases are the family of serine proteases that mainly process vasoconstrictor hormone angiotensin (Ang) II, by cleaving the Phe8-His9 peptide bond in Ang I, the precursor of Ang II, in the primate's heart and vessels [33]. Unlike chymotrypsin and rat chymase-1 that further cleave the Tyr4-Ile5 bond in Ang II, human chymase is highly specific for the Ang II formation [34, 35]. To clarify the evolutionary process of the human chymase, Chandrasekharan *et al.* reconstructed ancestral chymase and examined its activity [4]. The ancestral chymase was reconstructed by the MP method by using twenty-three leukocyte serine protease sequences including human chymase and rat chymase-1 (Figure 4). Kinetic studies revealed that the ancestral chymase was Ang II formation specific, that was, the enzyme could cleave the Phe8-His9 bond in Ang I to produce Ang II, but could not cleave the Tyr4-Ile5 bond in Ang I and II that was digested by extant rat chymase-1 and chymotrypsin. The results suggested that a part of chymases have acquired the additional function of the Ang II processing in the course of evolution.

a

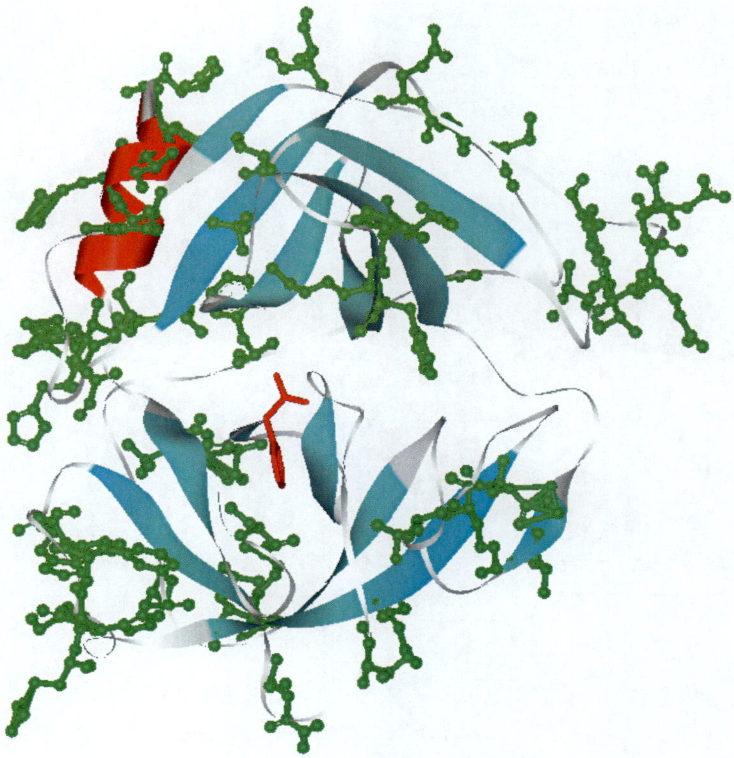

b

```
                   10         20         30         40         50         60         70
           ....|....| ....|....| ....|....| ....|....| ....|....| ....|....| ....|....|
human-Chy  IIGGTESKPH SRPYMAYLEI VTSNGPSKFC GGFLIRRNFV LTAAHCAGRS ITVTLGAHNI TEEDTWQKL  70
anc-Chy    ....V..... ........K. ..E..YKAR. .......... ......R... .......... NKK.P.Q..I 70

                   80         90        100        110        120        130        140
           ....|....| ....|....| ....|....| ....|....| ....|....| ....|....| ....|....|
human-Chy  EVIKQFRHPK YNTSTLHHDI MLLKLKEKAS LTLAVGTLPF PSQFNFVPPG RMCRVAGWGR TGVLKPGSDT 140
anc-Chy    K.E...P... ..D...FN.. .........K ..S..D...L ..P.D..K.. K......... ...NE..... 140

                  150        160        170        180        190        200        210
           ....|....| ....|....| ....|....| ....|....| ....|....| ....|....| ....|....|
human-Chy  LQEVKLRLMD PQACSHFRDF DHNLQLCVGN PRKTKSAFKG DSGGPLLCAG VAQGIVSYGR SDAKPPAVFT 210
anc-Chy    .R..E..I.. EK..KDY..Y .Y.......S ...L...... .......... ..H....... N......... 210

                  220
           ....|....| ....|.
human-Chy  RISHYRPWIN QILQAN 226
anc-Chy    .....L.... KV.KSK 226
```

Figure 4. a. Three dimensional structure of human chymase in ribbon model (PDB code 1klt). The amino acid residues differed from ancestral sequence (green) and bound PMFS (the protease inhibitor at the active site; red) are shown in ball and stick models. b. Amino acid sequence alignment of human chymase (human-Chy) and ancestral chymase (anc-Chy). The amino acid residues of anc-Chy are shown only for the sites different from the human chymase sequence.

a

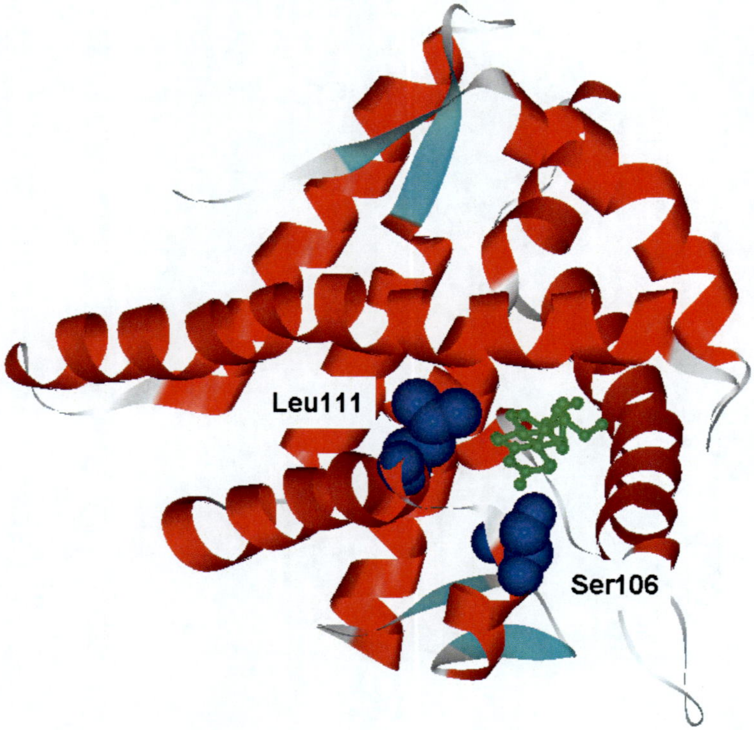

b

```
                  10         20         30         40         50         60         70
         ....|....| ....|....| ....|....| ....|....| ....|....| ....|....| ....|....|
AncCR    LISILEAIEP EVVYAGYDNS QPDTTNYLLS SLNRLAGKQM VSVVKWAKAL PGFRNLHLDD QMTLIQYSWM 70
humanGR  .V.L..V... ..L.....S. V..S.WRIMT T..M.G.R.V IAA......I .......... ....L..... 70
humanMR  PVMV..N... .I......S. K...AEN... T......... IQ......V. ...K..P.E. .I........ 70

                  80         90        100        110        120        130        140
         ....|....| ....|....| ....|....| ....|....| ....|....| ....|....| ....|....|
AncCR    SLMAFSLGWR SYKHTNGQML YFAPDLIFNE ERMQQSAMYD LCQGMRQISQ EFVRLQVTYE EFLCMKVLLL 140
humanGR  .....A.... ..RQSSANL. C......I.. Q..TLPC... Q.KH.LYV.S .LH....S.. .Y....T... 140
humanMR  C.LS.A.S.. ......S.F. ......V... .K.H.....E .....H...L Q.....L.F. .YTI...... 140

                 150        160        170        180        190        200        210
         ....|....| ....|....| ....|....| ....|....| ....|....| ....|....| ....|....|
AncCR    LSTVPKDGLK SQASFDEMRM NYIKELRRAI ARNENNSSQN WQRFYQLTKL LDSMHDLVGG LLQFCFYTFV 210
humanGR  ..S....... ..EL...I.. T.....GK.. VKR.G..... .......... .....EV.EN ..NY..Q..L 210
humanMR  ...I...... ...A.E...T .......KMV TKAP...G.S .......... ........SD ..E......R 210

                 220        230        240
         ....|....| ....|....| ....|....| ....|..
AncCR    QSQALSVEFP EMLVEIISDQ LPKVMAGMAK PLLFHK- 246
humanGR  D-KTM.I... ...A...TN. I..YSN.NI. K....QK 246
humanMR  E.H..K.... A......... ....ES.N.. ..Y..RK 247
```

Figure 5. a. Three dimensional structure of AncCR in complex with aldosterone (PDB code 2q1h). Bound aldosterone is shown in ball and stick model (green). The key residues for specificity shift of GR are shown in space filling models (blue). b. Amino acid sequence alignment of AncCR, human glucocorticoid receptor (humanGR), and human mineralocorticoid receptor (humanMR). The amino acid residues of humanGR and humanMR are shown only for the sites different from the AncCR sequence.

Thornton and colleagues have been studying steroid receptors in terms of their evolution in ligand specificities using ancestral protein reconstruction for nearly a decade [5, 36-43]. Vertebrates have six homologous nuclear receptors for steroid hormones: the estrogen receptors alpha and beta (ERα and ERβ), androgen receptor (AR), progesterone receptor (PR), glucocorticoid receptor (GR), and mineralocorticoid receptor (MR). Because it is thought that these proteins have evolved from a common ancestor through a series of gene duplications [44], the reconstruction of their ancestral proteins is useful for investigating their evolutionary process of ligand-specificity. Although GR and MR are close relatives, GR is activated only by the stress hormone cortisol in most vertebrates, while MR is activated by both aldosterone and cortisol [45, 46]. The amino acid substitutions responsible for the specificity of GR toward cortisol were identified by the reconstruction studies of the common ancestor of GRs and MRs using the ML methods [38-40]. Thornton and colleagues reconstructed the ancestral corticoid receptor (AncCR), which corresponded to the protein at the duplication event between the genes of GR and MR. Functional analysis showed that AncCR could be activated by both aldosterone and cortisol, suggesting that GR of vertebrates had lost aldosterone specificity in the evolutionary process. Furthermore, the site-direct mutagenesis and X-ray crystallographic studies of AncCR revealed that Ser106Pro and Leu111Gln were the key substitutions for the specificity shift of GR (Figure 5) [38-40]. AncCR is, at this point, the only complete domain ancestor (ligand-binding domain only was reconstructed in these studies) for which three-dimensional structure is determined.

Benner and colleagues reconstructed twelve candidates of ancestral alcohol dehydrogenases (Adh_A) that were the last common ancestors of alcohol dehydrogenase 1 (Adh1) and alcohol dehydrogenase 2 (Adh2) from yeast using the ML method [5]. Although Adh1 and Adh2 differ by only 24 out of a total 348 amino acids, the equilibrium between substrate and product is inverted between these enzymes. As a result, Adh1 catalyzed acetaldehyde to make ethanol, and Adh2 converted ethanol into acetaldehyde under the cellular concentration of these compounds [47].

In many cases, an ancestral sequence cannot be unambiguously determined, and several near equally probable amino acids might be assigned to a residue site. The case of Adh_A reconstruction provided a solution to avoid false conclusion due to such ambiguity. When the posterior probability of Adh_A sequence anticipated by the ML method was examined, those of three sites were found to be low. The sites 168, 211 and 236 had two (Met and Arg), three (Lys, Arg and Thr), and two (Asp and Asn) equally probable candidates as the ancestral residues, respectively (Figure 6). Therefore, they reproduced all possible combinations (2 x 3 x 2 = 12) of the candidates at the ambiguous sites. The kinetic properties of these ancestors rather resembled Adh1 than Adh2, and all Adh_As preferred acetaldehyde as the substrate and catalyzed it into ethanol. Thus, the results confirmed that acetaldehyde metabolism was the original function of Adh, and ancestral yeast could not consume ethanol. The function of ethanol metabolism was most likely acquired in the linage of Adh2 after the gene duplication.

a

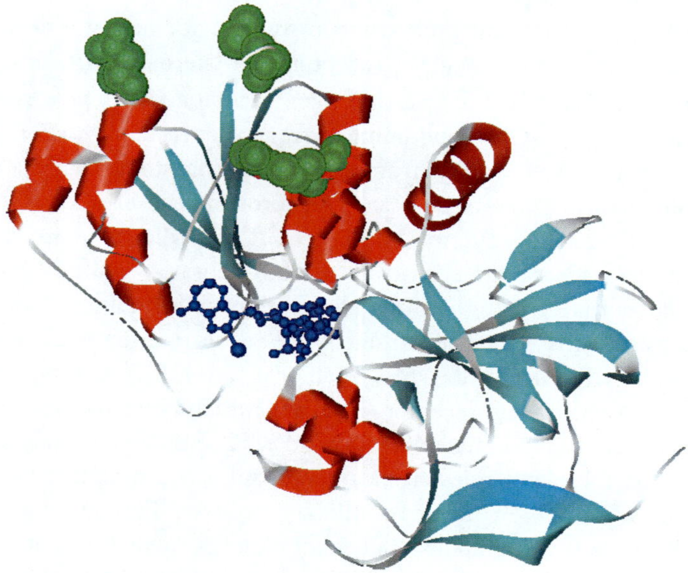

b

```
                10         20         30         40         50         60         70
        ....|....| ....|....| ....|....| ....|....| ....|....| ....|....| ....|....|
Adh1    SIPETQKGVI FYESHGKLEY KDIPVPKPKA NELLINVKYS GVCHTDLHAW HGDWPLPVKL PLVGGHEGAG  70
Adh2    .......AI. ....N....H .........P .......... .......... .......T.. ..........  70
AdhA    .......... ....N..... .........P .......... .......... .......T.. ..........  70

                80         90        100        110        120        130        140
        ....|....| ....|....| ....|....| ....|....| ....|....| ....|....| ....|....|
Adh1    VVVGMGENVK GWKIGDYAGI KWLNGSCMAC EYCELGNESN CPHADLSGYT HDGSFQQYAT ADAVQAAHIP  140
Adh2    .......... .......... .......... .......... .......... ......E... ..........  140
AdhA    ...A...... .......... .......... .......... .......... .......... ..........  140

               150        160        170        180        190        200        210
        ....|....| ....|....| ....|....| ....|....| ....|....| ....|....| ....|....|
Adh1    QGTDLAQVAP ILCAGITVYK ALKSANLMAG HWVAISGAAG GLGSLAVQYA KAMGYRVLGI DGGEGKEELF  210
Adh2    ......E... .......... .......R.. ..A....... .......... .......... ...P......  210
AdhA    ......E... .......... .......X.. .......... .......... .......... ..........  210
                                     [M,R]
               220        230        240        250        260        270        280
        ....|....| ....|....| ....|....| ....|....| ....|....| ....|....| ....|....|
Adh1    RSIGGEVFID FTKEKDIVGA VLKATDGGAH GVINVSVSEA AIEASTRYVR ANGTTVLVGM PAGAKCCSDV  280
Adh2    T.L....... ........S. .V...N.... .I........ ........C. ....V....L .......S..  280
AdhA    X.L....... .......... .I...X.... .......... ........C. ....V....L .......S..  280
        [K,R,T]                   [D,N]
               290        300        310        320        330        340
        ....|....| ....|....| ....|....| ....|....| ....|....| ....|....| ....|..
Adh1    FNQVVKSISI VGSYVGNRAD TREALDFFAR GLVKSPIKVV GLSTLPEIYE KMEKGQIVGR YVVDTSK 347
Adh2    ..H....... .......... .......... .......... ...S...... .......A.. ....... 347
AdhA    .......... .......... .......... .......... ...N...... .......... ....... 347
```

Figure 6. a. Three dimensional structure of Adh1 of yeast (PDB code 2hcy). Bound NADH is shown in ball and stick model (blue). Low posterior probability sites are shown in space filling models (green). b. Amino acid sequence alignment of Adh1, Adh2 and Adh_A. The amino acid residues of Adh2 and Adh_A are shown only for the sites different from the Adh1 sequence.

Reconstruction of Ancient Thermophilic Proteins

As mentioned above, the common ancestor of all organisms is hypothesized to be a hyperthromophile. Yamagishi and colleagues reported several ancestral protein reconstructions using isopropylmalate dehydrogenase (IPMDH), isocitrate dehydrogenase (ICDH), and Glycyl-tRNA synthetase families, and demonstrated that introductions of ancestral state amino acids into modern proteins frequently thermostabilize the molecules [27-30]. These studies demonstrated that the ancestral protein reconstruction was also useful strategy in thermostabilization protein engineering.

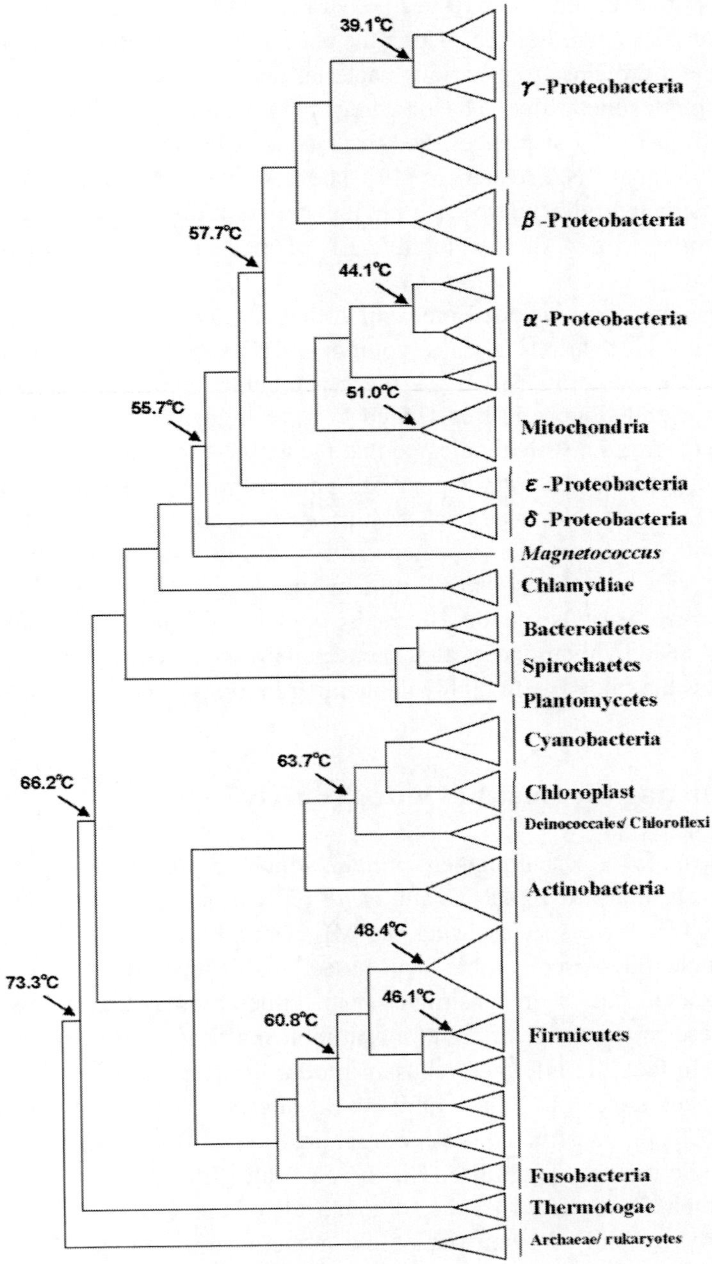

Figure 7. Schematic phylogenetic tree of EFs. The melting temperatures observed for ancestral EFs are indicated at the respective nodes.

Benner and colleagues also reconstructed ancestral elongation factors (EF) Tu predicted by the ML method to infer the physical environment surrounding ancient organisms [7]. Because EFs play a crucial role in protein synthesis in cells, the thermostability of EFs show a strong correlation with the optimal growth temperature of their host organisms. For example, the melting temperatures (T_m) of EFs from *Escherichia coli* and *Thermus thermophilus* (HB8) are 42.8°C and 76.7°C, respectively, and the optimal growth temperatures of their respective host are about 40°C and 74°C, respectively [48]. Thus, EFs are suitable to assess the ambient temperature for ancient life. To predict the ancestral sequences of EFs, amino acid sequences of fifty EF-Tu proteins from various bacterial lineages were used to construct two kinds of molecular phylogenetic tree. The first tree was constructed by using evolutionary distances evaluated from EF-Tu sequences, and the second one was based on the distances calculated from the ribosomal RNA sequences [49]. The last common ancestors of various bacterial EFs were reconstructed referring to each phylogeny. As a result, both ancestors had temperature profiles similar to that of thermophilic EF of modern *Thermus aquaticus*, supporting the hypothesis.

Furthermore, they included more microbial species into the analyses, and reconstructed the ancestral EFs at various depths (evolutionary distance from present time) in the phylogeny using the ML method [50, 51]. It was demonstrated that the ancestral EFs positioned closer to the root of the phylogenetic tree tended to have higher thermostabilities with a significant correlation (Figure 7), which indicated that the ambient temperature for ancient life should be higher than present. This study provided another way to overcome inherent ambiguity in ancestral sequences. Notably, because this study reconstructed the last common ancestral protein of all organisms, the expected error in the inferred sequence should be large, even though the relatively conservative protein was used. Therefore, they produced a lot of ancestors over various depths through phylogeny, and statistically compensated the uncertainty of individual ancestral sequences. However, testing many probable ancestors to the extent of this study is still highly demanding in terms of cost and time.

Reconstructing the Ancient Optic Property

Rhodopsin is the visual pigment protein, which absorbs light and converts its energy into ion gradient to transmit signal. Chang *et al.* reconstructed an ancestral archosaur rhodopsin from thirty vertebrate species using the ML method with three different types of models, namely, nucleotide-based, amino acid-based, and codon-based models [8]. Although an ancestral protein can be reconstructed with either model, the results might be different, because these models use different substitution matrices, which are not fully consistent to each other. In fact, the inferred archosaur rhodopsin using these models had the same amino acid sequences except for three amino acid sites (namely, positions 213, 217, and 218) (Figure 8). They reconstructed the ancestral protein, and four variants at the ambiguous sites, namely, single mutants Thr213Ile, Thr217Ala, Val218Ile, and the triple mutant of them, and measured their absorption spectra. Although all of the spectra of the ancestral rhodopsins were slightly red-shifted from that of the modern vertebrates (482-507 nm), they showed similar optical properties with apparent absorption maximum at 508-509 nm. These data indicated that the alternative ancestral amino acids predicted by the different likelihood models showed similar functional characteristics. So far, various wavelength-sensitive visual pigments have been also subjected to the ancestral reconstruction experiments, and the significant shift in absorption spectra has been detected in most of the ancestral proteins [53-56].

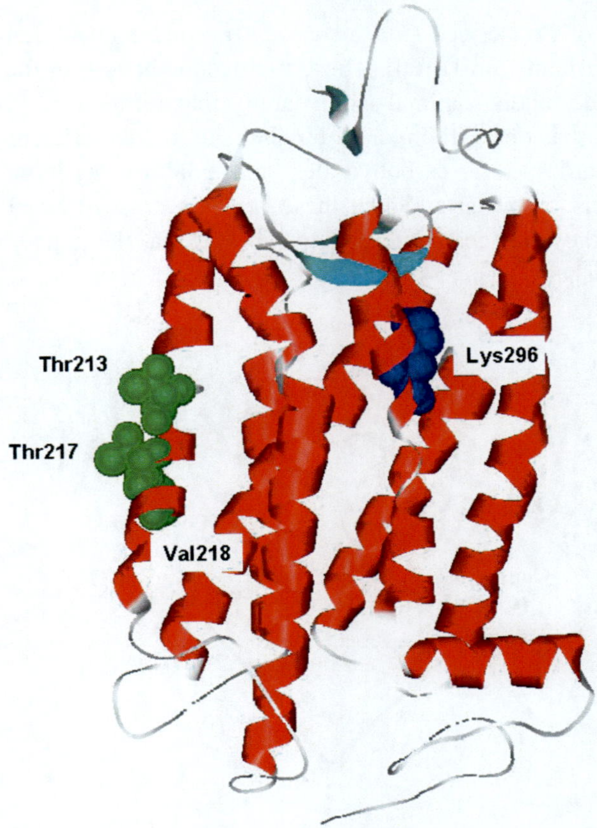

Figure 8. Three dimensional structures of an ancestral archosaur rhodopsin predicted by homology modeling using the bovine rhodopsin structure (PDB code 1u19) as a template. Homology modeling was performed at Swiss-model site [52]. Amino acid residues substituted in the inferred sensitive ancestral sequences and the 11-cis retinal binding lysine are shown in space filling models, and colored in green and blue, respectively.

The great star coral *Montastrea cavernosa* has several green fluorescent protein (GFP)-like proteins, which are classified into four paralogous groups based on their emission spectra: cyan (emission maximum at 480-495 nm), short wavelength green (500-510 nm), long wavelength green (515-525 nm), and red (575-585 nm) [57]. Matz and colleagues reconstructed the ancestral fluorescent proteins corresponding to the root of each color group, and the common ancestor of the groups by using the ML method [9, 58]. The analyses of the fluorescence spectra of the ancestral proteins depicted the evolutional process of the coral GFP-like proteins, in which the peak wavelength has shifted from green to red. Furthermore, they identified the amino acid substitutions responsible for the generation of recent cyan and red fluorescence proteins through the site-direct mutagenesis studies on the ancestral green fluorescent protein as a template [59].

THE CASE OF ANCESTRAL FISH GALECTIN

The Ancestor of the Proteins Differentiating under Natural Selection Pressure

The skin mucus of Conger eel, *Conger myriaster*, contains two lectins named congerin I and congerin II (ConI and ConII) [60]. These two lectins belong to the proto-type galectin, which is β-galactoside binding animal lectin family that forms a non-covalent homodimer. Because each subunit has one carbohydrate-binding site, the dimeric congerins have divalent binding activity toward specific carbohydrates, and exhibit cross-linking and agglutinating activities against cells or bacteria. Since these proteins are expressed in skin and upper-digestive tract, and they can coagulate pathogenic bacteria, the congerins are though to be defense proteins of fishes.

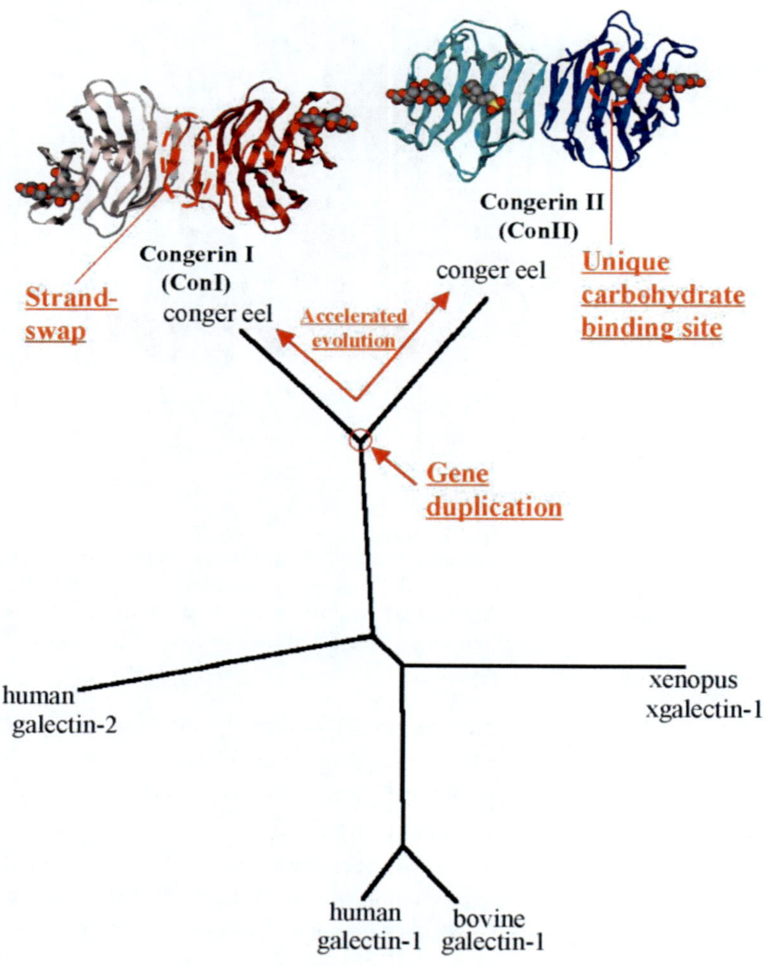

Figure 9. The molecular phylogenetic tree of galectins with the schematic representation of the dimer structures of ConI and ConII. Congerins have evolved by accelerating evolutionary rate after the gene duplication event.

Each subunit of ConI and ConII consists of 136 and 135 amino acid residues, respectively, and shares 47% amino acid identity [61, 62]. The previous studies of ConI and ConII based on the molecular evolutionary and X-ray crystallography analyses have revealed that these proteins have been evolving in high-speed under natural selection pressure. As a consequence of the selection pressure, ConI acquired a new strand-swap fold, and ConII acquired unique ligand-binding site (Figure 9) [63-68]. The new fold in ConI stabilizes its dimer structure, which is essential for the pathogen coagulating activity. The new binding site in ConII might facilitate a defense reaction against novel carbohydrates on pathogen. Therefore, the two different structural features observed in ConI and ConII are thought to be the reactions for natural selection pressure, which is most probably poured from invasions from the environment. The most interesting about congerins is that the two duplicated genes are reacting differently, therefore differentiating, against selection pressure. The congerins might provide an excellent model system in the study of functional differentiation of homologous proteins in natural evolution.

Recently, the authors of this chapter reconstructed the ancestral congerin, Con-anc, that had the putative amino acid sequence at the gene duplication between ConI and ConII genes in the molecular phylogenetic tree (Figure 9) in order to understand the mechanism of the rapid adaptive differentiation process of the proteins [11].

Ancestral Congerin Gene Construction

If the predicted sequence of an ancestral protein was close to the present protein, applying a series of site-directed mutagenesis to a cloned gene template should be the most feasible method to obtain ancestral protein. As already mentioned, however, this is rarely the case in ancestral protein construction, and a total gene synthesis or assembling partially mutated gene fragments is the only choice. In our case, the gene of Con-anc predicted by the MP method using the ODEN system [69] was constructed by assembling gene fragments generated by polymerase chain reaction (PCR) mutagenesis method, with the synthetic oligonucleotides bearing the ancestral sequence. Long synthetic oligonucleotides tend to form secondary structures, which might cause deletions or insertions in the PCR products. Therefore, we designed relatively short PCR primers to generate ancestral gene fragments, and ligated the gene fragments into a whole gene, step by step.

The overlapping oligonucleotides method is also used in constructing whole ancestral genes in the studies reviewed above [50, 70, 71]. In this method, the forward and reverse ancestral gene sequences are divided into complementally overlapping single-stranded fragments of 30-35 bases, which are chemically synthesized. Then, the fragments are ligated into a double-stranded gene though the overlapping regions including extra 5' terminal segments for PCR primer annealing. PCR amplification is finally performed using the ligation product to clone ancestral gene.

The Molecular Properties of Ancestral Congerin

The reconstructed ancestral gene of congerins was successfully cloned, and the recombinant Con-anc was expressed in soluble form in *Escherichia coli* expression system. To evaluate the structural stability of Con-anc, temperature- and urea-induced denaturations of the protein were measured, and their results were compared with those of ConI and ConII. Thermostability was assayed by residual hemagglutinating activity of the proteins after incubation at various temperatures for 30 min, and it was demonstrated that the temperatures, at which 50% of the initial hemagglutination activity was retained, were estimated to be 46.5°C for Con-anc and ConII, and 51°C for ConI.

The urea-induced denaturations estimated by the shift in far-UV circular dichroism (CD) spectra showed that the EC50 values for urea concentrations, under which 50% of total protein denatured, were found to be 3.58, 5.04 and 3.50 M for Con-anc, ConI, and ConII, respectively. These observations indicated that the stability of Con-anc was similar to that of ConII, and lower than that of ConI [11].

The carbohydrate-binding activity of Con-anc was characterized by the frontal affinity chromatography (FAC) [72-74] and the surface plasmon resonance (SPR) analyses [75]. The dissociation constants (K_d) for 34 kinds of pyridylaminated oligosaccharides with Con-anc determined by FAC showed that the sugar-binding profile of Con-anc closely resembled that of ConII except for the carbohydrates containing α2, 3-sialyl galactose such as GM3 and GD1a. Con-anc had generally weak binding activities as compared to ConI. The precise binding activity of Con-anc to α2, 3-sialyllactose was also analyzed by SPR. α2, 3-Sialyllactose is one of the characteristic carbohydrate moieties, which were recently found in marine bacteria. As a result, the overall character in carbohydrate-binding activities of Con-anc was similar to that of ConII in terms of thermostability and carbohydrate recognition specificity, although Con-anc shared a higher sequence similarity with ConI (76%) than with ConII (61%).

These analyses of Con-anc suggested functional evolutionary process of ConI and ConII: ConI has evolved from the ancestral congerin to increase the structural stability and the sugar-binding activity, whereas it has lost affinity to α2, 3-sialyl galactose. On the other hand, ConII has evolved to enhance the affinity to α2, 3-sialyl galactose (Figure 10). Thus, the results appeared to support the hypothesis of congerin evolution: ConI and ConII are differentiating under natural selection pressure. The selection pressure for ConI was thought to be structure stabilization. The lower stability of Con-anc (and ConII) than ConI was consistent with this assumption. The selection pressure, to which ConII reacted, was hypothesized to be a shift in carbohydrate affinity. The observed difference in α2, 3-sialyl galactose affinities between Con-anc and ConII was supportive for this hypothesis.

Ancestor reconstructions can also give useful information of the structure-activity relationship of proteins through a comparison between extant and ancestral sequences. In the case of ancestral congerin reconstruction, the candidates of amino acid residues responsible for the higher structural stability and carbohydrate-binding activity of ConI were reduced to 31 amino acid residues from a total of 71 residues, which were apparently different between ConI and ConII (Figure 10).

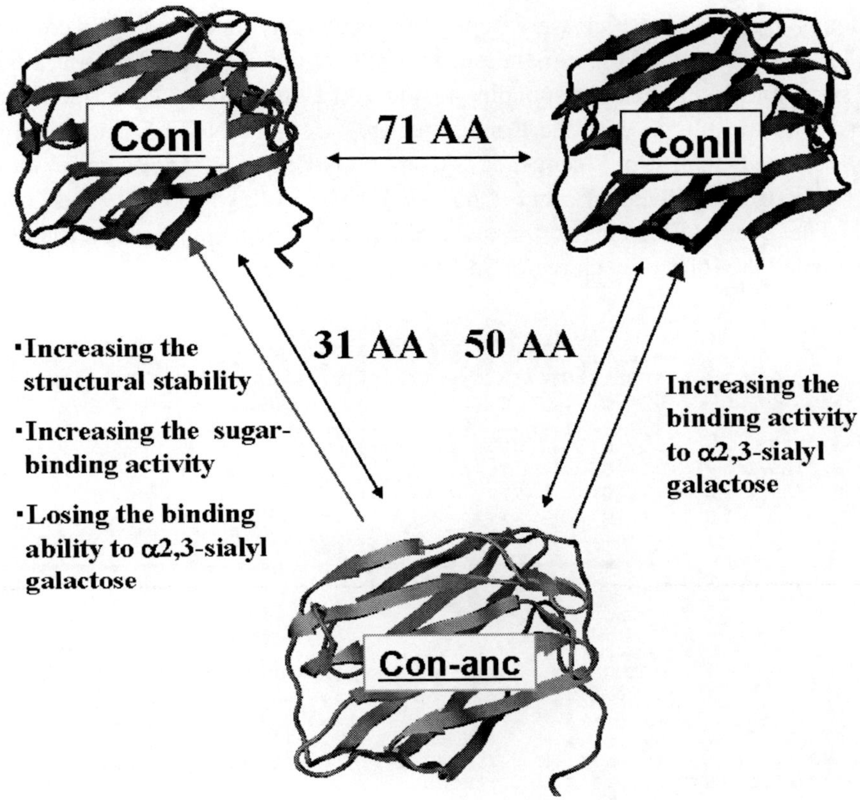

Figure 10. Functional evolutionary process predicted from the reconstruction and characterization of Con-anc. The numbers of different amino acid residues (AA) among these proteins are shown.

Molecular Evolution Tracing Analysis with Pseudo-Ancestors

Is there any way to study ancestral proteins in further detail? The inherent limitation of the ancestral reconstruction is that the current method can evaluate only the sequence at a node of phylogenetic tree. Although a series of ancestral sequences should have existed on each branch, they are not detected unless a molecular lineage that has branched out from the corresponding branch is discovered.

Therefore, to identify the residues responsible for the ConI properties, we next performed the molecular evolution tracing analysis by constructing pseudo-ancestral congerins [76]. This is a protein engineering approach that a part of amino acid residues of present protein are substituted with those of ancestor, to reveal the determinant residues of functional differences between present and ancestral proteins. Although these partially ancestral mutants still have a chance to be actual intermediate ancestors, because they have a part of substitutions between on-node ancestor (Con-anc) and leaf protein (present ConI), the probability of sequence is much lower as compared with the on-node ancestor. Therefore, they were called pseudo-ancestors.

The pseudo-Con-anc proteins focused on N- and C-terminals region and some loop regions (loops 3, 5 and 6), which were important for carbohydrate binding, were constructed (Figure 11). The constructed pseudo-ancestors were Con-anc-N/C (N- and C-termini of Con-anc were substituted with the corresponding residues of ConI), Con-anc-L5 (the loop 5 region of Con-anc was substituted with the ConI sequence), Con-anc-N/C/L5 (loop 5 of Con-anc-N/C was replaced with ConI sequence), Con-anc-N/C/L5/L6 (additionally loop 6 was replaced with ConI sequence), and Con-anc-N/C/L5/L6/L3 (loop 3 was additionally replaced). The characterization of thermostability and sugar-binding specificity of these mutants revealed the roles of each region in ConI evolution (Figure 12).

```
                              10         20         30         40         50
                      ....|....| ....|....| ....|....| ....|....| ....|....|
Con-anc               SGGLEVKNFD FKVGKFLTVG GVINNYAKRF SINVGESTNS LSLHLDHRFN  50
Con-anc-N/C           ....Q..... .T........ .......... .......... ..........  50
Con-anc-L5            .......... .......... .......... .......... ..........  50
Con-anc-N/C/L5        ....Q..... .T........ .......... .......... ..........  50
Con-anc-N/C/L5/L6     ....Q..... .T........ .......... .......... ..........  50
Con-anc-N/C/L5/L6/L3  ....Q..... .T........ .......... .......M.. ..........  50
ConI                  ....Q..... .T........ .F...SPQ.. .V.....M.. ..........  50
ConII                 .DRA..R.IP ..L.MY.... ..V.SN.T.. ........D. IAM.M....S  50
                         S1        L1        F2        L2     S3     L3    S4

                              60         70         80         90        100
                      ....|....| ....|....| ....|....| ....|....| ....|....|
Con-anc               YGADQNTIVL NSMVN--SGW ETEQRSKNFP FSAGEYFEIT ITFDTNKFYI  98
Con-anc-N/C           .......... .....--... .......... .......... ..........  98
Con-anc-L5            .........M ..TLKGDN.. .......... .......... .......... 100
Con-anc-N/C/L5        .........M ..TLKGDN.. .......... .......... .......... 100
Con-anc-N/C/L5/L6     .........M ..TLKGDN.. ......T..T L......... .......... 100
Con-anc-N/C/L5/L6/L3  .........M ..TLKGDN.. ......T..T L......... .......... 100
ConI                  .........M ..TLKGDN.. ......T..T L...Q..... LSY.I..... 100
ConII                 ......VL.. ..L.H-NV.. QQ.E...K.. .TK.DH.QT. .....HT...  99
                      L4    S5          L5     S6      L6        F3     L7  F4

                              110        120        130
                      ....|....| ....|....| ....|....| ....|.
Con-anc               ELLDGHKLEF PNRYKKEALN FLSLAGDARL TFVKLE 134
Con-anc-N/C           .......... .......F.P .......... .L.... 134
Con-anc-L5            .......... ......... .......... ...... 136
Con-anc-N/C/L5        .......... .......F.P .......... .L.... 136
Con-anc-N/C/L5/L6     .......... .......F.P .......... .L.... 136
Con-anc-N/C/L5/L6/L3  .......... .......F.P .......... .L.... 136
ConI                  DI...PN... ....S..F.P .......... .L.... 136
ConII                 Q.SN.ETV.. ...N.DA.F. LIY....... ...R.. 135
                         L8   F5      L9       S2   L10    F1
```

Figure 11. Amino acid sequences alignment of ConI, ConII, Con-anc, and pseudo-Con-anc mutants. The amino acid residues are shown only for the site substituted from Con-anc sequence. The positions of the strands (S1-S6, F1-F5) and loops (L1-L10) are indicated by arrows and thin horizontal lines, respectively.

Con-anc-N/C, Con-anc-L5, and Con-anc-N/C/L5 demonstrated higher thermostability compared with Con-anc. In particular, Con-anc-N/C/L5 had almost the same thermostability as ConI, whereas the carbohydrate-binding activity of the pseudo-ancestor was only 30–40% of ConI. The result suggested the substitutions at these three regions were responsible mainly for thermostability of ConI. Then, loops 3 and 6 of Con-anc-N/C/L5, which are located close to the carbohydrate-binding site (Figure 12), were additionally replaced with the corresponding sequences of ConI (Figure 11). The pseudo-ancestors, Con-anc-N/C/L5/L6 and Con-anc-N/C/L5/L6/L3, showed higher binding activities to almost all carbohydrates than Con-anc-N/C/L5 and the activities were comparable to that of ConI. However, the substitution of loop 3 decreased the binding-activity specifically to the lacto-*N*-fucopentaose-containing carbohydrates. These results indicated that the loops 3 and 6 played an important role in carbohydrate-binding activity and specificity.

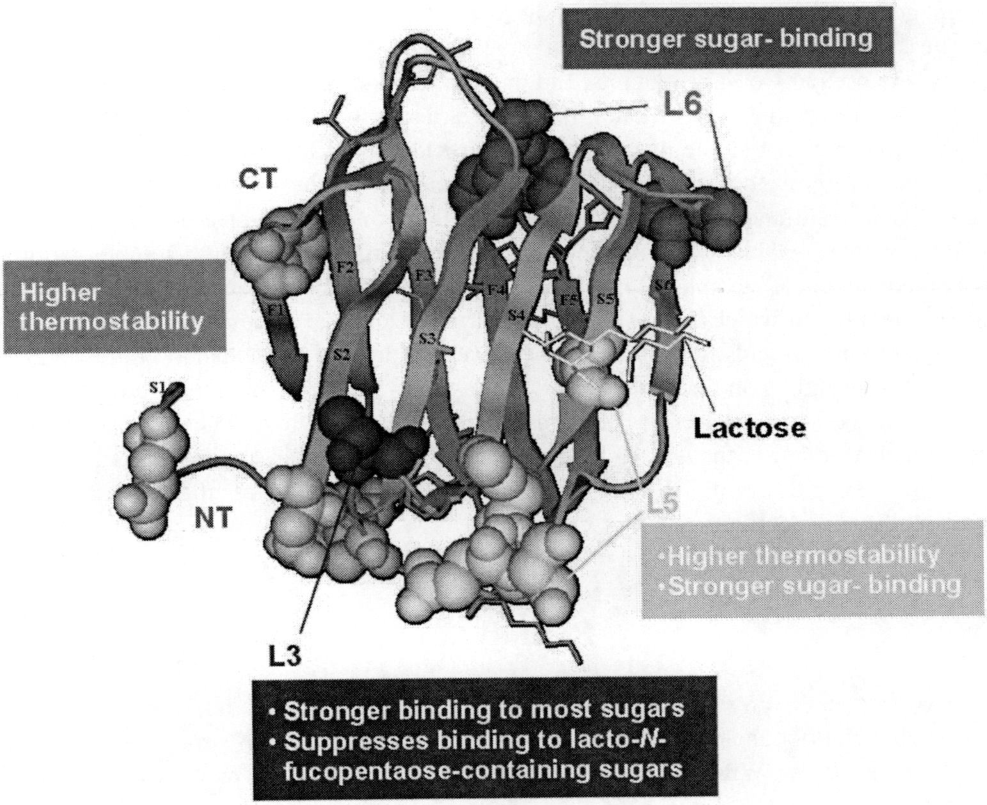

Figure 12. The functional roles revealed by the molecular evolution tracing analysis are indicated for the corresponding regions, on the three dimensional structure of ConI (PDB code 1c1l). The amino acid residues substituted from Con-anc to ConI in N- and C-termini (NT and CT, respectively) and loops 3, 5, and 6 (L3, L5, and L6, respectively) are shown in space-filling models.

The examined course of molecular evolution from Con-anc to ConI, bridged by these pseudo-ancestors in the order of Con-anc-N/C, N/C/L5, N/C/L5/L6, and N/C/L5/L6/L3, is only one of the huge amounts of possible courses. However, constructing such virtual evolutionary course of protein would facilitate the study of structure-function relationship of proteins. In the case of Con-anc, the pseudo-ancestors successfully elucidated the regions of the protein relevant to the adaptive features of ConI, namely, thermostability and higher carbohydrate-binding activity.

CONCLUSION

The ancestral protein reconstruction is a new and potentially useful method for molecular evolutionary study. This method can provide us with the experimental information of ancient genes and proteins, which cannot be obtained from fossil records or from simply constructing molecular phylogeny. Meanwhile, ancestral sequences have a major problem of uncertainty depending on the choice of evaluation method, evolutionary model, and sequences. We recently reconstructed another ancestral congerin by the ML method based on more sequences than the study introduced in this chapter, including recently determined fish galectin genes. The revised ancestor showed, to a certain extent, different sequence and properties from Con-anc, and it is under investigation to elucidate the relationship among Con-anc, the pseudo-Con-anc mutants, and the revised Con-anc.

The ancestral protein construction is also regarded to be a protein engineering technique. It is often the case that single or multiple site-directed mutation of a current protein leads to decrease in solubility, or even to a complete disruption of the function or structure of the protein. In the mentioned cases of Con-anc and other complete reconstruction studies, more than 20% of total residues were replaced at once, and still proteins were soluble and active. This is a considerable advantage in protein engineering. Actually, we also constructed some congerin mutants, in which the residues in N-terminus or loop regions were substituted with non-ancestral amino acids, and some of them were expressed as inclusion bodies in *E. coli*. On the other hand, Con-anc and all of the pseudo-ancestral mutants were successfully expressed in the same bacteria as the soluble form in high yield. Therefore, the ancestral protein reconstruction is thought to be a powerful tool in protein engineering, if the aim of engineering coincides with the expected ancestral properties of the protein, such as thermostability or specificity.

REFERENCES

[1] Stackhouse, J., Presnell, S. R., McGeehan, G. M., Nambiar, K. P. & Benner, S. A. (1990). The ribonuclease from an extinct bovid ruminant. *FEBS Lett.*, 262, 104-106.

[2] Malcolm, B. A., Wilson, K. P., Matthews, B. W., Kirsch, J. F. & Wilson, A. C. (1990). Ancestral lysozymes reconstructed, neutrality tested, and thermostability linked to hydrocarbon packing. *Nature*, 345, 86-89.

[3] Jermann, T. M., Opitz, J. G., Stackhouse, J. & Benner, S. A. (1995). Reconstructing the evolutionary history of the artiodactyl ribonuclease superfamily. *Nature*, 374, 57-59.

[4] Chandrasekharan, U. M., Sanker, S., Glynias, M. J., Karnik, S. S. & Husain, A. (1996). Angiotensin II-forming activity in a reconstructed ancestral chymase. *Science*, 271, 502-505.

[5] Thomson, J. M., Gaucher, E. A., Burgan, M. F., De Kee, D. W., Li, T., Aris, J. P. & Benner, S. A. (2005). Resurrecting ancestral alcohol dehydrogenases from yeast. *Nat. Genet.*, 37, 630-635.

[6] Watanabe, K., Ohkuri, T., Yokobori, S. & Yamagishi, A. (2006). Designing thermostable proteins: ancestral mutants of 3-isopropylmalate dehydrogenase designed by using a phylogenetic tree. *J. Mol. Biol.*, 355, 664-674.

[7] Gaucher, E. A., Thomson, J. M., Burgan, M. F. & Benner, S. A. (2003). Inferring the palaeoenvironment of ancient bacteria on the basis of resurrected proteins. *Nature*, 425, 285-288.

[8] Chang, B. S., Jonsson, K., Kazmi, M. A., Donoghue, M. J. & Sakmar, T. P. (2002). Recreating a functional ancestral archosaur visual pigment. *Mol. Biol. Evol.*, 19, 1483-1489.

[9] Ugalde, J. A., Chang, B. S., & Matz, M. V. (2004). Evolution of coral pigments recreated. *Science*, 305, 1433.

[10] Thornton, J. W. (2001). Evolution of vertebrate steroid receptors from an ancestral estrogen receptor by ligand exploitation and serial genome expansions. *Proc. Natl. Acad. Sci. USA*, 98, 5671-5676.

[11] Konno, A., Ogawa, T., Shirai, T. & Muramoto, K. (2007). Reconstruction of a probable ancestral form of conger eel galectins revealed their rapid adaptive evolution process for specific carbohydrate recognition. *Mol. Biol. Evol.*, 24, 2504-2514.

[12] Eck, R. V. & Dayhoff, M. O. (1966). Evolution of the structure of ferredoxin based on living relics of primitive amino acid sequences. *Science*, 152, 363-366.

[13] Felsenstein, J. (1981). Evolutionary trees from DNA sequences: a maximum likelihood approach. *J. Mol. Evol.*, 17, 368-376.

[14] Stewart, C. -B., Schilling, J. W. & Wilson, A. C. (1987). Adaptive evolution in the stomach lysozymes of foregut fermenters. *Nature*, 330, 401-404.

[15] Shirai, T., Suzuki, A., Yamane, T., Ashida, T., Kobayashi, T. & Ito, S. (1997). High-resolution crystal structure of M-protease: phylogeny aided analysis of the high-alkaline adaptation mechanism. *Protein Eng.*, 10, 191-108.

[16] Messier, W. & Stewart, C. -B. (1997). Episodic adaptive evolution of primate lysozymes. *Nature*, 385, 151-154.

[17] Yang, Z. (2007). PAML 4: phylogenetic analysis by maximum likelihood. *Mol. Biol. Evol.*, 24, 1586-1591.

[18] Pupko, T., Pe'er, I., Shamir, R. & Graur, D. (2000). A fast algorithm for joint reconstruction of ancestral amino acid sequences. *Mol. Biol. Evol.* 17, 890-896.

[19] Cai, W., Pei, J. & Grishin, N. V. (2004). Reconstruction of ancestral protein sequences and its applications. *BMC Evol. Biol.*, 4, 33.

[20] Edwards R. J. & Shields, D. C. (2004). GASP: Gapped ancestral sequence prediction for proteins. *BMC Bioinf.*, 5, 123.

[21] Beintema, J. J. (1987). Structure, properties and molecular evolution of pancreatic-type ribonucleases. *Life Chem. Rep.*, 4, 333-389.

[22] Carroll, R. L. (1988). Vertebrate Paleontology and Evolution, New York, Freeman.

[23] Blackburn, P. & Moore, S. (1984). In: (3rd Edn. ed.), *The Enzymes* Vol. 15, Academic Press, New York, pp. 317-436.

[24] Blake, C. C. F., Koenig, D. F., Mair, G. A., North, A. C. T., Phillips, D. C. & Sarma, V. R. (1965). Structure of Hen Egg-White Lysozyme: A Three-dimensional Fourier Synthesis at 2 Å Resolution. *Nature*, 206, 757-761.

[25] McCammon, J. A., Gelin, B. R., Karplus, M. & Wolynes, P. G. (1976). The hinge-bending mode in lysozyme. *Nature*, 262, 325-326.

[26] Jolles, P. & Jolles, J. (1984). What's new in lysozyme research? Always a model system, today as yesterday. *Mol. Cell. Biochem.*, 63, 165-189.

[27] Miyazaki, J., Nakaya, S., Suzuki, T., Tamakoshi, M., Oshima, T. & Yamagishi, A. (2001). Ancestral residues stabilizing 3-isopropylmalate dehydrogenase of an extreme thermophile: experimental evidence supporting the thermophilic common ancestor hypothesis. *J. Biochem.*, 129, 777-782.

[28] Iwabata, H., Watanabe, K., Ohkuri, T., Yokobori, S. & Yamagishi, A. (2005). Thermostability of ancestral mutants of Caldococcus noboribetus isocitrate dehydrogenase. *FEMS Microbiol. Lett.*, 243, 393-398.

[29] Watanabe, K. & Yamagishi, A. (2006). The effects of multiple ancestral residues on the Thermus thermophilus 3-isopropylmalate dehydrogenase. *FEBS Lett.*, 580, 3867-3871.

[30] Shimizu, H., Yokobori, S., Ohkuri, T., Yokogawa, T., Nishikawa, K. & Yamagishi, A. (2007). Extremely thermophilic translation system in the common ancestor commonote: ancestral mutants of Glycyl-tRNA synthetase from the extreme thermophile *Thermus thermophilus*. *J. Mol. Biol.*, 369, 1060-1069.

[31] Woese, C. R. (1987). Bacterial evolution. *Microbiol. Rev.*, 51, 221-271.

[32] Imada, K., Sato, M., Tanaka, N., Katsube, Y., Matsuura, Y. & Oshima, T. (1991). Three-dimensional structure of a highly thermostable enzyme, 3-isopropylmalate dehydrogenase of *Thermus thermophilus* at 2.2 A resolution. *J. Mol. Biol.*, 222, 725-738.

[33] Husain, A. (1993). The chymase-angiotensin system in humans. *J. Hypertens.*, 11, 1155-1159.

[34] Urata, H., Kinoshita, A., Misono, K.S., Bumpus, F.M. & Husain, A. (1990). Identification of a highly specific chymase as the major angiotensin II-forming enzyme in the human heart. *J. Biol. Chem.*, 265, 22348-22357.

[35] Le Trong, H., Neurath, H. & Woodbury, R. G. (1987). Substrate specificity of the chymotrypsin-like protease in secretory granules isolated from rat mast cells. *Proc. Natl. Acad. Sci. USA*, 84, 364-367.

[36] Thornton, J. W., Need, E. & Crews, D. (2003). Resurrecting the ancestral steroid receptor: ancient origin of estrogen signaling. *Science*, 301, 1714-1717.

[37] Thornton, J. W. (2004) Resurrecting ancient genes: experimental analysis of extinct molecules. *Nat. Rev. Genet.*, 5, 366-375.

[38] Bridgham, J. T., Carroll, S. M. & Thornton, J. W. (2006). Evolution of hormone-receptor complexity by molecular exploitation. *Science*, 312, 97-101.

[39] Dean, A. M. & Thornton, J. W. (2007) Mechanistic approaches to the study of evolution: the functional synthesis. *Nat. Rev. Genet.*, 8, 675-688.

[40] Ortlund, E. A., Bridgham, J. T., Redinbo, M. R. & Thornton, J. W. (2007) Crystal structure of an ancient protein: evolution by conformational epistasis. *Science*, 317, 1544-1548.

[41] Carroll, S. M., Bridgham, J. T. & Thornton, J. W. (2008) Evolution of hormone signaling in elasmobranchs by exploitation of promiscuous receptors. *Mol. Biol. Evol.*, 25, 2643-2652.

[42] Keay, J. & Thornton, J. W. (2009). Hormone-activated estrogen receptors in annelid invertebrates: implications for evolution and endocrine disruption. *Endocrinology*, 150, 1731-1738.

[43] Bridgham, J. T., Ortlund, E. A. & Thornton, J. W. (2009). An epistatic ratchet constrains the direction of glucocorticoid receptor evolution. *Nature*, 461:515-519.

[44] Thornton, J. W. & DeSalle, R. (2000). A new method to localize and test the significance of incongruence: detecting domain shuffling in the nuclear receptor superfamily. *Syst. Biol.*, 49, 183-201.

[45] Bentley, P. J. (1998). Comparative Vertebrate Endocrinology, Cambridge Univ. Press, Cambridge.

[46] Farman, N. & Rafestin-Oblin, M. E. (2001). Multiple aspects of mineralocorticoid selectivity. *Am. J. Physiol. Renal. Physiol.*, 280. F181-192.

[47] Wills, C. (1976). Production of yeast alcohol dehydrogenase isoenzymes by selection. *Nature*, 261, 26-29.

[48] Williams, R. A. D. & da Costa, M. S. (1992). The genus Thermus and related microorganisms. In Balows, A. Truper, H. G., Dworkin, M., Harder, W. & Schleifer, K. -H. (Eds.), *The Prokaryotes* (2nd edition, pp. 3745-3753). New York, Springer.

[49] Hugenholtz, P., Goebel, B. M. & Pace, N. R. (1998). Impact of culture-independent studies on the emerging phylogenetic view of bacterial diversity. *J. Bacteriol.*, 180, 4765-4774.

[50] Gaucher, E. A., Govindarajan, S., & Ganesh, O. K. (2008). Palaeotemperature trend for Precambrian life inferred from resurrected proteins. *Nature*, 451, 704-707.

[51] Gouy, M. & Chaussidon, M. (2008). Evolutionary biology: ancient bacteria liked it hot. *Nature*, 451, 635-636.

[52] Schwede, T., Kopp, J., Guex, N. & Peitsch, M. C. (2003). SWISS-MODEL: an automated protein homology-modeling server. *Nucleic Acids Res.*, 31, 3381-3385.

[53] Shi, Y. & Yokoyama, S. (2003). Molecular analysis of the evolutionary significance of ultraviolet vision in vertebrates. *Proc. Natl. Acad. Sci. USA.*, 100, 8308-8313.

[54] Yokoyama, S. & Takenaka, N. (2004). The molecular basis of adaptive evolution of squirrelfish rhodopsins. *Mol. Biol. Evol.*, 21, 2071-2078.

[55] Chinen, A., Matsumoto, Y. & Kawamura, S. (2005). Reconstitution of ancestral green visual pigments of zebrafish and molecular mechanism of their spectral differentiation. *Mol. Biol. Evol.*, 22, 1001-1010.

[56] Yokoyama, S., Tada, T., Zhang, H. & Britt, L. (2008). Elucidation of phenotypic adaptations: Molecular analyses of dim-light vision proteins in vertebrates. *Proc. Natl. Acad. Sci. USA.*, 105, 13480-13485.

[57] Kelmanson, I. V. & Matz, M. V. (2003). Molecular basis and evolutionary origins of color diversity in great star coral *Montastraea cavernosa* (Scleractinia: Faviida). *Mol. Biol. Evol.*, 20, 1125-1133.

[58] Chang, B. S., Ugalde, J. A. & Matz, M. V. (2005). Applications of ancestral protein reconstruction in understanding protein function: GFP-like proteins. *Methods Enzymol.*, 395, 652-670.

[59] Field, S. F., Bulina, M. Y., Kelmanson, I. V., Bielawski, J. P. & Matz, M. V. (2006). Adaptive evolution of multicolored fluorescent proteins in reef-building corals. *J. Mol. Evol.*, 62, 332-339.

[60] Kamiya, H., Muramoto, K. & Goto, R. (1988). Purification and properties of agglutinins from conger eel, *Conger myriaster* (Brevoort), skin mucus. *Dev. Comp. Immuno.*, 12, 309-318.

[61] Muramoto, K. & Kamiya, H. (1992). The amino-acid sequence of a lectin from conger eel, *Conger myriaster*, skin mucus. *Biochem. Biophys. Acta*, 1116, 129-136.

[62] Muramoto, K., Kagawa, D., Sato, T., Ogawa, T., Nishida, Y. & Kamiya, H. (1999). Functional and structural characterization of multiple galectins from the skin mucus of conger eel, *Conger myriaster*. *Comp. Biochem. Physiol. B*, 123, 33-45.

[63] Ogawa, T., Ishii, C., Kagawa, D., Muramoto, K. & Kamiya, H. (1999). Accelerated evolution in the protein-coding region of galectin cDNAs, congerin I and congerin II, from skin mucus of conger eel (*Conger myriaster*). *Biosci. Biotechnol. Biochem.*, 63, 1203-1208.

[64] Ogawa, T., Shirai, T., Shionyu-Mitsuyama, C., Yamane, T., Kamiya, H. & Muramoto, K. (2004). The speciation of conger eel galectins by rapid adaptive evolution. *Glycoconj. J.*, 19, 451-458.

[65] Ogawa, T. (2006). Molecular diversity of proteins in biological offense and defense systems. *Mol. Div.*, 10, 511-514.

[66] Shirai, T., Matsui, Y., Shionyu-Mitsuyama, C., Yamane, T., Kamiya, H., Ishii, C., Ogawa, T. & Muramoto, K. (2002). Crystal structure of a conger eel galectin (Congerin II) at 1.45 angstrom resolution: Implication for the accelerated evolution of a new ligand-binding site following gene duplication. *J. Mol. Biol.*, 321, 879-889.

[67] Shirai, T., Mitsuyama, C., Niwa, Y., Matsui, Y., Hotta, H., Yamane, T., Kamiya, H., Ishii, C., Ogawa, T. & Muramoto, K. (1999). High-resolution structure of the conger eel galectin, congerin I, in lactose-liganded and ligand-free forms: emergence of a new structure class by accelerated evolution. *Structure*, 7, 1223-1233.

[68] Shirai, T., Shionyu-Mitsuyama, C., Ogawa, T. & Muramoto, K. (2006). Structure based studies of the adaptive diversification process of congerins. *Mol. Div.*, 10, 567-573.

[69] Ina, Y. (1994). ODEN: a program package for molecular evolutionary analysis and database search of DNA and amino acid sequences. *Comput. Appl. Biosci.*, 10, 11-12.

[70] Ferretti, L., Karnik, S. S., Khorana, H. G., Nassal, M. & Oprian, D. D. (1986). Total synthesis of a gene for bovine rhodopsin. *Proc Natl Acad Sci USA.*, 83, 599-603.

[71] Chang, B. S., Kazmi, M. A. & Sakmar, T. P. (2002) Synthetic gene technology: applications to ancestral gene reconstruction and structure-function studies of receptors. *Methods Enzymol.*, 343, 274-294.

[72] Oda, Y., Kasai, K. & Ishii, S. (1981). Studies on the specific interaction of concanavalin A and saccharides by affinity chromatography. Application of quantitative affinity chromatography to a multivalent system. *J. Biochem.*, 89, 285-296.

[73] Hirabayashi, J., Hashidate, T., Arata, Y., Nishi, N., Nakamura, T., Hirashima, M., Urashima, T., Oka, T., Futai, M., Muller, W. E., Yagi, F. & Kasai, K. (2002). Oligosaccharide specificity of galectins: a search by frontal affinity chromatography. *Biochim. Biophys. Acta*, 1572, 232-254.

[74] Hirabayashi, J., Arata, Y. & Kasai K. (2003). Frontal affinity chromatography as a tool for elucidation of sugar recognition properties of lectins. *Methods Enzymol.*, 362, 353-368.

[75] Markey, F. (1999). What is SPR anyway? *BIA J.*, 6, 14-17.

[76] Konno, A., Yonemaru, S., Kitagawa, A., Muramoto, K., Shirai, T. & Ogawa, T. (2010). Protein engineering of conger eel galectins by tracing of molecular evolution using probable ancestral mutants. *BMC Evolutionary Biology,* in press.

In: Ribosomal Proteins and Protein Engineering… ISBN 978-1-60741-005-8
Editors: V. Ortendhal and H. Salchow, pp. 91-98

Chapter 6

Ribosomal Proteins in Important Tropical Diseases

Viroj Wiwanitkit
Wiwanitkit House, Bangkhae, Bangkok Thailand 10160

Abstract

The ribosome is an important cellular organelle that is necessary for the basic cellular process, protein synthesis. Ribosomal protein is a group of proteins that is presented with a focus on biomedical research. In this article, the author will review and present important data on the ribosomal proteins in important tropical diseases including malaria and tuberculosis.

Introduction to the Ribosome

The word "ribosome" is the combination of the two words "ribonucleic acid" and "soma". The meaning is the body of ribonucleic acid. Therefore, the correlation between ribosome and nucleic acid and consequent protein synthesis can be expected [1–2]. Ribosome is an important cellular organelle that is necessary for the basic cellular process, protein synthesis. Since the ribosome is an organelle in the cell, it therefore has a very small size and can be classified as a nanosize particle. Its size is about 20 nm. Functionally, the ribosome is central to protein synthesis and converts transcribed mRNAs into polypeptide chains [2]. The ribosome has been the focus of structural and biochemical studies for decades [2]. Structurally, peptide-bond formation is an important enzymatic activity of the ribosome [3-5]. Due to extensive X-ray crystallographic and NMR studies of the three-dimensional structure of ribosomal proteins, ribosomal RNA components and their complexes with ribosomal proteins and of several translation factors in different functional states lead to a better level of understanding of the mechanism of function of the protein synthesis [6]. Briefly, the ribosome's catalytic site is made up of ribosomal RNA, indicating that the ribosome is a ribozyme [3]. Concerning ribozyme, proteins are structural units that support and stabilize it [7].

The main reaction of protein synthesis, peptidyl transfer, is catalyzed in living organisms by the ribosome [6]. The translation is the well-known biological process of the ribosome. Translocation of the tRNA-mRNA complex is one of the most important tasks performed by the ribosome [8]. The translation can be divided into many steps: initiation, elongation, termination and recycling [9]. Initiation requires many more necessary accessory proteins, called initiation factors, than either elongation or termination [10]. A specific region of ribosomal 30S subunit is a ribosome binding site of regulatory domains of mRNA which locates on the non-translated 5'-end of the mRNA [11]. Marzi et al. reported that the existence of a universal stand-by site on the ribosome (the 30S platform) dedicated for binding regulatory 5' mRNA elements [12]. Marzi et al. also noted that different types of mRNA structures might be accommodated during translation preinitiation and regulated gene expression by transiently stalling the ribosome [12]. Yusupova et al. reported that the presence of the Shine-Dalgarno (SD) duplex caused strong anchoring of the 5'-end of mRNA onto the platform of the 30S subunit, with numerous interactions between mRNA and the ribosome in the initiation and post-initiation complexes [13]. Conversely, Yusupova et al. also reported that the 5' end of the 'elongator' mRNA lacking SD interactions was flexible, suggesting a different exit path for mRNA during elongation [13]. Culver also reported that elements of both 16S rRNA and ribosomal proteins could contribute to mRNA binding [14]. This work also revealed two tunnels that the mRNA passed through as it wrapped around the 30S subunit [14].

Ribosomal Protein: What Is It?

By definition, protein is a sequence of amino acids. Therefore, ribosomal protein is also a sequence of amino acids. Within the ribosome molecule, apart from nucleic acid, another important part is protein. By definition, a ribosomal protein is a protein in conjunction with rRNA that builds up ribosomal subunits involved in the cellular process of translation. There are many ribosomal proteins. Important ribosomal protein families are listed in Table 1.

Table 1. Important ribosomal protein families

Large subunit	Small subunit
L1p (L10Ae), L2p (L8e), L3p (L3e), L4p (L4e), L5p (L11e), L6p (L9e), L9p, L10p (P0), L11p (L12e), L12p (P1/P2), L13p (L13Ae), L14p (L23e), L15p (L27Ae), L16p, L17p, L18p (L5e), L19p, L20p, L21p L22p (L17e), L23p (L23Ae), L24p (L26e), L25p, L27p, L28p, L29p (L35e), L30p(L7e), L31p, L32p, L33p, L34p, L35p, L36p, L6e, L7Ae, L10e, L13e, L14e, L15e, L18e, L18Ae, L19e, L21e, L22e, L24e, L27e, L28e, L29e, L30e, L31e, L32e, L34e, L35Ae, L36e, L37e, L37Ae, L38e, L39e, L40e, L41e, L44e (L36Ae), L49em	S1p (S1e), S2p (SAe), S3p (S3e), S4p (S9e), S5p (S2e), S6p, S7p (S5e), S8p (S15Ae), S9p (S16e), S10p (S20e), S11p (S14e), S12p (S23e), S13p (S18e) S14p (S29e), S15p (S13e), S16p, S17p (S11e), S18p, S19p (S15e), S20p, S21p, S22p, S3Ae, S4e, S6e, S7e, S8e, S10e, S12e, S17e, S19e, S21e, S24e, S25e, S26e, S27e, S27Ae, S28e, S30e, S30Ae, S31e, var1

Ribosomal Proteins in Important Tropical Diseases

Ribosomal protein is a group of proteins that are presentedly the focus in biomedical research. In this article, the author will review and present important data on the ribosomal proteins in important tropical diseases focusing on malaria and tuberculosis.

A. Malaria

Malaria is an important tropical infection. It is classified as a vector borne disease. The mosquito is the specific vector for malaria. Development and approval of new knowledge are the hope for infectious control of the possible emerging pandemic of this pathogen. Despite decades of control success and a competent network of country-wide health infrastructure, malaria remains an important health threat in many countries. The correlation between malaria and ribosomal proteins has been proposed for a long time. A rapid, sensitive diagnosis of malaria based on ribosomal RNA is proposed [15]. According to the literature, it was mentioned that parasite and host ribosomes contain distinct ribosomal RNAs and ribosomal proteins [16]. Miller and Ilan studied the parasite and host ribosome in a rat model [16]. Miller and Ilan found that plasmodial small ribosomal subunits contained 30 proteins ranging in apparent molecular size from 11.7 to 40.7 kDa, while large subunits composed of 35 to 36 proteins ranging from 12.1 to 42.6 kDa and rat liver 40 S ribosomal subunit proteins numbered 30 and ranged from 9.2 to 37.5 kDa, while liver 60 S subunits contained 41 to 43 proteins with molecular sizes ranging from 10.3 to 45.2 kDa [16]. Miller and Ilan also reported that base composition studies showed *P. berghei* rRNA to be low in guanosine and cytosine content, as is the case for general protozoa [17].

There were several attempts to study the ribosomal protein and translation process in malaria. In 2007, Takebe et al. reported their success in purification of components of the translation elongation factor complex of *Plasmodium falciparum* by tandem affinity purification [18]. Takebe et al. reported that the translation elongation complex in *P.falciparum* contained at least two subunits of PfEF-1beta [18]. Indeed, PfEF-1beta has been shown to be phosphorylated for a few years and its phosphorylation is critical for optimal activity [19]. Mamoun and Goldberg found that PfEF-1beta was efficiently phosphorylated by protein kinase C and this phosphorylation could result in a four-fold increase in its nucleotide exchange activity [19]. Mamoun and Goldberg noted that the identification of a translation elongation component as substrate for serine/threonine protein phosphatase 2C (PP2C) suggested an important regulatory function for this enzyme and suggests that it may be a good target for drug design in the fight against malaria [19]. Vinkenoog et al. also reported that malaria parasites contained two identical copies of an EF-1 alpha gene [20]. Basically, EF-1alpha is a protein in eukaryotic cells, involved mainly in translation of mRNA on the ribosomes, and is frequently encoded by more than one gene [20]. Vinkenoog et al. also found the existence of a codon bias for highly expressed genes relating to EF-1 alpha [20].

Concerning the ribosomal protein of malaria, one of the most famous proteins is ribosomal phosphoprotein P0 of *P. falciparum* (PfP0). It is noted that antibodies against PfP0 have been detected in malaria-immune people [21]. Singh et al. reported that a large number of malaria immune persons from Eastern India were found to possess antibodies to the PfP0

[22]. There are many research studies in this area. Chatterjee et al. reported that antibodies against PfP0 of *P. falciparum* protected mice against challenge with *P. yoelii* [21]. They also demonstrated the cross-protection of the anti-PfP0 IgG and established parasite P0 protein as a target for invasion-blocking antibodies [21]. Chatterjee et al. also performed a study to characterize of domains of thePfP0 [23]. According to this work, they concluded PfP0 was a possible target for invasion-blocking antibodies [23]. Goswami et al. noted that this inhibition appeared to be mainly at the step of erythrocyte invasion by the *P. falciparum* [24]. Recently, Aruna et al. studied functional complementation of yeast ribosomal P0 protein with PfP0 [25]. According to this work, Aruna et al. found a close functional relationship of the PfP0 gene to fungal P0 genes [25]. Finally, there is another interesting report that PfP0 could be detected in the patients with systemic lupus erythrematosus (SLE) [22]. Singh et al. found that an SLE patient's serum could inhibit the parasite growth in vitro only during the phase in which anti-PfP0 antibodies were detected [22].

Apart from ribosomal protein of parasite and host, there are some interesting reports on malarial vector. Recently, Li and Fallon reported on recovery of cDNAs encoding ribosomal proteins S9 and L26 from *Aedes albopictus* mosquito cells and identification of their homologs in the malaria vector, *Anopheles gambiae* [26]. According to this report, in the malarial vector, the RpS9 coding region was distributed over 3 exons, spanning 2.6 kb, but the rpL26 protein coding region posed no introns [26]. Gentile et al. performed another work to isolate a fragment homologous to the rp49 constitutive gene of Drosophila, frequently used as an endogenous control in Drosophila gene expression experiments, in the malaria vector *An. Aquasalis* [27]. Gentile et al. concluded that rp49 could be used as a good constitutive control in gene expression studies [27].

B. Tuberculosis

Tuberculosis is an important respiratory infection. It is one of the most problematic infectious diseases of the world. Roles of ribosomal protein are widely mentioned in tuberculosis. There have been continuous research studies for years. The early reports were usually related to the preparation. In 1971, Youmans et al. reported on failure of synthetic polynucleotides to affect the immunogenicity of Mycobacterial ribonucleic acid and ribosomal protein preparations [28]. From this work, Youmans and Youmans concluded for the lack of correlation between conventional antibody formation and immunity against tuberculosis [28]. Youmans and Youmans noted that Mycobacterial RNA not incorporated in Freund's incomplete adjuvant was nonimmunogenic [29]. In 1972, Youmans and Youmans reported another additional work on effect of polybasic amines on the immunogenicity of mycobacterial ribonucleic acid [30]. According to this work, methylated bovine serum albumin, protamine sulfate, and neomycin sulfate could reduce the biological activity of mycobacterial RNA after complexing with it [30]. Youmans and Youmans noted that it was probable that these compounds could mask the immunogenic sites on the mycobacterial RNA structure [30]. In 1983, Portelance and Beaudet studied the biochemical characterization of three mycobacterial ribosomal fractions [31]. In this work, the induction of antituberculous immunity by crude ribosomal fractions isolated from *Mycobacterium tuberculosis* strain H37Ra, *M. bovis* strain BCG, and *M. smegmatis* was studied in CF-1 mice and it resulted that each purified preparation showed low antituberculous activity when injected separately, but when mixed together a high protection was determined [31].

There were also other reports on tuberculosis ribosomal protein after that. Important reports were on drug resistance. Medow et al studied to compare of ribosomes and ribosomal proteins of sensitive and resistant mycobacteria [32]. Medow et al concluded that streptomycin-resistant cells in comparison to sensitive cells have two additional 30S proteins [32]. Yamada et al also reported on pleiotropic antibiotic resistance mutations associated with ribosomes and ribosomal subunits in *M. smegmatis* viomycin-resistant strains [33]. McClatchy et al concluded that because of this variability in cross-resistance and the fact that kanamycin was a more toxic drug than capreomycin, therefore, capreomycin should be used in the first retreatment regimen for tuberculosis when streptomycin resistance had been demonstrated [34]. In 1992, Yamada et al performed genetic and biochemical analysis of ribosomal proteins of minocycline-susceptible and -resistant *M. smegmatis* and found that minocycline resistance could be mapped close to genes encoding the structure of ribosomal proteins in *M. smegmatis* [35].

The mutation research on ribosomal protein of tuberculosis is also widely mentioned. For sure, the implication is related to drug resistance. Doner et al. noted that specific mutations associated with resistance to streptomycin in Mycobacterium tuberculosis suggested themselves for its rapid prediction [36]. Meier et al. noted for ribosomal protein mutations associated with streptomycin resistance in *M. tuberculosis* [37]. In this report, a single point mutation at 16S rRNA position 491 (C-->T) or 512 (C-->T) was mentioned [37]. Okamoto et al. also recently reported that loss of a conserved 7-methylguanosine modification in 16S rRNA conferred low-level streptomycin resistance [38].

Finally, there are also other reports on the host responses to tuberculosis ribosomal proteins. Miyazaki et al. reported that the ribosomes from BCG strongly induced delayed type hypersensitivity (DTH) skin reactions in guinea pigs immunized with live BCG or heat killed *M. tuberculosis* H37Rv, and also induced lymphocyte proliferative response in mice immunized with ribosomes [39]. Kitaura et al. noted that ribosomal protein L7 included in tuberculin purified protein derivative (PPD) was a major heat-resistant protein inducing strong DTH [40]. The 16-kDa protein in the 50S subunit was mentioned as the important immune induced protein [41].

REFERENCES

[1] Robinson CV, Sali A, Baumeister W. The molecular sociology of the cell. *Nature.* 2007 Dec 13;450(7172):973-82.

[2] Steitz TA. A structural understanding of the dynamic ribosome machine. *Nat Rev. Mol. Cell Biol.* 2008 Mar;9(3):242-53.

[3] Rodnina MV, Beringer M, Bieling P. Ten remarks on peptide bond formation on the ribosome. *Biochem. Soc. Trans.* 2005 Jun;33(Pt 3):493-8.

[4] Fraser CS, Doudna JA. Quantitative studies of ribosome conformational dynamics. *Q Rev. Biophys*. 2007 May;40(2):163-89.

[5] Ramakrishnan V. Ribosome structure and the mechanism of translation. *Cell.* 2002 Feb 22;108(4):557-72.

[6] Al-Karadaghi S, Kristensen O, Liljas A. A decade of progress in understanding the structural basis of protein synthesis. *Prog. Biophys. Mol. Biol.* 2000;73(2-4):167-93.
[7] Cech TR. Structural biology. The ribosome is a ribozyme. *Science.* 2000 Aug 11;289(5481):878-9.
[8] Joseph S. After the ribosome structure: how does translocation work? RNA. 2003 Feb;9(2):160-4.
[9] Marintchev A, Wagner G. Translation initiation: structures, mechanisms and evolution. *Q. Rev. Biophys.* 2004 Aug-Nov;37(3-4):197-284.
[10] Gallie DR. Protein-protein interactions required during translation. *Plant. Mol. Biol.* 2002 Dec;50(6):949-70.
[11] Iusupova G, Jenner L, Iusupov M. Messenger RNA movement on the ribosome. *Mol. Biol.* (Mosk). 2007 Mar-Apr;41(2):274-83.
[12] Marzi S, Myasnikov AG, Serganov A, Ehresmann C, Romby P, Yusupov M, Klaholz BP.Structured mRNAs regulate translation initiation by binding to the platform of the ribosome. *Cell.* 2007 Sep 21;130(6):1019-31.
[13] Yusupova G, Jenner L, Rees B, Moras D, Yusupov M. Structural basis for messenger RNA movement on the ribosome. *Nature*. 2006 Nov 16;444(7117):391-4.
[14] Culver GM. Meanderings of the mRNA through the ribosome. *Structure.* 2001 Sep;9(9):751-8.
[15] Waters AP, McCutchan TF. Rapid, sensitive diagnosis of malaria based on ribosomal RNA. *Lancet.* 1989 Jun 17;1(8651):1343-6.
[16] Miller FW, Ilan J. Evidence for major differences in ribosomal subunit proteins from Plasmodium berghei and rat liver. *Mol. Biochem. Parasitol.* 1984 Jul;12(3):249-60.
[17] Miller FW, Ilan J. The ribosomes of Plasmodium berghei: isolation and ribosomal ribonucleic acid analysis. *Parasitology.* 1978 Dec;77(3):345-65.
[18] Takebe S, Witola WH, Schimanski B, Günzl A, Ben Mamoun C. Purification of components of the translation elongation factor complex of Plasmodium falciparum by tandem affinity purification. *Eukaryot. Cell.* 2007 Apr;6(4):584-91.
[19] Mamoun CB, Goldberg DE. Plasmodium protein phosphatase 2C dephosphorylates translation elongation factor 1beta and inhibits its PKC-mediated nucleotide exchange activity in vitro. *Mol Microbiol.* 2001 Feb;39(4):973-81.
[20] Vinkenoog R, Sperança MA, van Breemen O, Ramesar J, Williamson DH, Ross-MacDonald PB, Thomas AW, Janse CJ, del Portillo HA, Waters AP. Malaria parasites contain two identical copies of an elongation factor 1 alpha gene. *Mol. Biochem. Parasitol.* 1998 Jul 1;94(1):1-12.
[21] Chatterjee S, Singh S, Sohoni R, Singh NJ, Vaidya A, Long C, Sharma S. Antibodies against ribosomal phosphoprotein P0 of Plasmodium falciparum protect mice against challenge with Plasmodium yoelii. *Infect. Immun.* 2000 Jul;68(7):4312-8.
[22] Singh S, Chatterjee S, Sohoni R, Badakere S, Sharma S. Sera from lupus patients inhibit growth of P. falciparum in culture. *Autoimmunity*. 2001;33(4):253-63.
[23] Chatterjee S, Singh S, Sohoni R, Kattige V, Deshpande C, Chiplunkar S, Kumar N, Sharma S. Characterization of domains of the phosphoriboprotein P0 of Plasmodium falciparum. *Mol. Biochem. Parasitol.* 2000 Apr 15;107(2):143-54.
[24] Goswami A, Singh S, Redkar VD, Sharma S. Characterization of P0, a ribosomal phosphoprotein of Plasmodium falciparum. Antibody against amino-terminal domain inhibits parasite growth. J. *Biol. Chem.* 1997 May 2;272(18):12138-43.

[25] Aruna K, Chakraborty T, Rao PN, Santos C, Ballesta JP, Sharma S. Functional complementation of yeast ribosomal P0 protein with Plasmodium falciparum P0. *Gene.* 2005 Aug 29;357(1):9-17.

[26] Li L, Fallon AM. Recovery of cDNAs encoding ribosomal proteins S9 and L26 from Aedes albopictus mosquito cells and identification of their homologs in the malaria vector, Anopheles gambiae. *Arch. Insect. Biochem. Physiol.* 2005 Sep;60(1):44-53.

[27] Gentile C, Lima JB, Peixoto AA. Isolation of a fragment homologous to the rp49 constitutive gene of Drosophila in the Neotropical malaria vector Anopheles aquasalis (Diptera: Culicidae). *Mem. Inst Oswaldo Cruz.* 2005 Oct;100(6):545-7.

[28] Youmans AS, Youmans GP. Failure of Synthetic Polynucleotides to Affect the Immunogenicity of Mycobacterial Ribonucleic Acid and Ribosomal Protein Preparations. *Infect. Immun.* 1971 Jan;3(1):149-153.

[29] Youmans AS, Youmans GP. Effect of trypsin and ribonuclease on the immunogenic activity of ribosomes and ribonucleic acid isolated from Mycobacterium tuberculosis. *J. Bacteriol.* 1966 Jun;91(6):2146-54.

[30] Youmans AS, Youmans GP. Effect of polybasic amines on the immunogenicity of mycobacterial ribonucleic acid. *Infect. Immun.* 1972 Nov;6(5):798-804.

[31] Portelance V, Beaudet R. Biochemical characterization of three mycobacterial ribosomal fractions. *Can. J. Microbiol.* 1983 Feb;29(2):185-93.

[32] Medow I, Rüsch-Gerdes S, Schröder KH. Comparison of ribosomes and ribosomal proteins of sensitive and resistant mycobacteria. *Zentralbl. Bakteriol. Mikrobiol. Hyg.* [A]. 1987 Oct;266(3-4):359-69.

[33] Yamada T, Masuda K, Shoji K, Hori M. Pleiotropic antibiotic resistance mutations associated with ribosomes and ribosomal subunits in Mycobacterium smegmatis. *Antimicrob Agents Chemother.* 1974 Jul;6(1):46-53.

[34] McClatchy JK, Kanes W, Davidson PT, Moulding TS. Cross-resistance in M. tuberculosis to kanamycin, capreomycin and viomycin. *Tubercle.* 1977 Mar;58(1):29-34.

[35] Yamada T, Mizuguchi Y, Isono S, Isono K. Genetic and biochemical analysis of ribosomal proteins of minocycline-susceptible and -resistant Mycobacterium smegmatis. *Microbiol. Immunol.* 1992;36(2):139-48.

[36] Dobner P, Bretzel G, Rüsch-Gerdes S, Feldmann K, Rifai M, Löscher T, Rinder H. Geographic variation of the predictive values of genomic mutations associated with streptomycin resistance in Mycobacterium tuberculosis. *Mol. Cell Probes.* 1997 Apr;11(2):123-6.

[37] Meier A, Kirschner P, Bange FC, Vogel U, Böttger EC. Genetic alterations in streptomycin-resistant Mycobacterium tuberculosis: mapping of mutations conferring resistance. *Antimicrob. Agents Chemother.* 1994 Feb;38(2):228-33.

[38] Okamoto S, Tamaru A, Nakajima C, Nishimura K, Tanaka Y, Tokuyama S, Suzuki Y, Ochi K. Loss of a conserved 7-methylguanosine modification in 16S rRNA confers low-level streptomycin resistance in bacteria. *Mol. Microbiol.* 2007 Feb;63(4):1096-106.

[39] Miyazaki C, Ohara N, Yukitake H, Kinomoto M, Matsushita K, Matsumoto S, Mizuno A, Yamada T. Host immune responses to ribosome, ribosomal proteins, and RNA from Mycobacterium bovis bacille de Calmette-Gúerin. *Vaccine.* 1999 Jan 21;17(3):245-51.

[40] Kitaura H, Kinomoto M, Yamada T. Ribosomal protein L7 included in tuberculin purified protein derivative (PPD) is a major heat-resistant protein inducing strong delayed-type hypersensitivity. *Scand J. Immunol.* 1999 Dec;50(6):580-7.

[41] Tantimavanich S, Nagai S, Nomaguchi H, Kinomoto M, Ohara N, Yamada T. Immunological properties of ribosomal proteins from Mycobacterium bovis BCG. *Infect Immun.* 1993 Sep;61(9):4005-7.

In: Ribosomal Proteins and Protein Engineering… ISBN 978-1-60741-005-8
Editors: V. Ortendhal and H. Salchow, pp.99-102

Chapter 7

THREE CHALLENGES IN POST-PROTEOMICS TIME

Shaomin Yan[1] and Guang Wu[2,1]

[1]National Engineering Research Center for Non-food Biorefinery, Guangxi Academy of Sciences, 98 Daling Road, Nanning, Guangxi, 530007, China,

[2]Computational Mutation Project, DreamSciTech Consulting, 301, Building 12, Nanyou A-zone, Jiannan Road, Shenzhen, Guangdong, 518054, China

ABSTRACT

In this chapter, we discuss three challenges we face in explaining the life from genome and proteome viewpoint, that is, how to measure a living protein/RNA/DNA, how to restore the missing relation cross generations and how to correlate the structure with its function. Finally, we suggest the possible solutions for these three challenges.

History repeats in one way or another, similarly we would expect to see the scientific history parallels in one way or another. An important episode in scientific history is the time before the creation of relativity, then most physicists considered that the physics as a discipline had reached its final goal, and the next job was to use the physics to explain everything from the viewpoint of atom. The similar expectations were strongly advocated, i.e. we can explain everything from the viewpoint of genome prior genome time, and explain everything from the viewpoint of proteome prior proteome time. Likely these advocations would be intensified in post-proteome era.

Ironically, we are still not in the position to fully and convincingly explain the life beased on our knowledge on genome and proteome. More ironically our time is far more advanced than the time when physicists considered explaining everything from the viewpoint of atom, because we have a full spectrum of mathematical tools, supercomputers, amazing manpower, etc.

[1] Corresponding author, Tel: +86 771 2503 930, Fax: +86 771 2503 999, E-mail: hongguanglishibahao@yahoo.com.

Clearly, there are several challenges we are facing before accomplishing the task of explaining the life from viewpoints of genome and proteome.

Challenge 1. How to Measure Living Protein/RNA/DNA

The beginning of geometry came from the measurement of land in ancient time [1], and the beginning of relativity came from the measurement of light speed [2]. Thus, as Poincaré pointed out that the important thing is not to know what it is, but how to measure it [3].

However, we have not seen much development in the measurements that can measure the living protein/RNA/DNA in biomedical sciences. Currently, almost all of the measures we use in daily research borrow mainly from physics and chemistry. In other words, we have not seen the parallel development in genome and proteome research as we have evidenced in geometry and physics.

This would lead the difficulty in our mission of explaining the life from viewpoints of genome and proteome, because the measurements developed by physicists and chemists were originally designed for their own research purposes, which have different pretexts and circumstances from ours.

The theme of how to measure a living subject seemingly contains some supernatural spirit that is somewhat equivalent to the philosophical theme of what is life, however, it indeed has the real meanings in protein/RNA/DNA studies. For these, we suggest that the current measurements do not reflect the characteristics of living protein/RNA/DNA simply because we do not know if a protein/RNA/DNA is alive and if it functions well based on its physicochemical properties, for example.

By clear contrast, we argue that the measurements in clinical settings well reflect the life of patient. For example, the measurement of blood pressure can directly reflect if a patient is alive or not. In other words, a single number from blood pressue can tell us if a patient is alive or ill. A simlar rationale can easily be referred to the measurements of heart beats and breathes. Practically, any clinical measurement can specifically determine if the given organ works well.

However, we cannot use a simple number to indicate if a protein/RNA/DNA is alive or dead, or functions normally. We desprately need the meaurments that can work as the clinical meaurements to answer the basic questions on protein/RNA/DNA, which would have a great application in clinical settings.

The possible solution we suggest would be the measurements that must be subject to the protein length, amino-acid composition, amino-acid position, adjacent amino acids, etc. as the first step for the development to measure the living protein [4].

Challenge 2. How to Restore Missing Relation Cross Generations

Although we have accumulated an amazing amount of protein/RNA/DNA data, we still feel that something lacks, that is, we generally lack the data directly from generation to generation mainly due to the fact that nature frequently does not leave any trace for us.

Without these data, we really cannot determine the mutation relationship from parent to daughter, which reflects the actual and real evolution. Thus no matter of how hard we collect data, we still definitely cannot collect all of them, and definitely cannot determine all the actual and real parent-daughter relationship for all generations.

This imposes the extreme difficulty in predicting mutation, in tracing the pace of evolution, in designing the next generation of vaccines, etc., although we might have some data obtained from experiments.

This challenge requires us to find out a mathematical tool that can deduce the actual and real parent-daughter relationship with clear mutation path and steps. Ironically, although there are many, even too many mathematical tools or mathematical methods available, we could frequently hear such saying that there are no data available. At first glance, it seems to be strange because there are so many unused data, but it is a real problem that every researcher could encounter, because the available data are not compatible with the request of mathematical modeling.

The possible solution we suggest is to develop the method to prove that there are only limited number of pathways and steps, from which a protein to mutate to another one, and the pathway should be unique.

Challenge 3. How to Correlate Structure with Function

Although many efforts are made in the direction of structure-function relationship [5], we are still in the phase dealing with this relationship case by case.

There are several difficulties in this regard. For example, almost all the relationship between structure and function is based on phenomenological law rather than kinetic law, thus we use the regression or neural network to model the relationship between structure and function with various coefficients to balance different dimensions between structure and function. Still it seems that there is no point-to-point relationship between structure and function because each and every structure can correspond to many radically different functions.

However, the key point is if we can assume the final goal of a protein structure is the function we defined. Each protein would use different structures to reach the goal as we can use bike, car, or aircraft to reach point B from point A. In this view, the structure of bike, car or aircraft has no relationship with the goal, but has the relationship with the speed of reaching a goal. Also, the function in mathematics is always associated with time.

Thus, the possible solution we suggest is to correlate the structure with the speed of reaching the final function, that is, the relationship between structure and pathway reaching the goal.

References

[1] http://en.wikipedia.org/wiki/Geometry, 2010.
[2] http://en.wikipedia.org/wiki/Speed_of_light#Special_relativity. 2010

[3] Poincaré H (1902) Science and Hypothesis. Walter Scott, London and Newcastle-on-Cyne.

[4] Wu G, Yan S. Lecture Notes on Computational Mutation. New York, Nova Science Publishers, 2008.

[5] Chou KC. Structural bioinformatics and its impact to biomedical science. *Curr. Med. Chem.* 2004; 11: 2105-2134.

In: Ribosomal Proteins and Protein Engineering... ISBN 978-1-60741-005-8
Editors: V. Ortendhal and H. Salchow, pp. 103-119

Chapter 8

SELF-SIMILAR PROPERTY OF RIBOSOMAL PROTEINS

Chang-Yong Lee*
The Department of Industrial and Systems Engineering,
Kongju National University, Kongju, 314-701 South Korea

Abstract

Self-similar properties of the proteins in the ribosome, the ribosomal proteins, are investigated in terms of various types of fractal dimensions. We especially estimate both the chain fractal and capacity dimensions of backbones of the ribosomal proteins, together with the mass fractal dimension of the ribosomal proteins at the atomic level. The chain fractal and capacity dimensions of the ribosomal proteins are compared with that of the self-avoiding walk, which is a typical model of a polymer without interaction between monomers. These fractal dimensions of the ribosomal protein support the existence of the long and extended domain, which is hardly seen in the globular protein. We demonstrate that the mass fractal dimension of the ribosomal proteins is pertinent to explain their structural characteristics. The self-similarity also upholds the fact that the ribosomal proteins function primarily to stabilize the structure of the ribosome by both the long-extended domain of the protein penetrating into the inside of the RNA, and the globular domain interacting with the RNA on the exterior of it. These results partially, if not whole, unravel the structural characteristics of the proteins in the ribosome.

PACS 87.14.Gg, 87.15.-v, 61.43.Hv, 82.35.Lr

Keywords: Ribosome, Ribosomal proteins, self-similarity, fractal dimension,

1. Introduction

The ribosome [1], a large RNA-protein complex molecule, catalyzes protein synthesis in all organisms. The prokaryotic ribosome is a macromolecule of size about 250 $\mathring{A}$ in diameter, and sediments at 70S, where the unit "S" is the measure of the sedimentation rate. It is

*E-mail address: clee@kongju.ac.kr

mainly composed of two subunits: the 30S and 50S subunits. The 30S subunit, in turn, contains the 16S ribosomal RNA (rRNA) in addition to about 20 ribosomal proteins (r-proteins); whereas the 50S subunit consists of two rRNAs, the 5S and 23S rRNAs, together with about 30 r-proteins, each of which is a polypeptide chain that is made up of residues of 20 different amino acids linked by peptide bonds.

Since the ribosome, as a tightly packed macromolecule, was considered too large for a high-resolution structural analysis, quantitative studies toward an understanding of the structure proved difficult until recent progress in the high resolution crystallography has been made. The 2.4 Å resolution of the 50S subunit from the *Haloarcula marismortuii* [2, 3] and the 3.05 Å resolution of the 30S subunit from the *Thermus thermophilus* [4, 5] uncover the first detail structure of both subunits. So far, these subunits are the largest asymmetric molecules that have been resolved at the atomic level by the x-ray crystallography. In addition, the 3.5 Å resolution of the intact 70S from the *Escherichia coli* [6] reveals the features of the inter-subunit bridges. These results provide not only the high-resolution structure of the rRNAs and r-proteins, but also the first detail pictures of how the rRNAs and r-protein components are interacted to form the ribosome.

The structure of a biopolymer is important since it is widely believed that the structure dictates its biological function. Thus, finding characteristics of the three-dimensional shape of biomolecules serve for a better understanding of their biological functions and associated applications in medicine. It is known that the main functional role of r-proteins is to shape and stabilize the ribosome by interacting with the rRNA even deep inside of it [3, 5]. This is in contrast to the case of the globular proteins in that they independently play a role of an enzyme. From this, we can infer that the conformational feature of the r-proteins may differ from that of globular proteins.

Due to considerable advances in the structural conformation of the ribosome, it is now possible to investigate quantitative characteristics of r-proteins. Also, the quantitative approach may at least serve as a complement to experimental achievements. One quantitative approach toward the understanding of the structure of the r-proteins is the self-similar property via the fractal dimension [7]. It, particularly, may help to get the insight of the local and global structures of r-proteins in the ribosome, especially from the statistical physics perspective.

The quantitative investigation on the structure of a r-protein has been relatively less studied than that of a global protein, mainly due to the difficulty in obtaining the highly resolved structure conformation. In contrast to r-proteins, the geometric and self-similar properties of the globular proteins have been studied extensively [8, 9, 10, 11, 12, 13, 14, 15, 16, 17]. It was reported that the relation between the average radius and the mass of the protein chains can be described by a fractal dimension [14]. The mass fractal dimension of the protein was shown to lie near three, suggesting a compact three-dimensional object [10]. However, more recent and extensive studies, including a statistical analysis in estimating the dimension, argue smaller values that are consistent with the result of the vibrational analysis of proteins [15, 16]. These results suggest that the fractal dimension of less than three may be intrinsic and universal characteristics of the protein chain.

The chain fractal dimension of a globular protein has been conventionally estimated by two different approaches. The first method is based on the scaling exponent of the contour length, which is proportional to the number of monomers with respect to the end-to-end length [9]. The second method is based on the length of a line segment along the chain measured with a scale of the number of monomers in the segment [8, 11]. Although these two methods are essentially the same, we believe that the latter is easier to implement than the former. Using the latter, called the internal line segment method, it was found that the chain fractal dimension of the globular proteins depended on the coarseness of the scale m, which is nothing but the number of residues included in a straight line segment. From this, it was argued that the folding mechanism of the local and global backbones could be separately characterized by the different chain fractal dimensions. Besides these methods, the fractal dimension of the globular proteins is estimated from the method of the radius of gyration for the protein compactness [12, 14].

With this structural information of r-proteins at the atomic level, in this paper, we investigate the self-similar property of r-proteins and their biological implication. The fractal dimension analysis of r-proteins may not only unravel the intrinsic structural characteristics of them in the ribosome but disclose the structural difference between the globular and ribosomal proteins in conjunction with its biological function. We especially focus on the scale invariance, by estimating various fractal dimensions, of the r-proteins in *Thermus thermophilus* 30S and *Haloarcula marismortui* 50S subunit. We estimate the fractal dimensions of about 20 and 30 r-proteins in the 30S and 50S subunits, respectively. The structural details of r-proteins are deposited in the Protein Data Bank (PDB) [18].

From the notion that r-proteins are long chains of amino acids with multiple interactions between them, we focus on two different types of the fractal dimensions: the chain fractal dimension [10] and the capacity dimension [19]. The former takes into account the fact that a linear polymer can be described by successive "walks" between monomers along the chain; while the latter considers only the geometrical configurations of the backbone without reflecting the ordering of the monomers along the chain. In addition, it has been known that the protein synthesis occurs in the context of the intact ribosome and the moving part of the ribosome enables the dynamic process of the translation. In this sense, the ribosome function is closely related to its spatial conformation in the physiological medium. Thus, the mass fractal dimension may help to reveal any characteristics of the r-proteins in the ribosome.

This paper is organized as follows. In section II, we describe methods for estimating various fractal dimensions and their feasibility. This section is followed by results of the estimates and their biological implications; the last section is devoted to the summary and conclusion.

2. Methods for Estimating Fractal Dimensions

One reason that the fractal concept attracts researchers' attention is its compact representation, in terms of a single fractal dimension, of seemingly complex inhomogeneous biomolecules, such as proteins [7, 10]. Although the fractal analysis averaged out the detail structural conformations of a bio-molecule, a fractal description may serve as a first step to characterize its strucutre as well as function.

A mathematically rigorous definition of the fractal dimension is given by F. Hausdorff [20]. It, however, is impractical especially for the numerical estimation. To practically estimate a fractal dimension, one often looks for a power law relation of the form

$$y \propto x^{\gamma} , \tag{1}$$

where y is a property, x is the variable, and γ is a not-integer which is related to a fractal dimension. The existence of such a power law suggests a symmetry under a change of the scale of the variable. This property is known as the scale invariance, which is closely associated with the self-similarity. The self-similarity implies there is no characteric scale (natural or preferred scale) for the system under consideration. That is, under the transformation $x \longrightarrow bx$, the property y remains unchanged up to the normalization constant.

$$y \propto (bx)^{\gamma} \propto x^{\gamma} . \tag{2}$$

In the case of protein structures, it can be expected that proteins have an intrinsic self-similarity with regard to the compactness and packing of their structures. Thus, depending on specifically desired properties, there are various aspects of measuring and describing the fractality of proteins. Note, of course, that due to the finite-size of the molecules, there are both upper and lower size limits beyond which the biopolymers are no longer fractal.

2.1. The chain fractal dimension

A chain fractal dimension D_c of a biopolymer chain is conventionally defined, in terms of the root mean square of the end-to-end distance R, as

$$R \propto N^{1/D_c} \propto N^{\nu} , \tag{3}$$

where N is the number of monomers along the chain, and ν is the Flory exponent. It is known that, in three dimension, $D_c = 2$ for the Gaussian walk (or Θ solvent), and $D_c = 5/3$ for the self-avoiding walk (SAW), which includes the excluded volume effect [21]. Note that since the SAW does not take into account any attractive force, the deviation from $D_c = 5/3$ is expected when there are interactions between monomers which are far apart along the chain but close in space [10]. Thus, more realistic polymers are likely to have chain fractal dimensions different from that of the SAW.

Practically, the end-to-end distance is measured by the radius of gyration R_g [12]. It is a measure of the size of an object, and is known to be proportional to the root mean square distance R [22]. Note that to get the scaling behavior of Eq. (3), this method requires R_g's for polymers of different numbers of monomers N. Thus, all polymers that participated in the estimation have the *same* fractal dimension. Therefore, this method is valid under the condition that the radius of gyration of a set of polymers depends only on their number of monomers, not their structural characteristics. When the structural variation among polymers is anticipated, this method may yield unreliable result. To illustrate this point, we calculate the radius of gyrations for all r-proteins in the ribosome, and plot R_g versus N in Fig 1. As shown in the figure, R_g for different r-proteins deviates rather sharply from the lease square fit (the dotted straight line). This suggests that R_g may depend not solely on the number of monomers, thus the structural variation among r-proteins is expected.

Incidentally, ignoring the deviation from the least square fit, the scaling behavior of the radius of gyration versus the number of monomers is best described by the exponent

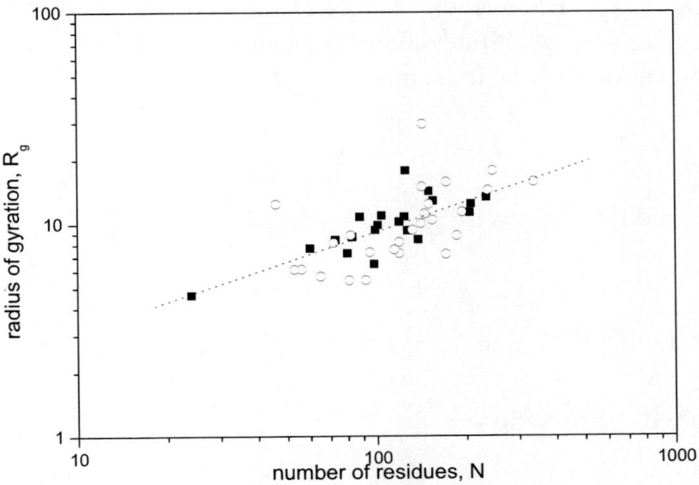

Figure 1. A log-log plot of the radius of gyration versus the number of residues for r-proteins in the 30S (■) and 50S (○) subunits. The dotted line is the least square fit to all proteins yielding the slope $\nu = 0.48 \pm 0.09$.

$\nu = 0.48 \pm 0.09$, which is very close to $\nu = 0.5$. This suggests that r-proteins are in an ideal or so called Θ solvent in a sense that r-proteins interact with the solvent with almost the same strength as they interact with themselves [12].

The chain fractal dimension of a biopolymer is also often estimated by the internal line segment method [8, 11]. It is based on the scaling behavior of the segmental length $L(m)$ of the backbone which is defined as the sum of stepwise connections of straight lines measured for different interval of m monomers. That is,

$$L(m) = \ell(m) + \frac{k}{m+1} \langle \ell(m) \rangle \, , \tag{4}$$

where k ($k < m$) is the number of monomers left unconnected, and the second term of the right hand side of Eq. (4) is a correction term which takes into account the contribution from the monomers left unconnected. $\langle \ell(m) \rangle$ is the mean length of the $\ell(m)$, and it is the average value of the end-to-end distance of m monomers. $\ell(m)$ is, in turn, defined as

$$\ell(m) = \sum_{i=0}^{N/m-1} |\vec{x}_{(i+1)m} - \vec{x}_{im}| \, , \tag{5}$$

where $\vec{x}_j$ is the coordinate of the monomer j, and N is the total number of monomers in the biopolymer.

Using the backbone segment length of a biopolymer, another the chain fractal dimension D_L is defined [8, 11] as

$$L(m) \propto m^{1-D_L} \, . \tag{6}$$

Thus, the chain fractal dimension D_L is equal to one plus the slope obtained from the plot of $\log L(m)$ versus $\log m$. The relation between D_c and D_L can be found as follow. When, similar to Eq. (3), the average of internal line segments scales $\langle \ell(m) \rangle \propto m^{\nu}$, Eq. (4) can be rewritten, ignoring the correction term, as

$$L(m) \propto \frac{N}{m} \langle \ell(m) \rangle \propto m^{\nu-1} . \tag{7}$$

Equating Eq. (6) and Eq. (7), and using Eq. (3), we get

$$D_L = 2 - 1/D_c . \tag{8}$$

Thus, $D_L = 1.40$ for the SAW, and $D_L = 1.50$ for a Gaussian polymer [8, 13].

2.2. The capacity dimension

A rigorous definition of the fractal dimension of a solid object was first proposed by Hausdorff, known as the Hausdorff dimension [19]. It is, however, not suitable for the numerical estimation. By relaxing the requirement of the infimum over all covering in the Hausdorff dimension and by choosing simply a fixed-size grid as a covering, we obtain an upper bound on the Hausdorff dimension. This is called the box-counting method, and the corresponding dimension is referred to as the capacity (or box-counting) dimension. It is known that the capacity and the Hausdorff dimensions are the same for most fractal sets [23]. The box-counting method, however, has practical limitations particularly at high embedding dimensions. It was argued that a direct attempt to estimate the fractal dimension using the box-counting method is almost impractical when the embedding dimension is greater than two [24, 25]. Considering that the biopolymers are embedded in the three-dimensional Euclidean space, the application of the box-counting method to them may yield inaccurate results.

An alternative is the correlation integral method [25]. Since the correlation integral method estimates the fractal dimension entirely based on the statistics of pair wise distance, it is much less affected by the embedding dimension. Furthermore, due to the statistics of pair wise distance, it has a dynamic range of $O(N^2)$ for a given data of N points, which is twice that available to the box-counting method. This wider range is another advantage that the method has over the box-counting method [19]. The correlation integral $C_N(r)$ of the distance r is defined as

$$C_N(r) = \frac{1}{N(N-1)} \sum_{i \neq j}^{N} \Theta(r - |\vec{x_i} - \vec{x}_j|) , \tag{9}$$

where $|\vec{x}_i - \vec{x}_j|$ is the distance between a pair of monomers i and j, and $\Theta(x)$ is the Heaviside function. That is,

$$\Theta(x) = \begin{cases} 1 & \text{if } x > 0 \\ 0 & \text{otherwise} . \end{cases} \tag{10}$$

From this, the capacity dimension D_H can be obtained by the behavior of $C_N(r)$ as a power of r:

$$C_N(r) \propto r^{D_H} , \tag{11}$$

or

$$D_H = \lim_{r\to 0} \frac{\log C_N(r)}{\log r} \, . \tag{12}$$

It was shown that D_H is, in most cases, very close to the actual fractal dimension [25]. In this paper, we adopt the correlation integral method to estimate the capacity dimension of the biopolymer backbones.

Whether the correlation integral or the box-counting method is applied to a biopolymer backbone, it does not take into account its sequence information; instead, it considers only the geometrical configuration of the backbone in the Euclidean space. In this perspective, the capacity dimension concerns the spatial distribution of monomers independent of their connectivity, thus emphasizes different aspect of the self-similar property of the biopolymer from the chain fractal dimension.

2.3. The mass fractal dimension

The mass fractal dimension, which can be used as a measure of the compactness, is defined as the number of elements (atoms in our case), N, enclosed in a sphere of radius r. When a molecule has a fractal structure, it is expected that

$$N \propto r^{D_M} \, , \tag{13}$$

where D_M is the mass fractal dimension. It can be estimated by plotting the number of all atoms contained inside concentric spheres of varying radius r on a log-log scale. The mass fractal dimension exhibits the molecule's space-filling ability: the larger is D_M, the more atoms are in the sphere. When the mass fractal dimension is less than three, the structure has "empty" or "void" space.

In estimating the mass fractal dimension of r-proteins at the atomic level, we ignore hydrogen atoms since the x-ray crystal structures do not contain the geometric information of hydrogen atoms, which cannot be seen but in very high resolution. Incidentally, this is also true for the global protein case. Thus, most of descriptions of the r-proteins focus on the position of the heavy atoms, such as C, N, O, and P. Note that, for a comparison, we exclude water molecules from the estimation of the mass fractal dimension, since the PDB data for the 30S subunit does not contain water molecules whereas that for the 50S does.

There is an attempt to estimate the mass fractal dimension by considering only C^{α} atoms of the protein backbone [26], ignoring the contribution from the side-chains. Even though it simplifies the calculation and C^{α} atoms are important in the polypeptide backbone, this approach may give rise to inaccurate estimation due to the large differences in the side-chain masses. Considering that proteins have twenty different possible amino acids, these monomers may differ from side-chain to side-chain.

3. Results and Discussion

Detail accounts on the atomic level structure of r-proteins in the 30S and 50S subunits are used to extract the fractal dimensions. There are 20 and 29 r-proteins in the 30S and 50S subunits, respectively; out of which 19 and 26 r-proteins are resolved well enough to estimate the dimensions [27]. The result of the estimated fractal dimensions for each r-protein in the 30S and 50S subunits are shown in Table 1 and 2, respectively.

Table 1. Fractal dimensions of r-protein backbones in the 30S subunit with corresponding least square fit errors in the parentheses. N stands for the number of modeled residues, D_M, for the mass fractal dimension, and D_H, for the capacity dimension. D_L represents the chain fractal dimension; $*$ and $**$ of D_L value stand for the chain fractal dimension in the short and long m ranges, respectively. Note that the number of modeled residues may not necessarily be the same as the number of residues in each protein.

Name	N	D_M	D_L	D_H
S2	234	3.03 (0.07)	1.35 (0.02)* 1.75 (0.03)**	2.35 (0.04)
S3	206	-	1.30 (0.02)* 1.63 (0.03)**	2.32 (0.03)
S4	208	2.89 (0.05)	1.39 (0.02)* 1.93 (0.04)**	2.35 (0.04)
S5	150	-	1.27 (0.02)	2.15 (0.03)
S6	101	2.69 (0.05)	1.28 (0.02)	2.18 (0.04)
S7	155	2.46 (0.03)	1.39 (0.02)	2.15 (0.03)
S8	138	-	1.28 (0.03)* 1.70 (0.04)**	2.26 (0.03)
S9	127	2.39 (0.03)	1.32 (0.02)* 1.79 (0.06)**	2.20 (0.03)
S10	98	2.63 (0.04)	1.29 (0.03)	2.23 (0.05)
S11	119	2.64 (0.04)	1.36 (0.03)* 1.80 (0.04)**	2.24 (0.04)
S12	124	2.49 (0.05)	1.36 (0.02)* 1.52 (0.03)**	1.97 (0.03)
S13	125	-	1.51 (0.02)	1.95 (0.04)
S14	60	2.63 (0.06)	1.45 (0.04)	1.98 (0.04)
S15	88	2.43 (0.03)	1.32 (0.03)	2.18 (0.03)
S16	83	3.06 (0.05)	1.37 (0.03)	2.07 (0.04)
S17	104	2.47 (0.05)	1.31 (0.03)	2.16 (0.05)
S18	73	2.89 (0.03)	1.42 (0.03)	2.30 (0.03)
S19	80	-	1.36 (0.04)	2.03 (0.04)
S20	99	3.08 (0.06)	1.26 (0.02)	1.79 (0.07)

3.1. Chain fractal dimensions of the r-protein backbones

In estimating the chain fractal dimension of r-proteins, we regard the connection of straight lines between the C^{α} atoms of the protein backbones as a protein chain, where the C^{α} atom is the carbon atom along the polypeptide backbone at which amino acid residues attach. A typical scaling behavior of $L(m)$ versus m is plotted in Fig. 2. A prominent feature in the chain fractal dimension analysis is that the scaling behavior of $L(m)$ for some r-proteins, if not all, depends on the nucleotide interval m. This is manifested by the existence of the cross over, indicated by arrows in Fig. 3. Note that 7 out of 19 r-proteins in the 30S subunit and 11 out of 26 r-proteins in the 50S subunit have their $L(m)$ being scaled differently with respect to m in two ranges with the cross over at $7 \lesssim m \lesssim 15$. The existence of the cross over suggests that characteristics of the local and global conformation of r-proteins are different.

From Table 1 and 2, for r-proteins with the cross over, we find that their D_L in the short m range smaller than that in the long m range. Furthermore, D_L in the short m range are less than that of the SAW ($D_L < 1.40$). Since the SAW is obtained from a mean field approach, in which any attractive force between monomers are ignored, deviation from the SAW is

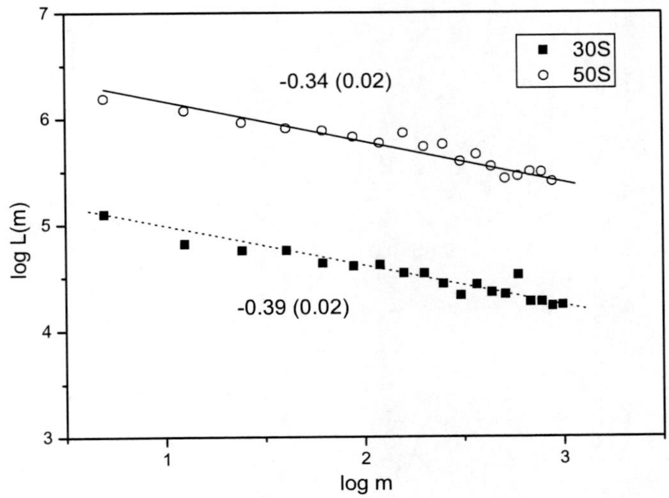

Figure 2. Plots of $\log L(m)$ versus $\log m$ for the backbone of the r-protein S7 (■) in the 30S subunit, and the backbone of L10e (○) in the 50S subunit. The least square fit to S7 (dotted line) and L10e (solid line) yields the slope -0.34 ± 0.02 and -0.39 ± 0.02, respectively. Note that, for the display purpose, the plot for S7 is shifted downward by the factor of 1.2.

expected if there are cross-links (e.g. hydrogen bonds) between constituents of the amino acids. Considering that the SAW is a typical model for a polymer with the excluded volume effect [21], this suggests that the local conformation of these r-proteins are more extended or swollen than that of both the SAW and the global conformation.

In the long m range, on the other hand, the chain fractal dimensions of these r-proteins extend $1.52 \lesssim D_L \lesssim 1.93$, greater than that of the SAW. This finding supports the existence of the long-range interaction in forming the tertiary structure among resides, thus making the molecule more globular shape. This suggests that the global conformation of r-proteins is more compact than that of the SAW. This compactness of r-proteins is supposed to be dictated by the predominant long-range interaction among the residues which are far apart along the chain but close in space. Furthermore, most of these r-proteins consist of relatively large number of residues. This is related, at least partly, to the fact that an r-protein having more residues is more likely to form a globular domain outside the rRNA surface. This result coincides with the finding that r-proteins having small number of residues are often embedded by rRNA, whereas those having large number of residues tend to have independent globular domains that lie on the outside of the subunit [3, 5].

The existence of the cross over in the chain fractal dimension is closely related to the extended structure in r-proteins. The extended structure is a long stretch of folded polypeptide that emanates from otherwise normal globular domain, and it is the single most striking feature of r-proteins. In the case of the 50S subunit, we find that 7 out of 11 r-proteins having their extensions exhibit the cross over. Furthermore, the largest and most conserved r-proteins (namely, L2, L3, L4, L22, and L24) all have both the extension and the cross

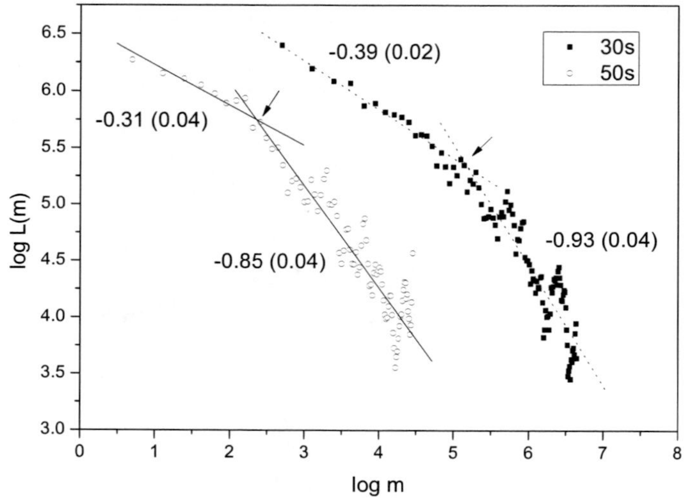

Figure 3. Plots of $\log L(m)$ versus $\log m$ for the backbone of the r-protein S4 (■) in the 30S subunit, and the backbone of L6 (○) in the 50S subunit. For each r-protein, there exhibits a cross over, with which the two scaling is divided (indicated by the arrow). The least square fit to S4 (dotted line) gives the slope -0.31 ± 0.04 and -0.85 ± 0.04 for the short and the long m ranges, respectively; to L6 (solid line) yields the slope -0.39 ± 0.02 and -0.93 ± 0.04 for the short and the long m ranges, respectively. Note that, for the display purpose, the plot for S4 is shifted horizontally to the right by the factor of 2.

over. It was reported [3, 5] that besides the globular domain found in the exterior of both rRNAs, many r-proteins have long and extended regions (either in the terminal or in internal loops) penetrating into the interior of the rRNAs. The extended domain makes extensive contact with the rRNA, and helps to stabilize the rRNA tertiary structure. Thus, the chain fractal dimension being less than that of the SAW supports the presence the extended domain which stabilizes the rRNA.

It is also known that the extension lacks hydrophobic cores which are typical of globular proteins and their structures depend on interactions with rRNA [3]. The fact that the extension lacks hydrophobic cores suggests that the extension is more likely composed of hydrophilic amino acids. In the case of r-proteins in the 50S subunit, it was reported [3] that the most abundant amino acids in extensions are arginine (15.9%), glycine (13.7%), lysine (12.7%), and serine (7.1%); these amino acids consist of about 50% of all amino acids in extensions. This result is consistent with the finding that the loss of the solvent-accessible surface area of individual amino acid and its hydrophobicity are closely correlated and the above amino acids are proven to have hydrophilic side chains [28, 29].

In the case of r-proteins in the 30S subunit, it was reported [5] that arginine and lysine, typical amino acids having hydrophilic side chains, are more abundant in the extended region than the globular domain. That is, these amino acids consist of 26.8% in the extended region; whereas 18.8% in the globular domain. In addition, the cross over in the chain fractal dimension is more likely to occur to relatively large proteins as can be seen in Table 1. This is supported by the finding in Ref. [5] that smaller r-proteins usually have a greater fraction of their surface area buried in RNA, while the larger proteins tend to have more independent globular domains that lies on the outside of the subunit.

Table 2. Fractal dimensions of r-protein backbones in the 50S subunits. Other inscriptions are the same as Table 1.

Name	N	D_M	D_L	D_H
L37e	56	2.50 (0.03)	1.42 (0.02)	1.73 (0.05)
L39e	46	2.73 (0.07)	1.27 (0.03)	1.66 (0.03)
L44e	92	2.76 (0.05)	1.28 (0.03)	1.93 (0.04)
L2	237	-	1.38 (0.03)* 1.54 (0.02)**	2.26 (0.03)
L3	337	2.73 (0.04)	1.30 (0.02)* 1.71 (0.02)**	2.48 (0.03)
L4	246	3.13 (0.06)	1.30 (0.02)* 1.54 (0.03)**	2.38 (0.04)
L6	172	2.28 (0.03)	1.31 (0.04)* 1.85 (0.04)**	2.45 (0.05)
L7Ae	119	2.75 (0.04)	1.36 (0.02)	2.50 (0.05)
L10e	171	3.24 (0.04)	1.34 (0.02)	2.43 (0.04)
L13	141	2.75 (0.04)	1.32 (0.02)	2.44 (0.05)
L14	132	2.92 (0.02)	1.34 (0.05)* 1.91 (0.08)**	2.51 (0.05)
L15	145	-	1.46 (0.02)	2.12 (0.04)
L15e	194	2.84 (0.05)	1.39 (0.02)	2.34 (0.04)
L18	186	2.63 (0.06)	1.36 (0.03)* 1.68 (0.02)**	2.44 (0.04)
L18e	115	2.70 (0.02)	1.33 (0.05)* 1.77 (0.05)**	2.48 (0.07)
L19e	143	-	1.25 (0.02)	2.14 (0.08)
L21e	95	2.57 (0.04)	1.38 (0.05)* 1.84 (0.06)**	2.34 (0.04)
L22	150	2.92 (0.05)	1.30 (0.02)* 1.78 (0.04)**	2.39 (0.05)
L23	81	2.79 (0.06)	1.30 (0.04)	2.39 (0.06)
L24	119	3.00 (0.06)	1.42 (0.03)* 1.62 (0.05)**	2.35 (0.06)
L24e	53	2.75 (0.09)	1.40 (0.04)	2.25 (0.09)
L29	65	2.89 (0.05)	1.19 (0.03)	2.16 (0.08)
L30	154	2.93 (0.05)	1.37 (0.02)* 1.92 (0.05)**	2.47 (0.05)
L31e	82	2.97 (0.09)	1.36 (0.04)	2.40 (0.07)
L32e	142	3.10 (0.07)	1.36 (0.03)	2.33 (0.05)
L37Ae	72	2.61 (0.05)	1.35 (0.05)	2.24 (0.09)

3.2. Capacity dimensions of the r-protein backbones

A typical result of the capacity dimension of r-proteins is shown in Fig. 4. Due to structural characteristics of the extension, it is plausible to expect that the capacity dimension of the globular domain is greater than that of the extended domain. Thus, the capacity dimension of a r-protein having extension is, in general, smaller than that of a globular r-protein. In addition, it was demonstrated that the primary function of a r-protein is to stabilize and maintain the subunit's structure via various types of protein-RNA and/or protein-protein interactions [3]. In particular, the protein-RNA interaction is mediated by the extensive size of the buried surface area of the r-protein. Large surface area buried for r-proteins is in part, if not entirely, due to r-protein extensions penetrating into the rRNA core. Since the role of extension is mainly for interaction with the rRNA, we expect that r-proteins with extensions have higher percentage of surface area buried by the rRNA and less percentage of protein-protein interaction. This tendency is shown quantitatively in Table 3 for r-proteins in the 50S subunit.

Table 3. Comparison of various average quantities between r-proteins with and without the extension. The average values of buried surface areas from the protein-protein interaction and protein-RNA interaction are estimated from Table II and VIII of Ref. [3]. The capacity dimension of r-proteins with the extension is smaller by 10% on the average than those without extension.

Interaction	With extensions	Without extension
protein-protein (%)	10.3 (5.5)	30.6 (13.0)
protein-rRNA (%)	47.1 (7.5)	28.1 (7.8)
D_H	2.19 (0.25)	2.40 (0.11)

R-proteins in the 30S subunit also contain extended structures in internal loops and/or long N or C-terminal that is not seen in the isolated structures [5]. It was reported [5] that almost all r-proteins (18 out of 20 r-proteins) have their extensions; furthermore, they are either less than about 10% or greater than about 30% of their surface packed against the rRNA through the protein-RNA interaction. We thus classify r-proteins into two groups according to the above percentage of packed surface, and estimate the average value of the capacity dimension for both groups. The average value of the capacity dimension of the group having less than 10% of their surface packed against the RNA is 2.27 ± 0.07, while that of the second group is 2.10 ± 0.14. This supports an assertion that large capacity dimension of r-proteins implies their shape being more spherical, thus they are less adaptive for the protein-RNA interaction.

The capacity dimensions of r-proteins are less than their embedding dimension, three; furthermore, they are, in general, less than that of the rRNAs [30]. This supports a structural characteristic of the ribosome as the RNA-protein complex macromolecule. The rRNAs are main constituents of the subunits with some empty space; the extended and globular domains of r-proteins fill the empty space in the interior and exterior of the rRNAs, respectively. A plot of the capacity dimensions versus the number of residues computed for r-proteins in both subunits is shown in Fig. 5. R-proteins with large number of residues are referred to as r-proteins of high capacity dimensions. The converse, however, is not true. This is in contrast with the case of the chain fractal dimension, where we find that the chain fractal dimensions of r-proteins are almost independent of the number of their residues. These characteristics are supported by the estimation of the cross correlation coefficient. For r-proteins in both subunits, the cross correlation between the number of residues and the chain fractal dimension is 0.01 and 0.11 for the 50S and 30S subunits, respectively; while the cross correlation rises to 0.49 and 0.51 for the 50S and 30S, respectively between the number of residues and the capacity dimension.

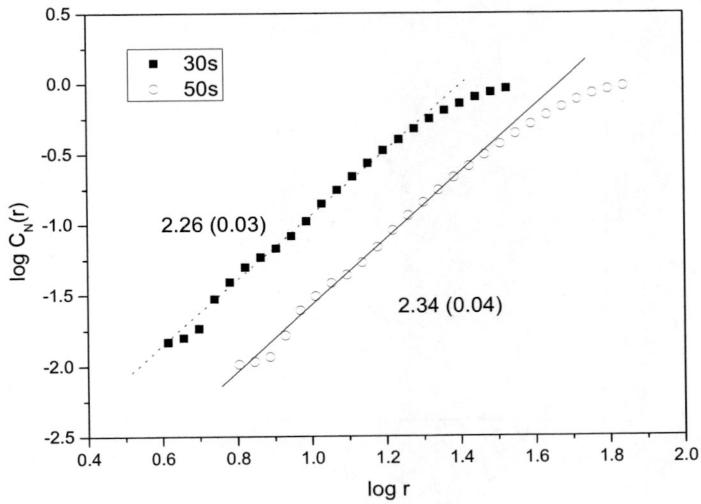

Figure 4. A typical plot of $\log C_N(r)$ versus $\log r$ for the S8 (■) and the L15e (○) backbones in the 30S and 50S subunits, respectively. The least-square fits on the slope of the S8 (solid line) and the L15e (dotted line) yield $D_H = 2.26 \pm 0.03$ and $D_H = 2.34 \pm 0.04$, respectively. The plot for the 23S rRNA is shifted to the right by the factor of 0.2 for the display purpose.

3.3. Mass fractal dimensions of r-proteins

The mass fractal dimension exhibits the molecule's space-filling ability: the larger is D_M, the more atoms are in the sphere. When fractal dimension is less than three, the structure has "empty" or "void" space. From a typical behavior shown in Fig. 6 and the results in Tables 1 and 2, the mass fractal dimensions of 11 and 16 r-proteins in the 30S and 50S subunits respectively are less than three. This indicates that the mass inside a radius r does not increase with the Euclidean dimension as an exponent but with some lesser power. Note that 5 and 3 r-proteins in the 30S and 50S subunits respectively do not show a scaling behavior. Also, 3 and 7 r-proteins in the 30S and 50S subunits respectively are compact three-dimensional collapsed polymers having their mass fractal dimension close to three.

Table 4. Average value of the mass fractal and the capacity dimensions over r-proteins in the 50S and 30S subunits. We estimated the mass fractal dimension for two levels: atom and backbone levels. D_M and D_H stand for the mass fractal and the capacity dimensions, respectively; the corresponding standard deviations are in the parentheses.

Subunit	D_M (atom)	D_M (backbone)	D_H
50S	2.80 (0.21)	2.39 (0.39)	2.30 (0.22)
30S	2.70 (0.24)	2.41 (0.35)	2.16 (0.15)

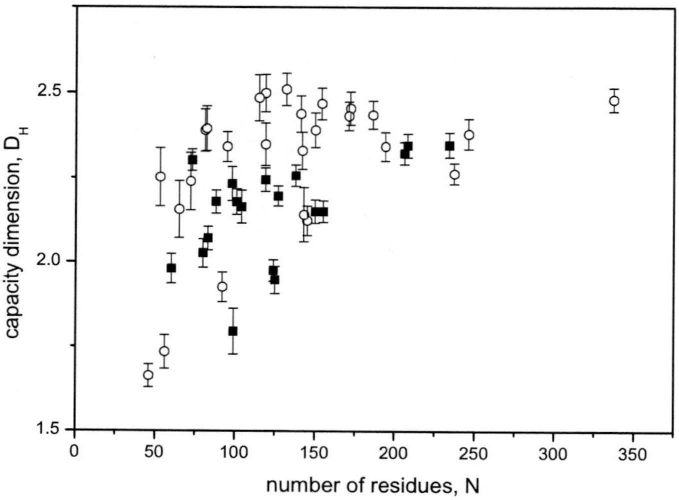

Figure 5. Plots of the capacity dimension versus the number of residues in r-proteins in the 30S (■) and 50S subunits (◯) with errorbars.

As stated in Table 4, the average mass fractal dimension of the r-protein in the 30S subunit is 2.70 ± 0.24; whereas those in the 50S subunit is 2.80 ± 0.21. Thus, the average fractal dimension of r-proteins in the 30S subunit is smaller than that in the 50S subunit. This result is consistent with the previous finding that the mass fractal dimension of the 30S subunit is smaller than that of the 50S subunit implying a compact three dimensional macromolecule [31]. This is probably closely related to the fact that almost all r-proteins in the 30S subunit have extensions, while about a half of r-proteins in the 50S subunit have extensions. Since any extended structure is less globular, thus has less mass fractal dimension, it can be expected that the mass fractal dimensions of r-proteins in the 30S subunit are expected less than those in the 50S subunit. In the case of the 50S subunit, most of conserved r-proteins have their mass fractal dimension close to three, implying they are dominated by the compact three dimensional structure. Furthermore, all r-proteins that do not exhibit a scaling behavior have extensions. The converse, however, is not true.

For the completeness, we estimate average values of the mass fractal and capacity dimensions over r-protein backbones in both subunits, and the result is shown in Table 4. We find that the average mass fractal dimension at the backbone level is less than that at the atomic level. Furthermore, at the backbone level, the average capacity dimension is similar to the average mass fractal dimension. This is justified by the fact that, for self-similar and deterministic fractals, both capacity and the mass fractal dimensions give the same result [7]. These results can be also compared with the result of the mass fractal dimension of globular proteins. From Refs. [16, 17], we find that the average mass fractal dimension for all globular proteins is 2.49 ± 0.09. At the atomic level, we find that mass fractal dimensions of r-proteins are, on the average, slightly greater than those of globular proteins. This suggests that r-proteins are in general more compact than the globular protein, which

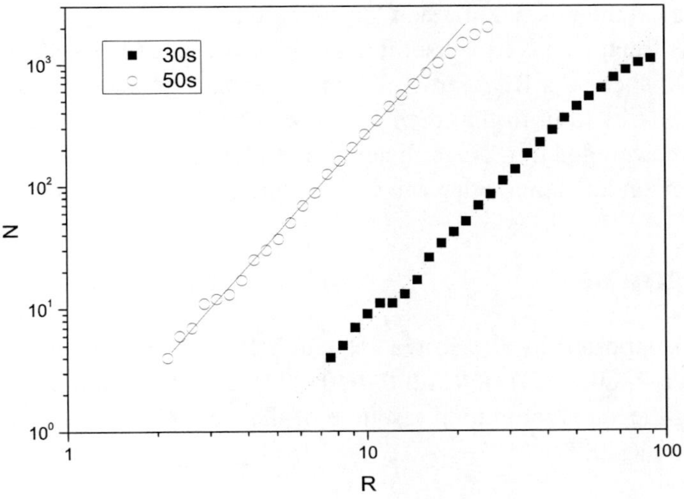

Figure 6. A typical log-log plot of the number of atoms versus the radius for the r-protein S7 (■) in the 30S subunit and r-proteins L3 (○) in the 50S subunit. The solid and dotted lines are least square fits to the mass fractal dimensions of L3 and S7, yielding the slope $D_M = 2.73 \pm 0.04$ and $D_M = 2.45 \pm 0.03$, respectively. The plot for the r-protein S7 is shifted to the right by the factor of 4 for the display purpose.

is probably due to the interaction with the rRNA.

4. Conclusion

In this paper, we investigated the self-similarity of r-proteins to unravel intrinsic structural characteristics of the biopolymers in conjunction with its biological function. To this end, we estimate the chain fractal and capacity dimensions as well as the mass fractal dimension of r-proteins with and without regarding them as successive walks between monomers along the chain. We, in particular, adopt the line segment and correlation integral methods to estimate the fractal dimensions.

The chain fractal dimension of r-proteins is less than that of the SAW in the short m range explains the existence of the long-extended domain in the r-protein structure. This is unusual to globular proteins. Although not all r-proteins have their chain fractal dimension in the long m range, r-proteins of relatively large number of residues are more likely to have the chain fractal dimension in large m range. The chain fractal dimension greater than that of the SAW in the large m range is a consequence of the formation of the globular domain that exists the exterior of the rRNA. These structural characteristics are found to be different from that of normal globular proteins, and can be explained in part by the chain fractal dimension. The capacity dimension of r-proteins is less than that of the rRNA, from which we may infer the structural role of r-proteins: r-proteins fill the empty space left by the rRNA. The average mass fractal dimension of the r-protein in the 30S subunit is smaller than those in the 50S subunit. This explains partly, if not entire, that r-proteins in the 30S subunit are more extended structure than those in the 50S.

Although the fractal analysis is useful to get intrinsic characteristics of highly inhomogeneous systems, such as a RNA-protein complex macromolecule, it is also true that the fractal analysis has its limit in that their detail structural conformations which determine their function are averaged out. Nevertheless, the fractal property of r-proteins provides an evidence of the extended domain that can not, in general, be seen in usual globular proteins.

Acknowledgment

This work was supported by the Korea Research Foundation Grant funded by the Korean Government (MOEHRD) (KRF-2006-H00007), in which main calculations were performed by using the supercomputing resource of the Korea Institute of Science and Technology Information(KISTI).

References

[1] For a general reference of the ribosome and its function, see, for example, Nierhaus, K. H.; Wilson, D. N. *Protein Synthesis and Ribosome Structure: Translating the Genome* ; Wiley-VCH, Weinheim, 2004; Garrett, R.; Douthwaite, S.; Liljas, A.; Matheson, A.; Moore, P.; Noller, H. *The Ribosome: Structure, Function, Antibiotics, and Cellular Interactions*; ASM Press: Washington, DC, 2000.

[2] Ban, N.; Nissen, P.; Hansen, J.; Moore, P.; Steitz, T. *Science* 2000, *289*, 905-920.

[3] Klein, D.; Moore, P.; Steitz, T. *J Mol Biol* 2004, *340*, 141-177.

[4] Wimberly, B.; Brodersen, D.; Claemons, W.; Morgan-Warren, R.; Carter, A.; Vonhein, C.; Hartsch, T.; Ramakrishnan, V. *Nature* 2000, *407*, 327-339.

[5] Brodersen, D.; Clemons, W.; Carter, A.; Wimberly, B,; Ramakrishnan, V. *J Mol Biol* 2002, *316*, 725-768.

[6] Schuwirth, B.; Borovinskaya, M.; Hau, C.; Zhang, W.; Vila-Sanjurjo, A.; Holton, J.; Cate, J. *Science* 2005, *310*, 827-834.

[7] Dewey, T. *Fractals in Molecular Biophysics* ; Oxford University Press: Oxford, 1998.

[8] Isogai, Y.; Itoh, T. *J Phys Soc Jpn* 1984, *53*, 2162-2171.

[9] Colvin, J.; Stapleton, H. *J Chem Phys* 1985, *82*, 4699-4706.

[10] Elber, R. In *The Fractal Approach to Heterogeneous Chemistry* ; Avnir, D.; Ed.; John Wiley & Sons: New York, NY, 1989; pp 407-423.

[11] Wang, C. X.; Shi, Y. Y.; Huang, F. H. *Phys Rev A* 1990, *41*, 7043-7048.

[12] Dewey, T. *J Chem Phys* 1993, *98*, 2250-2257.

[13] Xiao, Y. *Phys Rev E* 1994, *49*, 5903-5905.

[14] Moret, M. A.; Miranda, J. G. V.; Nogueira, Jr., E.; Santana, M. C.; Zebende, G. F. *Phys Rev E* 2005, *71*, 012901.

[15] Yu X.; Leitner, D. M. *J Chem Phys* 2003, *119*, 12673-12679.

[16] Enright, M. B.; Leitner, D. M. *Phys Rev E* 2005, *71*, 011912.

[17] Enright, M. B.; Yu, X.; Leitner, D. M. *Phys Rev E* 2006, *73*, 051905.

[18] The structural information for the subunits can be found in the Protein Data Bank (PDB) at http://www.rcsb.org/pdb/. The access numbers are 1FJF for *Thermus thermophilus* 30S subunit including the 16S rRNA, and 1S72 for *Haloarcula marismortui* 50S including the 23S rRNA. There are 21 and 29 r-proteins modeled in the 30S and 50S subunit, respectively. These are the highest resolution results for each subunit. Anything worse than about 3.5 Å would normally not be possible to construct an accurate model of a macromolecule.

[19] See, for example, Theiler, J. *J Opt Soc Am A* 1990, *7*, 1055-1073.

[20] Hausdorff, F. *Math Annalen* 1919, *79*, 157-179.

[21] See, for example, de Gennes, P. *Scaling concepts in polymer physics*; Cornell University Press: Ithaca, NY 1996.

[22] See, for instance, Doi, M.; Edwards, S. *The Theory of Polymer Dynamics*; Clarendon Press, Oxford, 1986, and references therein.

[23] Farmer, J.; Ott, E.; Yorke, J. *Physica D* 1983, *7*, 153-180.

[24] Greenside, H. S.; Wolf, A.; Swift, J.; Pignataro, T. *Phys Rev A* 1982, *25*, 3453-3456.

[25] Grassberger P,; Procaccia, I *Phys Rev Lett* 1983, *50*, 346-349.

[26] Colvin, J.; Stapleton, H. *J Chem Phys* 1985, *82*, 4699-4706.

[27] There are 20 r-proteins in the 30S subunit. We exclude r-proteins called THX, composed on 27 residues, since it contains too small number of residue to get a reliable statistics. There are 29 r-proteins in the 50S subunit. We exclude r-proteins, called L5 and L10*, since the coordinate information of than 10 residues in the middle of their amino acid sequence are missed. We also exclude the r-protein L1, since the coordinate information for atoms of the protein was not provided in the PDB.

[28] Creamer, T. P.; Srinivasan, R.; Rose, G. D. *Biochemistry* 1995, *34*, 16245-16250.

[29] Moret, M. A.; Zebende, G. F. *Phys Rev E* 2007, *75*, 011920.

[30] Lee, C.-Y. *Physica A* 2008, *387*, 4871-4880.

[31] Lee, C.-Y. *Phys Rev E* 2006, *73*, 042901.

In: Ribosomal Proteins and Protein Engineering... ISBN 978-1-60741-005-8
Editors: V. Ortendhal and H. Salchow, pp.121-141

Chapter 9

SYNTHESIZING MAMMALIAN CELL ADHESIVE SURFACES USING SELF-ASSEMBLING FIBRONECTIN DOMAINS

***Patricia Pereira*[*1] *and Christopher F. van der Walle*[†2]**
[1]Universidade de Lisboa, Lisboa, Portugal
[2]University of Strathclyde, Scotland

ABSTRACT

For tissue growth, organisation and repair, interaction between cells and their extracellular matrix (ECM) is critical. Fibronectin ECM renders a polyvalent ligand display for cell adhesion through binding to cell membrane integrin receptors. Mimicking this polyvalent display and designing ligands which bind specific integrin receptor subtypes will be discussed in this chapter. The central peptide motif for integrin binding is RGD (arginine-glycine-aspartic acid) but the integrin $\alpha5\beta1$ subtype requires the synergy site on the 9th type III fibronectin domain (FIII) in addition to RGD on the 10th FIII domain. Polyvalent integrin $\alpha5\beta1$ ligands have been synthesised using a protein engineering approach, designing chimeras consisting of a N-terminal 9th-10th FIII domain pair, IgG-derived hinge and leucine zipper-derived helix; the latter mutated to yield di-, tri- and tetrameric coiled coils and thus self-assembling, multimeric integrin $\alpha5\beta1$ ligands. A unique C-terminal cysteine was appended to the helix to facilitate 'anchoring' of the chimeras with a defined orientation on a surface. Careful characterization of such self-assembling chimeras is required and can be achieved through the use of complementary techniques including size exclusion chromatography, analytical ultracentrifugation and circular dichroism. Surface adsorption and specific immobilization can then be unambiguously characterized by neutron reflectivity, relating the determined models of the adsorbed protein layers to their biological function. Application of these engineered proteins and cell adhesive surfaces as scaffolds or other

* Patricia Pereira, PhD, iMed.UL - Research Institute for Medicines and Pharmaceutical Sciences, Faculdade de Farmácia, Edificio E, Universidade de Lisboa, Av. Prof. Gama Pinto, 1649-003 Lisboa, Portugal. Tel: +351-217946400 (Ext: 245), Fax: +351-217946470
† C. F. van der Walle, PhD, Strathclyde Institute of Pharmacy and Biomedical Sciences, University of Strathclyde, 27 Taylor St, Glasgow, G4 0NR, Scotland. Tel: +44 (0) 141 548 5755, Fax: +44 (0) 141 552 2562

biomedical devices has the potential to promote tissue repair and regeneration, for a wide variety of tissues including bone and skin. These applications will be discussed in the context of the future outlook for these ECM protein engineering approaches.

INTRODUCTION

The extracellular matrix (ECM) component involved in integrin mediated cell adhesion is dependent upon the integrin subtype, for example, osteoclasts bind collagen via $\alpha 1\beta 1$ and $\alpha 2\beta 1$ integrins whereas fibroblasts bind fibronectin (FN) via $\alpha 5\beta 1$ integrin [1]. The primary integrin binding motif, arginine-glycine-aspartic acid (RGD), is found in several ECM proteins [2]. However, while RGD binds $\alpha v\beta 3$ integrin with the same affinity as its ECM ligand, vitronectin, this is not true for all integrins, especially $\alpha 5\beta 1$ integrin which requires the 'synergy site' (PHSRN) on the 9th type III FN domain (FIII9) in addition to RGD on the 10th type III FN domain (FIII10) [3, 4]. This is often overlooked because cells will adhere to surfaces furnished with RGD [4], including cells expressing both $\alpha v\beta 3$ and $\alpha 5\beta 1$ integrins since RGD favours $\alpha v\beta 3$ integrin binding over $\alpha 5\beta 1$ integrin [4]. This hides the limitation that, for fibroblasts binding via $\alpha 5\beta 1$ integrin, cell spreading and signalling cascades are not fully invoked [5]. Peptide amphiphiles harboring PHSRN and RGD motifs have been reported but demonstrate very weak synergistic effect [6], corroborating direct evidence that the spatial relationship of the FIII9-10 domain pair (Figure 1) is critical to $\alpha 5\beta 1$ integrin binding [7, 8].

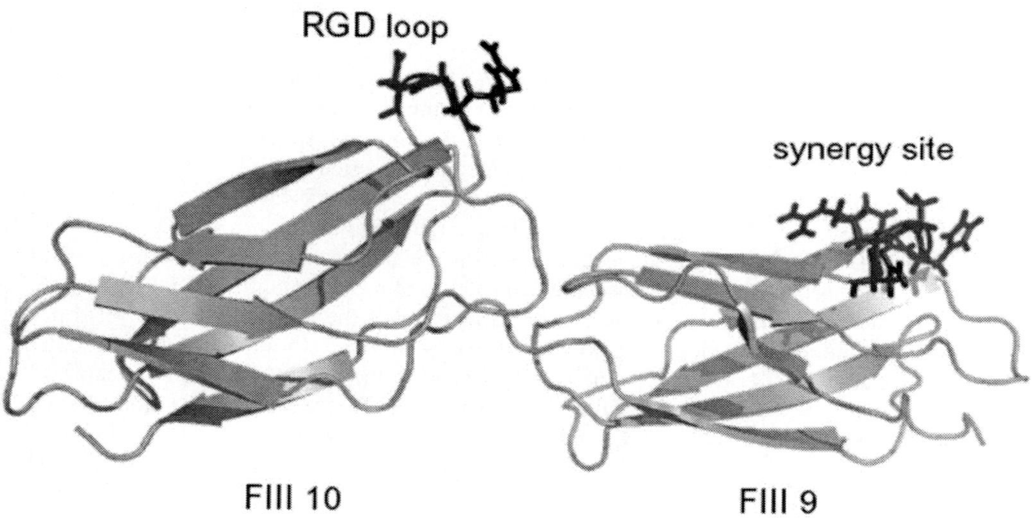

Figure 1. Ribbon diagram of the 3-dimensional structure of FIII9-10. β-strands are shown as grey ribbons, with the integrin binding domains shown in black as stick diagrams.

With RGD being a poor mimic of FN ECM, one option would be to coat surfaces with FN. While simple, this does not address the issues of adsorption induced denaturation and elasticity of the ECM. Nor does the size of FN permit controlled cross-linking with polymeric supports on account of steric hindrance and multiple functional sites. Protein immobilisation is not facile; while adsorption is fast and essentially irreversible, denaturation and random orientations result [9]. For RGD-peptides at least, the development of comb-like polymers and self-assembling monolayers (SAMs) has permitted calculation of ligand densities [10, 11]. Application of SAMs to larger protein domains remains unclear since the cross-linking of proteins to the gold/alkanethiolate brush border appears to leave significant proportions of the immobilised protein non-specifically adsorbed [12]. Similarly, one report for cross-linking FIII7-10 to adsorbed BSA did not address random ligand orientation or adsorption-induced unfolding [13]. A 'carrier' facilitating uniform ligand orientation upon immobilisation is therefore required.

The ECM is not static: soluble FN polymerises rendering a polyvalent display wherein the fibrils may extend and contract [14]. Integrin-FN binding triggers the clustering of integrins to focal adhesion complexes and cytoskeletal reorganisation [15]. Fibroblasts probe the elasticity of their supporting matrix, responding to greater elasticity with increased motility and protruding lamellipodia [16]. The importance of polyvalent ligand presentation to NR6 fibroblasts expressing both $\alpha v\beta 3$ and $\alpha 5\beta 1$ integrins was explored using RGD on comb-like arrays, which increased cell migration speeds over random RGD arrays [17]. Further, actin stress fibre formation was 2-3 fold higher and cell 'adhesion reinforcement' (a strengthening of the links between integrins and the cytoskeleton [18]) was observed for RGD-cluster sizes between 3.6 to 5.4 [19]. These studies now raise questions over previous results for minimal RGD densities required to support cell adhesion and spreading because these were based on random ligand distributions [4]. Therefore, concerns over ligand density, orientation and polyvalency cannot be ignored if biomimetic supports are to realise their full potential.

THE DESIGN CONCEPT

Directed Mutation of the 9th-10th Type III Fibronectin Domain Pair

A previous report [20] demonstrated that the mouse FIII9-10 domain pair has increased conformational stability compared to the human FIII9-10 domain pair [21]. Mouse FIII9-10 was also sufficiently stable for resolution of its 3-dimensional (3-D) solution structure by NMR. The sequence identity between mouse and human FIII9 is high (83%) and a series of mutant human FIII9-10 proteins were generated by QuikChange™ (Stratagene, UK) site directed mutagensis. The mutations were made with the aim of identifying specific residues that could confer thermodynamic stability on FIII9 in the FIII9-10 pair, and that have a key function in the folding of FIII9. Alignment of mouse and human FIII9-10 primary sequences highlighted 5 non-conservative and 10 semi-conservative or conservative amino acid mismatches that exist between the mouse and human FIII9 modules (Figure 2). Minimal changes, involving semi conservative amino acid substitutions and Ser-Val for Phe-Ser^{1367} were ignored, focussing instead on three non-conservative amino acid substitutions: Ser for Phe^{1335}, Ile for Arg^{1358}, and Pro for Leu^{1408}, introduced either alone or in combination [22].

```
                    1335      1345      1355      1365      1375      1385
                     .         .         .         .         .         .
mouse-FIII9 GLDSPTGFDSSDITANSFTVHWVAPRAPITGYIIRHHAEHSVGRPRQDRVPPSRNSITLT
            :::::::.: ::::::::::::.::::.:::: ::::.::  ::::.::::.::::::::
human-FIII9 GLDSPTGIDFSDITANSFTVHWIAPRATITGYRIRHHPEHFSGRPREDRVPHSRNSITLT

                    1395      1405      1415
                     .         .         .
mouse-FIII9 NLNPGTEYVVSIIAVNGREESPPLIGQQAT
            ::.:::::::::.:.::::::: :::::.:
human-FIII9 NLTPGTEYVVSIVALNGREESPLLIGQQST
```

Figure 2. FASTA sequence alignment of the murine and human FIII9 modules; colons indicate identical amino acid pairs, with stops indicating conservative or semiconservative amino acid pairs and spaces indicating non-conservative pairs. Every tenth amino acid is noted with an upper dot along with the corresponding sequence number (corresponding to its position in FN [23]). β-Strands are underlined and the L^{1408} substitution is shown in bold. Adapted from [22].

FIII9-10 proteins are readily expressed in *Escherichia coli* strain BL21 with good yield (10-20 mg/L), appended at their N-terminus with a hexahistidine tag (cloning into pRSET-A – Invitrogen, UK) to facilitate purification by Ni^{2+} chelation chromatography. The conformational stabilities of the FIII9-10 were characterised by equilibrium chemical denaturation, monitoring the change in fluorescence of the aromatic residues. It was found that only the L1408P mutant had a greatly improved stability over the wild type FIII9-10. The reason for this is thought to involve the location of the proline (Pro) residue, which lies at the boundary between the β-strand G and the F-G loop [23], i.e. at the beginning of a predicted β-strand. Similar locations for other Pro residues, and Pro-Pro pairs exist in the other type III FN domains. In both cases, the local flexibility of the protein backbone will be restricted because of the narrow range of dihedral angles allowed by proline. The rapid refolding of the human FIII10 module has also been explained with reference to its numerous Pro residues found at key locations in the secondary structure [24]. Thus, the L1408P substitution is likely to contribute to the increased conformational stability, and increased aqueous solubility, by adopting and maintaining a favourable turn in FIII9. An added consequence of the improved structural stability was a concomitant improvement in the biological activity of the L1408P mutant, measured as a function of cell adhesion and spreading [22]. No interaction between Pro^{1408} and the integrin α5β1 binding site is predicted since Pro^{1408} lies on the opposite side of the module from the synergy site. Nor does Pro^{1408} alter the FIII9/FIII10 interface, which has been shown to affect the domain pair's biological function [7], and is a target of mutation is later studies described below. The generation of the L1408P mutant (hereafter termed FIII9'10) therefore provides an example of how a single residue substitution can markedly alter a protein’s solution properties and biological activity through consideration of conformation [22].

Design of Self-Assembling Integrin A5β1 Ligands

Given the work described above, we employed FIII9'10 as the ligand to support integrin α5β1-mediated cell adhesion and spreading. To mimic the polyvalent nature of the FN ECM, is was necessary to achieve clustering (multivalency) of FIII9'10, avoiding chemical conjugation protocols which may result in protein denaturation. The route chosen was to oligomerise FIII9'10 in a controlled fashion via fusion with self-assembling coiled coil domains. The envisaged multivalent integrin α5β1 ligand displays were hypothesised to combine the cell spreading/adhesion reinforcement seen for multivalent RGD-integrin αvβ3 binding with strong activation of Rho signalling-differentiation cascade seen for FN-integrin α5β1 binding; i.e. it was hoped that cell supports developed on this basis would best mimic the FN ECM *in vivo*. With regards to tissue engineering, ligand displays were initially developed on 2-D surfaces to enable physical characterisation and preliminary cell function assays.

Coiled coil domains are stable parallel or antiparallel oligomers of α-helical peptides which self-assemble with left- or right-handedness according to their amino acid sequence [25], and have been used for tetramerisation of miniantibodies [26]. The work described below built on experience gained with tetrameric coiled coil tetrabrachion fragments extended with the GRGDS pentapeptide [27]. However, we wanted greater flexibility in the number of FIII9'10 ligands per cluster, since RGD cluster size was previously shown to be important for cell function [17]. For this reason, smaller clusters of 2 and 3 were constructed using a modified leucine zipper domain of the GCN4 protein [28]. The leucine zipper domain provides the archetypal parallel left-handed coiled coil based on the heptad repeat, wherein mutation of residues ‘a’ and ‘d’ will afford dimers, trimers and tetramers (Figure 3). The use of parallel peptide sequences in the coiled coil ensured that all FIII9'10 ligands were presented in one face to bound cells. The range of cluster sizes (2 to 4) that can be achieved with mutants of the GCN4 leucine zipper corresponds to the comb-like RGD-polymers and simulation data for occupancy of α5β1 integrin populations [11, 29].

An important consideration was the potential affect of steric hindrance between neighbouring, bound integrin receptors. A ‘spacer’ between the coiled coil domain and FIII9'10 ligand was therefore required and based on the immunoglobulin (IgG) ‘hinge’ region. In the crystallographic 3-D structure for IgG2A, the 24 amino acid structural hinge is predicted to be highly flexible and extend over 50 Å [30]. While one 10 amino acid IgG hinge facilitated polyvalent binding for tetravalent miniantibodies, for an integrin receptor with a diameter of around 100 Å the IgG2A hinge may still be too short to prevent steric clashes. Instead, the 62 amino acid IgG3 hinge region proved useful since it represents the quadruplicated ‘primitive IgG hinge’ - a highly conserved 15 amino acid sequence [31]. Lengthening of the hinge simply involved insertion of extra 15 amino acid cassettes (Figure 3). On the basis that the 24 amino acid IgG2A hinge extends over 50Å it was not unreasonable to begin with a duplicated, 30 amino acid, hinge. When fusing FIII9'10 with other domains it is necessary to avoid inclusion peptide motifs with inherent integrin binding activity: in this case neither the coiled-coils nor hinge had any intrinsic cell-binding activity. Conversely, self-association has not been reported for the FIII9-10 domain pair.

Construct,	N-His$_6$-FIII9'10,	IgG-like hinge,	helix for coiled coil, C-term. Cys
Dimer	FIII9'10-	(EPKSSDTPPGSPRSP)$_2$	GR**I**KQ**L**EDK**I**EE**L**LSK**I**YH**L**ENE**I**AR**L**KKL**I**GE**L**EDK**I**EN**L**GC
Trimer	FIII9'10-	(EPKSSDTPPGSPRSP)$_2$	GR**I**KQ**I**EDK**I**EE**I**LSK**I**YH**I**ENE**I**AR**I**KKL**I**GE**I**EDK**I**EN**I**GC
Tetramer	FIII9'10-	(EPKSSDTPPGSPRSP)$_2$	GR**L**KQ**I**EDK**L**EE**I**LSK**L**YH**I**ENE**L**AR**I**KKL**L**GE**I**EDK**L**EN**I**GC
Heptad position			g**a**bc**d**efg**a**bc**d**efg**a**bc**d**efg**a**bc**d**efg**a**bc**d**efg**a**bc**d**

Figure 3. Amino acid sequences in single letter code for the hinge-coiled coils appended to FIII9'10. The N- to C-terminal order follows: N-FIII9'10-hinge-coiled coil-C; cysteines substituted for serines in the (duplicated) wild-type IgG hinge are underlined; the ‘a’ and ‘d’ positions in the coiled coil heptad repeat are shown in bold.

During the design process, all cysteine residues present in the wild-type IgG hinge were mutated to serine, so that a sole cysteine could be appended to the C-terminus of the protein. This was intended to facilitate sulfhydryl-specific biotinylation of the protein C-terminus, by reaction with biotin-maleimide, affording immobilisation to streptavidin-coated surfaces in a specific orientation (Figure 4). While non-covalent bonds may be considered too weak to support cell-detachment, the strength of the biotin-streptavidin bond has been directly measured as 3-170 pN for an increasing loading rate of 0.05-100 pN/s [32]. In the cell detachment protocol described by Koo et al. [19] the maximum detachment force was 1000 pN/cell (cell volume 500 μm^3, cell density 1.07 g/ml, medium density 1.00 g/ml), using a Sorvall centrifuge which would have typically reached the required RCF (~30 g) in around 10 s. The loading rate would therefore have been around 100 pN/s, predicting a biotin-streptavidin bond force of 50 pN [32]; clearly too weak to support a detachment force of 1000 pN. However, there are estimated to be 100,000 integrin molecules per cell [29] and this order of multiplication gives a detachment force >> 1000pN.

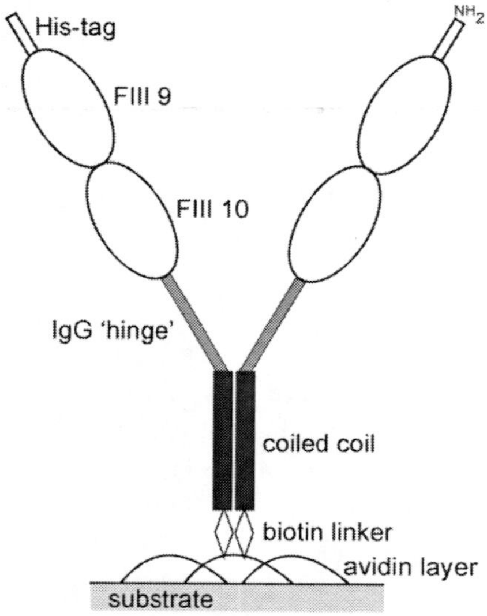

Figure 4. Schematic of the FIII9'10-dimer and proposed attachment to a support via biotin-streptavidin. A similar scheme is envisaged for the trimer and tetramer. The orientation and angle between the domain pair is not experimentally determined.

Although approximate lengths for FIII9-10 and the coiled-coil are known, until experimental data is available it is not possible to predict how far apart the FIII 9-10 ligands may be in (b) due to peptide chain mobility (especially at the hinge). The design therefore aimed to overcompensate, using the tetravalent miniantibodies as a precedent [26].

CHARACTERISATION OF OLIGOMERIC PROTEINS

Fusion proteins generally have an altered expression yield compared to the wild type. In some cases, this can be advantageous, for example, maltose binding protein and glutathione-S-transferase are used as carrier proteins to increase yield in commercially available cloning and expression systems. However, the converse is also true but generally under-reported in the literature. In the example here, fusion of an IgG-derived hinge and coiled coil to the C-terminus of FIII9'10 did not significantly effect the yield of expressed recombinant protein from *E. coli*, though yields for the tetramer and trimer were less than for the dimer. The secondary structure of the coiled coil domain (α-helical) contrasts against the entirely β-sandwich FIII9'10 domain pair, and a composite spectrum was observed in the far ultra-violet (UV) region by circular dichroism (CD) spectroscopy [33]. CD is commonly used to rapidly define protein structure as part of a quality control or exploratory investigation [34]. It is generally useful to have a reference spectrum against which changes in the protein secondary structure can readily be determined. For example, wild type fibronectin has distinctive positive CD band centred around 226 nm, as a consequence of aromatic contributions in the far UV [35].

It is of course important to address the question of oligomerisation, in the example here, had the coiled coil domains self-assembled into dimers, trimers and tetramers as predicted? Characterisation of this property has direct consequences for integrin clustering and cell adhesion response. In the engineering of novel antibody scaffolds through the use of coiled coil, the same question arises too [36]. Through a shift in size of the oligomeric proteins, simple size-exclusion chromatography may be sufficient to determine the oligomerisation state [27]. However, while isolated coiled coils are known to form exclusively dimers, trimers or tetramers [28], when fused with large ligands it has been shown that some 'plasticity' may occur. For example, the dimeric construct designed here (FIII9'10-dimer) formed a mixture of dimers and trimers. Furthermore, the ratio of dimer:trimer appeared to increase with increasing ionic strength of the buffer. This type of behaviour has previously been reported for some coiled coil structures [37-39] but not for the GCN4 leucine zipper domain. The hallmark of coiled coils is the distinctive packing of amino acid side chains in the hydrophobic core of the helix bundles, called 'knobs-into-holes' packing, which presents the major stabilizing force [28]. However, ionic interactions between the side chains of neighbouring helices affect stability, orientation and stoichiometry of coiled coils [40]. Accordingly, inter- and intra-helical salt bridges between polar residues in positions g and e of the heptad repeat are particularly important, with higher salt concentration favouring the higher oligomerisation state [37, 40]. This involves higher ionic strengths reducing the contribution of ionic interactions to coiled coil stability, due to increased shielding of charged side chains, while increasing the hydrophobic interaction. Therefore, the dimer-trimer transition for FIII9'-10-dimer was possibly due to improved shielding of the hydrophobic side chains from the very polar environment in a trimeric coiled coil compared with a dimeric coiled coil although the packing of the side chains would favour the dimeric structure.

One known issue with the use of size-exclusion chromatography is that the estimated and theoretical M_r generally do not agree [41], particularly with irregularly shaped proteins as would be expected for the constructs here. To better characterise the oligomerisation of a particular protein, analytical ultracentrifuge should be employed, to perform sedimentation velocity (SV) experiments. For the example of the self-assembling FIII9'10 ligands, under reducing conditions the FIII9'10-trimer sedimented as one major species ($s^0_{20,w}$ = 3.69 S; where $s^0_{20,w}$ is the measured sedimentation coefficient at 20 °C in water with units of S, the Svedberg), in agreement with the theoretical sedimentation coefficient of 3.9 S. In contrast, under non-reducing conditions the main species observed at $s^{0}_{20,w}$ = 5.44 S was suggestive of a hexamer composed of two trimers linked via a disulphide bond (oxidation of the C-terminal cysteine residues). Figure 5 shows the theoretical models used in the theoretical calculation of the sedimentation coefficient for comparison to the experimental data [42]. For the FIII9'10-dimer, a dimeric species was observed under reducing conditions but under non-reducing conditions the hexameric species observed for the FIII9'10-trimer was also observed. Thus, the SV data can be used to support size-exclusion data, in this example, that the FIII9'10-dimer can undergo oligomerisation to a trimer under specific buffer conditions depending on the ionic concentration and redox state [42].

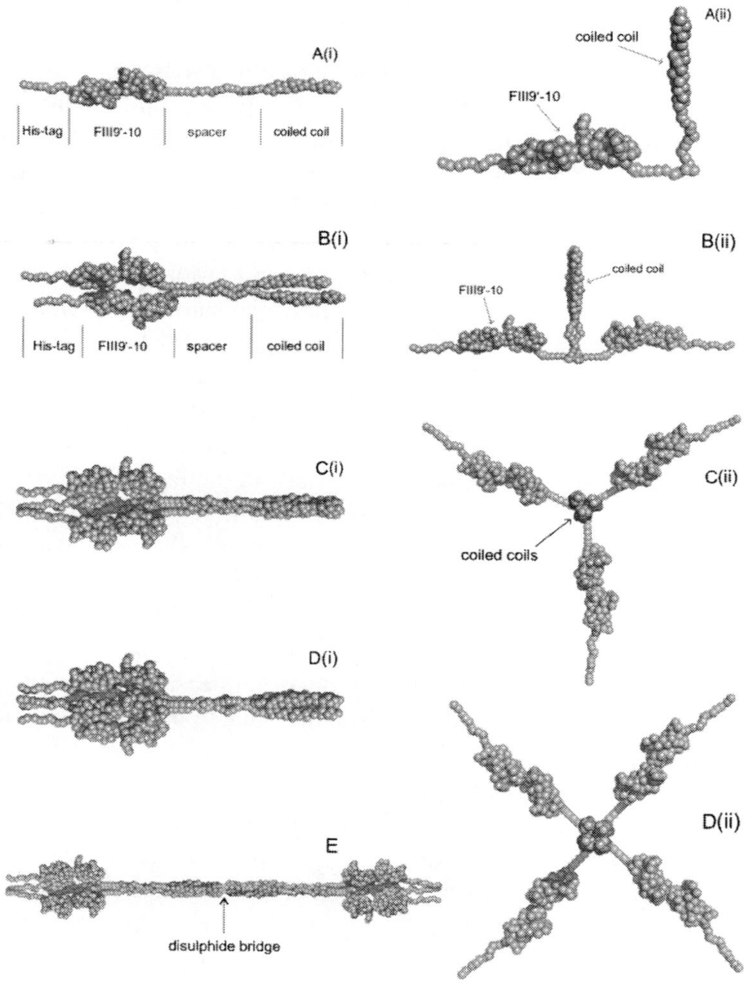

Figure 5. Straight' (i) and 'bent' (ii) bead models for the FIII9'-10-monomer (A), FIII9'-10-dimer (B), FIII9'-10-trimer (C) and FIII9'-10-tetramer (D). The disulphide-linked dimerised trimer was modelled as the straight form only (E). The individual peptide domains (N-terminal polyhistidine tag, FIII9'-10 domain pair, IgG-derived spacer and coiled coil) are indicated by demarcation in A(i) and B(i). The same peptide domains were used to build the bent model, which has a 90° turn at the centre of the spacer and the other straight models. Images C(ii) and D(ii) are shown as the elevated plan. Adapted from [42].

Related sedimentation equilibrium (SE) experiments can be used to experimentally determine the dissociation constant for (non-covalent) oligomeric proteins [43]. Note that the term dissociation constant needs defining carefully since the values are sometimes better interpreted as the concentration at which monomer and oligomer are equimolar. Similarly, if no single oligomeric species can be observed with confidence then this K_d value cannot be calculated. In the case of the FIII9'10-trimer, the dissociation constant ($K_{d(1,3)}$ = [molar concentration of monomer]^3/[molar concentration of trimer]) was 0.03 μM^2 and for the FIII9'10-tetramer the $K_{d(1,4)}$ was 0.3 μM^3; such that monomer and oligomer were equimolar at ca. 0.2 μM and 0.7 μM respectively. Other techniques such as functional or biological assays have been used to indirectly estimate the K_d of oligomers, using the assumption that a loss in activity indicates a shift from polyvalent to monovalent receptor-ligand binding [27]. Others have noted that determination of the dissociation constant for coiled coils is not trivial, partly due to mixed oligomers [39], and a wide range of values from 0.03 μM to < 10 μM have been reported for GCN4 leucine zipper domains [44, 45]. What is most important with respect to cell adhesion and differentiation is that self-assembling integrin ligands remain in their oligomeric form over the range of concentrations used in cell function assays [33].

ENGINEERING OF CELL ADHESIVE SURFACES

A vast amount of data has been acquired describing the adsorption of whole FN to various solid/liquid and air/liquid surfaces, which is beyond detailed examination in this book section; a review focussed to tissue engineering is provided by Wilson *et al.* [46]. However, it is important to note that FN is strongly surface active. It has been demonstrated that FN will displace albumin from an oil/water interface [47], the molecular basis of which was suggested to be due to the domain-domain flexibility of modular proteins such as FN. FN is well known to adopt extended and compact conformations for the same reason Direct imaging of these conformations is possible using atomic force microscopy, and indirectly using FRET (fluorescence resonance energy transfer), with extended FN conformations adopted at hydrophilic surfaces, and compact conformations at hydrophobic surfaces [48, 49]. The same conformations influence cell adhesion behaviour, with extended FN (on glass) promoting greater cell adhesion than compact FN (on fluorosilane), for bulk FN concentrations of 1 μg/ml [49]. In all cases, desorption from a surface is possible and engineering approaches have been used to address this issue. Calonder et al. overcame protein desorption by covalent coupling strategies which formed oriented FN layers at a surface [50]. These FN oriented FN layers led to the suggestion that the FN molecules were vertically aligned at the surface, yielding "thicker and denser" FN layers. However, the extent of vertical alignment over the whole FN molecule may be limited by its large size and modular nature.

It should be remembered that protein adsorption to an interface may involve: i) a change in conformation [51], ii) partial desorption and reorientation [52, 53], iii) orientation/alignment as a function of surface coverage [54]. Protein adsorption-unfolding has been studied using FTIR (Fourier transform infrared spectroscopy), surface pressure and neutron reflection [54-56]. More recently 'solid state' circular dichroism (CD) developed by Ganesan *et al.* [57] has been used in conjunction with TIRF (Total internal reflection fluorescence) to look at adsorption of FIII9-10 and related mutants to hydrophilic and hydrophobic surfaces [58]. The solid state CD methodology involves an empirical method wherein the CD spectra acquired are corrected for distortion through 'absorption flattening' of closely packed chromophores at the surface [59, 60]. It was found that the wild type FIII9-10 domain pair did not adsorb as strongly, and desorbed more completely, than the more stable FIII9'10 mutant. In addition, unfolding of FIII9-10 was more extensive than for FIII9'10, which maintained the typical FN type III fold on adsorption to silica. Such data give insight to cell behaviour on cell adhesive surfaces since FIII9-10 interaction with integrin α5β1 is sensitive to the relative spatial arrangement of the RGD (FIII10) and PHSRN (FIII9) loops [21]. Thus, the extensive unfolding of FIII9-10 compared to FIII9'10 may help explain the observed increased in cell adhesion to the latter [22].

However, the surface analytical techniques described above are unable to resolve the adsorbed protein layer coverage (surface fraction occupied by protein) and layer thickness to the accuracy possible with neutron scattering techniques [61, 62]. Recent analysis of FIII9-10 and related mutants has corroborated evidence from CD and TIRF but also yielding insight to the various stages of adsorption of constrained FN type III domain pairs [63]. FIII9-10 adsorbed 'side-on', that is, with the long axis of the domain pair aligning horizontally with the silica substrate surface. Its surface coverage was also found to have an upper limit of ~22%, compared to ~31% for FIII9'10 and ~100% for FIII9'10-CC (a mutant of FIII9'10 with highly constricted interdomain flexibility [8]). What was particularly interesting was the observed re-orientation of FIII9'10-CC away from the horizontal to the vertical (Figure 6). This observation is reminiscent of the data acquired by Calonder et al. [50]. The hypothesis put forward to explain the observed re-orientation, was to form a simple analogy with the surface pressure area isotherm of a surfactant at the air/liquid interface, where the average orientation of the surfactant molecule changes as a function of the available surface area [64]. Further engineering of FN surfaces is likely to become of more interest to researchers who require tailored surfaces for tissue engineering, particularly when defining the relationship between ligand density/presentation, and integrin clustering/cell behaviour. Such applications are described in the next section of this chapter.

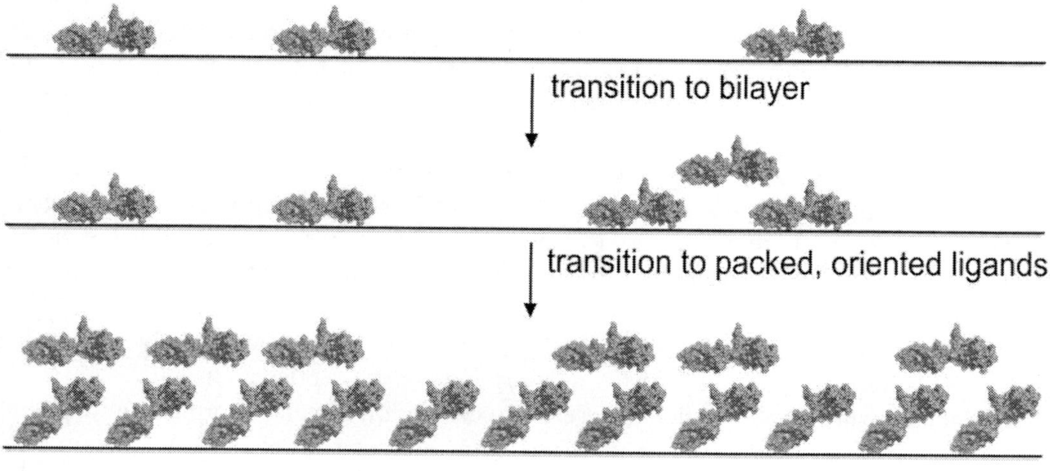

Figure 6. Cartoon of the three transitions observed for adsorption of FIII9'10-CC from bulk concentrations of 10 mg/L (top), 30 mg/L (middle) and 400 mg/L (bottom). The SiO_2 surface is shown as a black line with space filling models for FIII9'10-CC. Adapted from [63].

MEDICAL APPLICATIONS OF CELL ADHESIVE LIGANDS AND SURFACES

The constant increasing medical needs in the field of tissue repair and regeneration have led to the necessity of investigating different cell adhesive ligands and surfaces. In FN-mimetic surfaces, the use of recombinant fragments instead of the entire molecule offers several advantages: i) flexibility in engineering specific characteristics to increase immobilisation and activity [13]; ii) reduction in antigenicity; and iii) elimination of domains that may cause adverse reactions, such as the collagen-binding domain which otherwise may up-regulate expression of metalloproteinases by articular chondrocytes that specifically degrade collagen and aggrecan [65]. Furthermore, the importance of having the RGD and PHSRN motifs adequately spaced in such surfaces has proven successful in applications where α5β1 integrin has to be specifically activated [66] or blocked (e.g. certain cancers) over the remaining cell surface receptors; key examples are described below.

Bone Regeneration

Interactions between FN and integrin α5β1 are required for bone morphogenesis and osteoblast differentiation [67], essential processes for the repair of bony defects. A study conducted in 2001 using microplate dishes coated with 1 μM of entire FN and FIII9-10, found no significant differences in the activity of the two molecules as regards to human osteogenic sarcoma (HOS) cells adhesion, proliferation and differentiation; thus demonstrating that FIII9-10 can be used as a substitute for FN [68]. Peptides harbouring both the RGD and PHSRN motifs and separated by a polyglycine sequence (mimicking the native distance between the two motifs) have been coupled to acrylated PEG polymerised with PEG-diacrylate to form a hydrogel with a well defined surface density (Figure 7). These hydrogels supported a simultaneous increase in the adhesion, spreading, focal contact formation, proliferation and metabolic activity of osteoblasts isolated from neonatal rat calvaria, when compared to gels containing a scrambled co-localised peptide sequence or RGD and PEG alone [69].

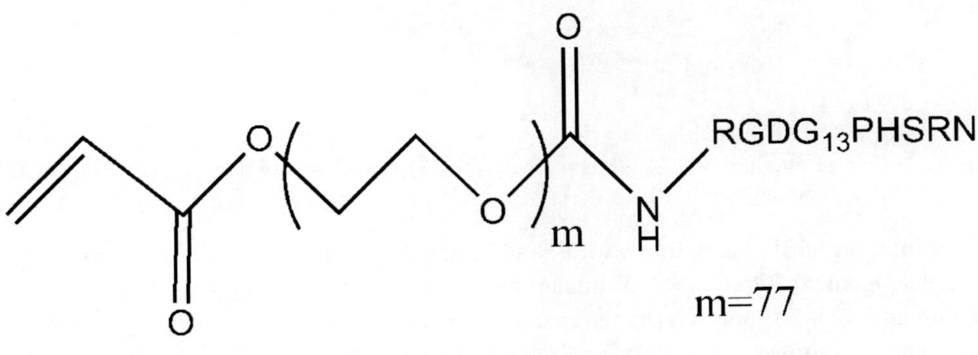

Figure 7. Chemical structure of acryloyl-PEG-RGDG$_{13}$PHSRN. Adapted from [69].

Clinical-grade titanium implants grafted with a non-fouling polymer coating, functionalised with a controlled density (0.9 pmol/cm^2) of FIII7-10 (encompassing the 7th-10th FIII domains, Figure 8) or RGD were tested *in vitro* and *in vivo*. *In vitro*, the FIII7-10 tethered surfaces promoted rat bone marrow stromal cells adhesion via integrin α5β1, **whereas the RGD-tethered surfaces primarily bound αvβ3 integrin** [70]. This differential integrin specificity was used to explain the raised levels of Cbfa1 (a transcription factor essential for osteoblastic differentiation and bone formation) and matrix mineralization observed for FIII7-10 surfaces when compared to the RGD-tethered and unmodified titanium counterparts. *In vivo*, the FIII7-10 functionalized implants placed in artificially created rat tibia cortical defects, exhibited a 70% increase in the bone-implant contact area, also in comparison to the RGD-tethered and unmodified titanium implants.

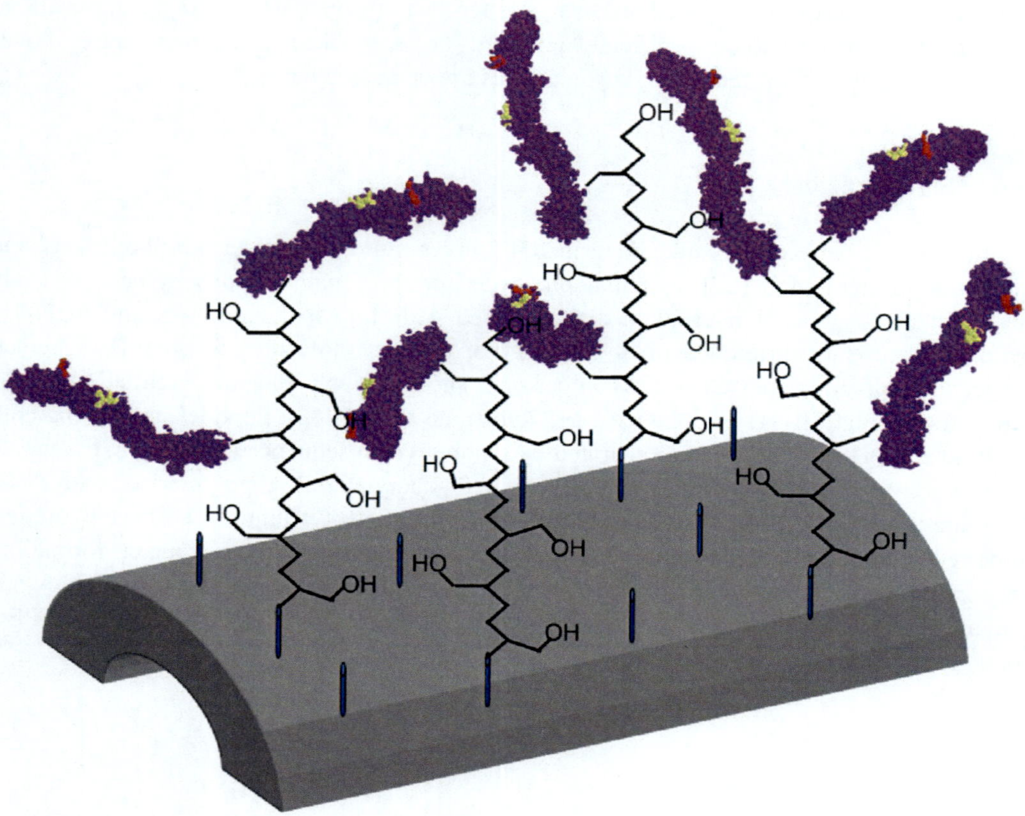

Figure 8. Cartoon of titanium FIII7-10-tethered poly(OEGMA) system. Purple: FIII7-10, Red: RGD, Yellow: PHSRN. Adapted from [70].

The integrin **α5β1 interaction** with FN and related FN fragments is also preponderant in driving osteogenic differentiation of human mesenchymal stem cells (MSCs) in 2-D and 3-D environments [71]. For polystyrene tissue culture plates coated in equimolar concentrations of integin binding motifs: 0.5 μM FN or 2 μM FIII9-10/FIII9'10/FIII10, inhibition of integrin α5β1-mediated MSC adhesion by functional blocking antibodies (FBA) was strong for FIII9'10 (43.4 ± 3.7%) and FIII9-10 (57.4 ± 4.7%) but less pronounced for FN and FIII10.

MSC proliferation on these surfaces was found to be significantly greater for FN, followed by FIII9'10 and FIII9-10, whilst FIII10 proliferation levels were significantly lower. In contrast, MSC proliferation on these surfaces but in the presence of integrin α5β1 FBAs was significantly less for FIII9'10 and FIII9-10 but not FN or FIII10, thus proving the importance of integrin specificity in the MSC proliferation response. PCR analysis for specific osteoblastic gene expression also showed FIII9'10 to be the most potent adhesion molecule (of the four tested) for driving osteogenic differentiation in 2-D environments, with these effects reduced by integrin α5β1 FBA. For 3-D functionalised fibrin matrices, similar observations were made [71].

Wound Healing

The wound healing process is complex and depends on the simultaneous orchestration of different events. Rapid induction of keratinocyte and fibroblast migration to the site of injury is necessary for tissue repair. FN is known to induce migration of fibroblasts [72], monocytes [73], epithelial [68] and endothelial cells [68] through its interaction with integrin α5β1. Some studies have demonstrated the potential therapeutic use of FN and FN-mimicking peptides in wound healing *in vitro* and *in vivo*. On full-thickness wounds surgically created in male Wistar rats, a single topical application of plasma FN significantly enhanced the wound closure rate, macrophage and fibroblast migration, collagen regeneration and epithelialisation (14 days post-wounding), compared to an equivalent dose of human plasma albumin used as the control [74]. Similar results were found for a single application of FN in full-thickness wounds inflicted on streptozotocin diabetic induced Male Sprague-Dawley rats, again with human serum albumin as the control (at 21 days post-wounding) [75]. Acetylated and amidated PHSRN peptide (Ac-PHSRN-NH_2) applied in artificially created dermal wounds in obese diabetic C57BL6/KsJ *db/db* mice stimulated wound contraction and re-epithelialisation in a concentration dependent manner, in contrast with randomised peptide (Ac-HSPNR-NH2) or normal saline treated wounds [76]. Ac-PHSRN-NH_2 at 200 and 2000 nM concentrations has been shown to have a significantly higher activity in promoting corneal epithelial migration in a rabbit organ system than FN at 100 nM; the randomised acetylated peptide (Ac-NRSHP) had no significant activity in all concentrations tested. The clinical outcome of the PHSRN peptides may be considered surprising given the conventional wisdom that PHSRN acts in synergy with RGD. In a similar study, corneal epithelial wounds were created in rabbit eyes; Ac-PHSRN-NH_2 at 200 μM and hFN 1 μM induced complete resurfacing of the created defect after 36 hours (Figure 9), a significant effect when compared to Ac-NRSHP (200 μM) or PBS as control [77]. Despite the apparent progresses demonstrated by these results, more studies are needed in order to adequately compare the clinical effects of FN, FN fragments (e.g. FIII9-10 and FIII7-10) and PHSRN/RGD peptides in wound healing. The focus should then also distinguish between the different roles of the integrin subtypes, particularly α5β1 and αvβ3 in this context.

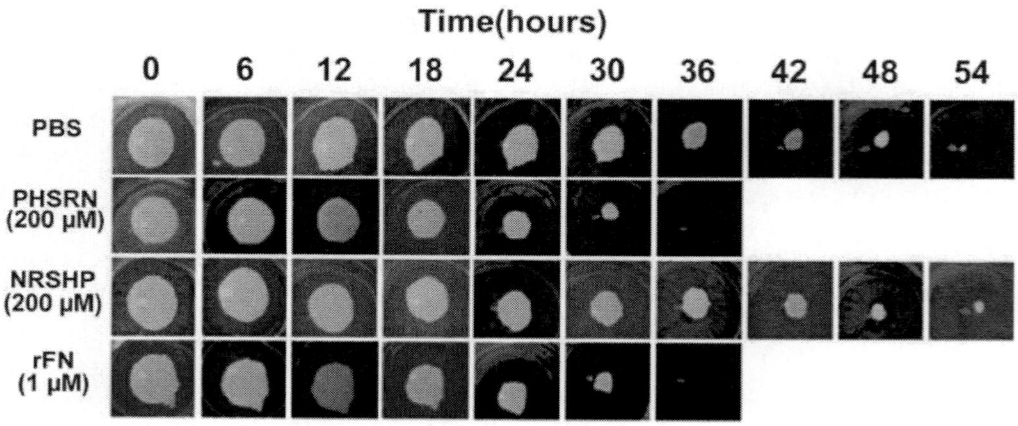

Figure 9. Effect of a PHSRN peptide on the healing of corneal epithelial wounds *in vivo*. Corneal epithelial defects were stained with fluorescein and monitored for closure after treatment. Adapted from [77].

Cancer

The potential for the use of FN, FN fragments and PHSRN/RGD peptides in the fight against cancer is huge. Livant and co-workers have shown that plasma FN is necessary to stimulate the invasive behaviour of human (DU145) and metastatic MATLyLu (MLL) rat prostate carcinoma cell, through interaction of PHSRN with integrin α5β1 [78]. Based on this, substitution of arginine for cysteine in the PHSRN peptide generated an efficient, competitive invasion inhibitor, PHSCN. Moreover, the acetylated and amidated peptide, Ac-PHSCN-NH_2, showed a 30 fold increase in inhibition of invasive activity compared to PHSCN; scrambled Ac-HSPNC-NH_2 had no activity *in vitro. In vivo*, MLL cells were injected in the right hind legs of Copenhagen rats to create tumours and 1 mg of Ac-PHSCN-NH_2 was injected subcutaneously three times per week (13 injections in total). After 5 injections, the tumours in the treated rats displayed a mean diameter of < 0.5 mm, whereas in untreated rats the tumour mean diameter reached 1.8 cm. Furthermore, the density of blood vessels was 8- to 10-fold lower in Ac-PHSCN-NH_2 treated rats and no lung metastases were found [78].

Ac-PHSCN-NH_2 (also termed ATN-161) activity in colorectal liver metastases was also investigated, using ATN-161 alone or in combination with the chemotherapeutic agent 5-fluorouracil (5-FU) [79]. CT-26 murine colon carcinoma cells were injected into the spleens of BALB/c mice to produce liver metastases. Treatment with ATN-161 (100 mg/Kg) or saline (control) was started 4 days after CT-26 cells injection and the administration was done every third day thereafter by intraperitoneal injection. One week later, osmotic pumps containing 5-FU (100 mg/Kg/2 weeks) or saline were implanted into the mice intraperitoneal cavity. Combination therapy of ATN-161 + 5-FU reduced the number of liver metastases (Figure 10) and tumour vascularisation and proliferation, whilst increasing the number of tumour apoptotic cells, in relation to the control group [79]. In another study, an amino acid modified PEG (aaPEG) was used as spacer/carrier for RGD and PHSRN peptides, the overall polymer hybrid being PHSRN-aaPEG-RGD in order to mimic the physiological distance between the two adhesion motifs in FN. The polymer hybrid exhibited a poorer Baby Hamster Kidney (BHK) cell spreading response than native FN but had a potent anti-metastatic activity in

C57BL/6 mice induced melanoma when compared to control (saline), aaPEG, PHSRN and RGD alone. However, the effect of the polymer hybrid on metastasis development was not significantly different from aaPEG-RGD [80] and further studies are required.

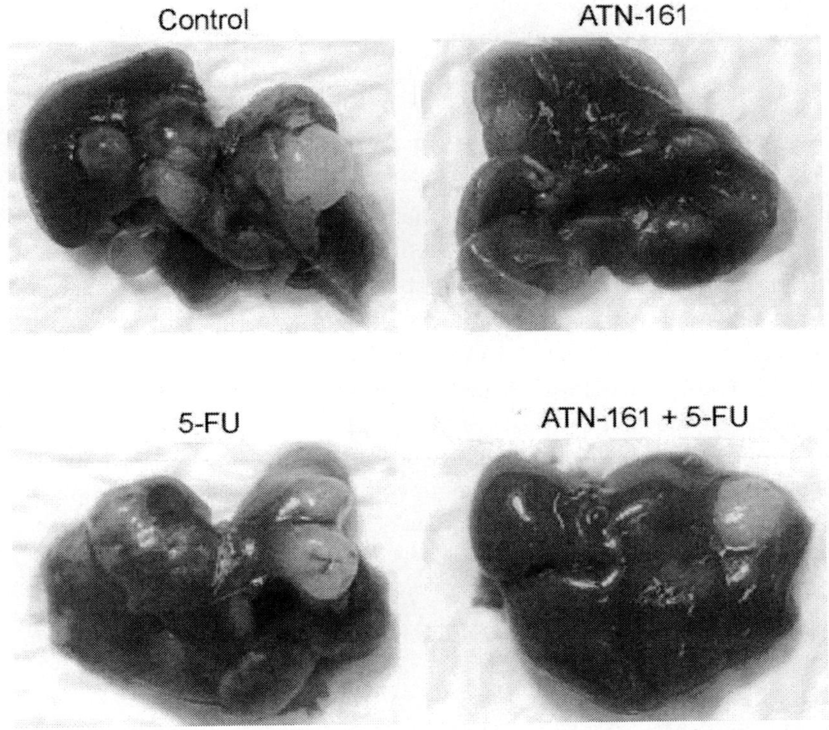

Figure 10. Effect of ATN-161 and 5-FU alone or in combination therapy on the formation of colorectal hepatic metastasis formation on BALB/c mice. Adapted from [79].

A near-native spacing between RGD and PHSRN motifs has also motivated the development of a peptide head group (composed of PHSRN, a $(SG)_4$ linker and RGDSP) for inclusion into an amphiphilic structure comproside of C_{16} dialkyl ester tails, a glutamic acid linker and a $(CH_2)_2$ spacer: the overall construct then being $(C_{16})_2$-Glu-$(CH_2)_2$-KSSPHSRNSGSGSGSGSGRGDSP, termed PR_b and which forms liposomes [81]. PR_b liposomal formulations for specific delivery of drugs to cancer cells have recently been investigated. Liposomes functionalised with PR_b alone or PR_b plus PEG 2000 were assayed for prostate cancer therapy *in vitro*. When PR_b concentrations were 2.5 and 5.3 mol%, irrespective of the PEG 2000 concentration, liposomal binding to LNCap cells was enhanced. The internalization of modified PR_b + PEG liposomes by LNCap cells was also higher when compared to PEG 2000 or GRGDSP-PEG 2000 modified liposomes. This explains the higher level of LNCap cell apoptosis when PR_b+ PEG targeted liposomes are loaded with tumour necrosis factor-α (TNF-α) [82]. Liposomes containing PR_b and two different molecular weight PEGs (750 and 2000) were also investigated for the binding of integrin α5β1 expressing colon cancer cells, namely CT26.WT and HCT116 cells. It was found that liposomes containing 2.2–2.6 mol. percent. PR_b and low PEG concentration for both PEG 750 and PEG 2000 displayed a maximum binding efficiency. When 5-FU loaded liposomes were used, 1.6–1.9 mol. percent. PR_b PEG 2000 liposomes improved the cytotoxicity of 5-FU for CT26.WT cells, compared to 2.9–3.8 mol. percent. GRGDSP PEG 2000 liposomes. [83]. These data demonstrate that in attaining a higher specificity and affinity for integrin α5β1 by using peptide harbouring both PHSRN and RGD motifs, tumours can be effectively targeted where their growth is dependent on integrin α5β1-mediated adhesion; this includes angiogenic and metastic phenomena.

CONCLUSION

The awareness of the importance of integrin α5β1 targeting in specific pathways is growing. Experimental progress towards a clinical candidate which includes FN mimetic peptides or domains has been, so far, remarkable. However, an improvement in methods for the characterisation of these peptides and proteins *in situ* (cf. in solution or adsorbed to surfaces) is needed, in order that the physiological response(s) can be understood at the molecular level. Thus, when making the transition from the laboratory to the clinic, an underlying molecular mechanism for a particular peptide/protein system can be defined and related to the appropriate clinical interventions. It is clear that protein engineering will continue to be a vital tool for both developing and testing hypotheses relating to FN-integrin interactions of utility in biomedical applications.

REFERENCES

[1] Clark K, Newham P, Burrows L, Askari JA, Humphries MJ. Production of recombinant soluble human integrin alpha4beta1. *FEBS Lett.* 2000 Apr 14;471(2-3):182-6.

[2] Pierschbacher MD, Ruoslahti E. Cell attachment activity of fibronectin can be duplicated by small synthetic fragments of the molecule. *Nature.* 1984 May 3-9;309(5963):30-3.

[3] Mardon HJ, Grant KE. The role of the ninth and tenth type III domains of human fibronectin in cell adhesion. *FEBS Lett.* 1994 Mar 7;340(3):197-201.

[4] Massia SP, Hubbell JA. An RGD spacing of 440 nm is sufficient for integrin alpha V beta 3-mediated fibroblast spreading and 140 nm for focal contact and stress fiber formation. *J Cell Biol.* 1991 Sep;114(5):1089-100.

[5] Hotchin NA, Kidd AG, Altroff H, Mardon HJ. Differential activation of focal adhesion kinase, Rho and Rac by the ninth and tenth FIII domains of fibronectin. *J Cell Sci.* 1999 Sep;112 (Pt 17):2937-46.

[6] Mardilovich A, Kokkoli E. Biomimetic Peptide-Amphiphiles for Functional Biomaterials: The Role of GRGDSP and PHSRN. *Biomacromolecules.* 2004 May-Jun;5(3):950-7.

[7] Grant RP, Spitzfaden C, Altroff H, Campbell ID, Mardon HJ. Structural requirements for biological activity of the ninth and tenth FIII domains of human fibronectin. *J Biol Chem.* 1997 Mar 7;272(10):6159-66.

[8] Altroff H, Schlinkert R, van der Walle CF, Bernini A, Campbell ID, Werner JM, Mardon HJ. Interdomain tilt angle determines integrin-dependent function of the ninth

and tenth FIII domains of human fibronectin. *J Biol Chem.* 2004 Dec 31;279(53):55995-6003.
[9] Norde W. Adsorption of proteins from solution at the solid-liquid interface. *Adv Colloid Interface Sci.* 1986 Sep;25(4):267-340.
[10] Mrksich M, Whitesides GM. Using self-assembled monolayers to understand the interactions of man-made surfaces with proteins and cells. *Annu Rev Biophys Biomol Struct*. 1996;25:55-78.
[11] Irvine DJ, Ruzette AV, Mayes AM, Griffith LG. Nanoscale Clustering of RGD Peptides at Surfaces Using Comb Polymers. 2. Surface Segregation of Comb Polymers in Polylactide. *Biomacromolecules*. 2001 Summer;2(2):545-56.
[12] Griko Y, Sreerama N, Osumi-Davis P, Woody RW, Woody AY. Thermal and urea-induced unfolding in T7 RNA polymerase: calorimetry, circular dichroism and fluorescence study. *Protein Sci.* 2001 Apr;10(4):845-53.
[13] Cutler SM, Garcia AJ. Engineering cell adhesive surfaces that direct integrin alpha5beta1 binding using a recombinant fragment of fibronectin. *Biomaterials*. 2003 May;24(10):1759-70.
[14] Ohashi T, Kiehart DP, Erickson HP. Dynamics and elasticity of the fibronectin matrix in living cell culture visualized by fibronectin-green fluorescent protein. *Proc Natl Acad Sci U S A.* 1999 Mar 2;96(5):2153-8.
[15] Critchley DR. Focal adhesions - the cytoskeletal connection. *Curr Opin Cell Biol.* 2000 Feb;12(1):133-9.
[16] Pelham RJ, Jr., Wang Y. Cell locomotion and focal adhesions are regulated by substrate flexibility. *Proc Natl Acad Sci U S A.* 1997 Dec 9;94(25):13661-5.
[17] Maheshwari G, Brown G, Lauffenburger DA, Wells A, Griffith LG. Cell adhesion and motility depend on nanoscale RGD clustering. *J Cell Sci.* 2000 May;113 (Pt 10):1677-86.
[18] Coussen F, Choquet D, Sheetz MP, Erickson HP. Trimers of the fibronectin cell adhesion domain localize to actin filament bundles and undergo rearward translocation. *J Cell Sci*. 2002 Jun 15;115(Pt 12):2581-90.
[19] van de Weert M, Hennink WE, Jiskoot W. Protein instability in poly(lactic-co-glycolic acid) microparticles. *Pharm Res.* 2000 Oct;17(10):1159-67.
[20] Copie V, Tomita Y, Akiyama SK, Aota S, Yamada KM, Venable RM, Pastor RW, Krueger S, Torchia DA. Solution structure and dynamics of linked cell attachment modules of mouse fibronectin containing the RGD and synergy regions: comparison with the human fibronectin crystal structure. *J Mol Biol*. 1998 Apr 3;277(3):663-82.
[21] Spitzfaden C, Grant RP, Mardon HJ, Campbell ID. Module-module interactions in the cell binding region of fibronectin: stability, flexibility and specificity. *J Mol Biol*. 1997 Feb 7;265(5):565-79.
[22] van der Walle CF, Altroff H, Mardon HJ. Novel mutant human fibronectin FIII9-10 domain pair with increased conformational stability and biological activity. Protein Eng. 2002 Dec;15(12):1021-4.
[23] Leahy DJ, Aukhil I, Erickson HP. 2.0 A crystal structure of a four-domain segment of human fibronectin encompassing the RGD loop and synergy region. *Cell*. 1996 Jan 12;84(1):155-64.

[24] Plaxco KW, Spitzfaden C, Campbell ID, Dobson CM. Rapid refolding of a proline-rich all-beta-sheet fibronectin type III module. *Proc Natl Acad Sci U S A.* 1996 Oct 1;93(20):10703-6.

[25] Harbury PB, Plecs JJ, Tidor B, Alber T, Kim PS. High-resolution protein design with backbone freedom. *Science.* 1998 Nov 20;282(5393):1462-7.

[26] Pack P, Muller K, Zahn R, Pluckthun A. Tetravalent miniantibodies with high avidity assembling in Escherichia coli. *J Mol Biol.* 1995 Feb 10;246(1):28-34.

[27] Watson N, Duncan G, Annan WS, van der Walle CF. A tetravalent RGD ligand for integrin-mediated cell adhesion. *J Pharm Pharmacol.* 2006 Jul;58(7):959-66.

[28] Harbury PB, Zhang T, Kim PS, Alber T. A switch between two-, three-, and four-stranded coiled coils in GCN4 leucine zipper mutants. *Science.* 1993 Nov 26;262(5138):1401-7.

[29] Irvine DJ, Hue KA, Mayes AM, Griffith LG. Simulations of cell-surface integrin binding to nanoscale-clustered adhesion ligands. *Biophys J.* 2002 Jan;82(1):120-32.

[30] Harris LJ, Larson SB, Hasel KW, McPherson A. Refined structure of an intact IgG2a monoclonal antibody. *Biochemistry.* 1997 Feb 18;36(7):1581-97.

[31] Michaelsen TE, Frangione B, Franklin EC. Primary structure of the "hinge" region of human IgG3. Probable quadruplication of a 15-amino acid residue basic unit. *J Biol Chem.* 1977 Feb 10;252(3):883-9.

[32] Merkel R, Nassoy P, Leung A, Ritchie K, Evans E. Energy landscapes of receptor-ligand bonds explored with dynamic force spectroscopy. *Nature.* 1999 Jan 7;397(6714):50-3.

[33] Kreiner M, Li Z, Beattie J, Kelly SM, Mardon HJ, van der Walle CF. Self-assembling multimeric integrin alpha5beta1 ligands for cell attachment and spreading. *Protein Eng Des Sel.* 2008 Sep;21(9):553-60.

[34] Kelly SM, Jess TJ, Price NC. How to study proteins by circular dichroism. *Biochim Biophys Acta.* 2005 Aug 10;1751(2):119-39.

[35] Khan MY, Villanueva G, Newman SA. On the origin of the positive band in the far-ultraviolet circular dichroic spectrum of fibronectin. *J Biol Chem.* 1989 Feb 5;264(4):2139-42.

[36] Pack P, Pluckthun A. Miniantibodies: use of amphipathic helices to produce functional, flexibly linked dimeric FV fragments with high avidity in Escherichia coli. *Biochemistry.* 1992 Feb 18;31(6):1579-84.

[37] Burkhard P, Meier M, Lustig A. Design of a minimal protein oligomerization domain by a structural approach. *Protein Sci.* 2000 Dec;9(12):2294-301.

[38] Erb EM, Tangemann K, Bohrmann B, Muller B, Engel J. Integrin alphaIIb beta3 reconstituted into lipid bilayers is nonclustered in its activated state but clusters after fibrinogen binding. *Biochemistry.* 1997 Jun 17;36(24):7395-402.

[39] Thomas RM, Zampieri A, Jumel K, Harding SE. A trimeric, alpha-helical, coiled coil peptide: association stoichiometry and interaction strength by analytical ultracentrifugation. *Eur Biophys J.* 1997;25(5-6):405-10.

[40] Burkhard P, Ivaninskii S, Lustig A. Improving coiled-coil stability by optimizing ionic interactions. *J Mol Biol.* 2002 May 3;318(3):901-10.

[41] Cabre F, Canela EI, Canela MA. Accuracy and precision in the determination of Stokes radii and molecular masses of proteins by gel filtration chromatography. *J Chromatogr.* 1989 Jun 23;472(2):347-56.

[42] Kreiner M, Byron O, Domingues D, van der Walle CF. Oligomerisation and thermal stability of polyvalent integrin alpha5beta1 ligands. *Biophys Chem.* 2009 Jun;142(1-3):34-9.

[43] Howlett GJ, Minton AP, Rivas G. Analytical ultracentrifugation for the study of protein association and assembly. *Curr Opin Chem Biol.* 2006 Oct;10(5):430-6.

[44] O'Shea EK, Rutkowski R, Kim PS. Evidence that the leucine zipper is a coiled coil. *Science*. 1989 Jan 27;243(4890):538-42.

[45] O'Shea EK, Lumb KJ, Kim PS. Peptide 'Velcro': design of a heterodimeric coiled coil. *Curr Biol.* 1993 Oct 1;3(10):658-67.

[46] Wilson K, Stuart SJ, Garcia A, Latour RA, Jr. A molecular modeling study of the effect of surface chemistry on the adsorption of a fibronectin fragment spanning the 7-10th type III repeats. *J Biomed Mater Res A.* 2004 Jun 15;69(4):686-98.

[47] Vaidya SS, Ofoli RY. Adsorption and interaction of fibronectin and human serum albumin at the liquid-liquid interface. *Langmuir*. 2005 Jun 21;21(13):5852-8.

[48] Bergkvist M, Carlsson J, Oscarsson S. Surface-dependent conformations of human plasma fibronectin adsorbed to silica, mica, and hydrophobic surfaces, studied with use of Atomic Force Microscopy. *J Biomed Mater Res A.* 2003 Feb 1;64(2):349-56.

[49] Baugh L, Vogel V. Structural changes of fibronectin adsorbed to model surfaces probed by fluorescence resonance energy transfer. *J Biomed Mater Res A.* 2004 Jun 1;69(3):525-34.

[50] Calonder C, Matthew HW, Van Tassel PR. Adsorbed layers of oriented fibronectin: a strategy to control cell-surface interactions. *J Biomed Mater Res A.* 2005 Nov 1;75(2):316-23.

[51] Norde W, Macritchie F, Nowicka G, Lyklema J. Protein Adsorption at Solid Liquid Interfaces - Reversibility and Conformation Aspects. *Journal of Colloid and Interface Science.* 1986 Aug;112(2):447-56.

[52] Wertz CF, Santore MM. Adsorption and relaxation kinetics of albumin and fibrinogen on hydrophobic surfaces: Single-species and competitive behavior. *Langmuir*. 1999 Dec;15(26):8884-94.

[53] Wertz CF, Santore MM. Adsorption and reorientation kinetics of lysozyme on hydrophobic surfaces. *Langmuir*. 2002 Feb;18(4):1190-9.

[54] Krammer A, Lu H, Isralewitz B, Schulten K, Vogel V. Forced unfolding of the fibronectin type III module reveals a tensile molecular recognition switch. *Proc Natl Acad Sci U S A.* 1999 Feb 16;96(4):1351-6.

[55] Chittur KK. FTIR/ATR for protein adsorption to biomaterial surfaces. *Biomaterials*. 1998 Mar;19(4-5):357-69.

[56] Lad MD, Birembaut F, Matthew JM, Frazier RA, Green RJ. The adsorbed conformation of globular proteins at the air/water interface. *Physical Chemistry Chemical Physics.* 2006 May;8(18):2179-86.

[57] Ganesan A, Price NC, Kelly SM, Petry I, Moore BD, Halling PJ. Circular dichroism studies of subtilisin Carlsberg immobilised on micron sized silica particles. *Biochimica Et Biophysica Acta-Proteins and Proteomics*. 2006 Jun;1764(6):1119-25.

[58] Pereira P, Kelly SM, Gellert PR, van der Walle CF. Interdomain mobility and conformational stability of type III fibronectin domain pairs control surface adsorption, desorption and unfolding. *Colloids and Surfaces B: Biointerfaces*. 2008;64(1):1-9.

[59] Duysens LN. The flattening of the absorption spectrum of suspensions, as compared to that of solutions. *Biochim Biophys Acta.* 1956 Jan;19(1):1-12.

[60] Wallace BA, Teeters CL. Differential absorption flattening optical effects are significant in the circular dichroism spectra of large membrane fragments. *Biochemistry*. 1987 Jan 13;26(1):65-70.

[61] Su TJ, Lu JR, Thomas RK, Cui ZF. Effect of pH on the adsorption of bovine serum albumin at the silica water interface studied by neutron reflection. *Journal of Physical Chemistry B.* 1999 May;103(18):3727-36.

[62] Holt SA, Le Brun AP, Majkrzak CF, McGillivray DJ, Heinrich F, Lösche M, Lakey JH. An ion channel containing model membrane: structural determination by magnetic contrast neutron reflectometry. *Soft Matter.* 2009:DOI:10.1039/b822411k.

[63] Kreiner M, Chillakuri CR, Pereira P, Fairhead M, Li Z, Mardon HJ, Holt SA, F. van der Walle CF. Orientation and surface coverage of adsorbed fibronectin cell binding domains and bound integrin α5β1 receptors. *Soft Matter.* 2009;DOI:10.1039/B908706K.

[64] Harvey RD, Heenan RK, Barlow DJ, Lawrence MJ. Effect of cholesterol and phospholipid on the behavior of dialkyl polyoxyethylene ether surfactant (2C18E12) monolayers and bilayers. *Langmuir*. 2004 Oct 12;20(21):9282-90.

[65] Stanton H, Ung L, Fosang AJ. The 45 kDa collagen-binding fragment of fibronectin induces matrix metalloproteinase-13 synthesis by chondrocytes and aggrecan degradation by aggrecanases. *The Biochemical journal.* 2002 May 15;364(Pt 1):181-90.

[66] Petrie TA, Capadona JR, Reyes CD, Garcia AJ. Integrin specificity and enhanced cellular activities associated with surfaces presenting a recombinant fibronectin fragment compared to RGD supports. *Biomaterials*. 2006 Nov;27(31):5459-70.

[67] Moursi AM, Globus RK, Damsky CH. Interactions between integrin receptors and fibronectin are required for calvarial osteoblast differentiation in vitro. *J Cell Sci.* 1997 Sep;110 (Pt 18):2187-96.

[68] Kim S, Bell K, Mousa SA, Varner JA. Regulation of angiogenesis in vivo by ligation of integrin alpha5beta1 with the central cell-binding domain of fibronectin. *Am J Pathol.* 2000 Apr;156(4):1345-62.

[69] Benoit DS, Anseth KS. The effect on osteoblast function of colocalized RGD and PHSRN epitopes on PEG surfaces. *Biomaterials*. 2005 Sep;26(25):5209-20.

[70] Petrie TA, Raynor JE, Reyes CD, Burns KL, Collard DM, Garcia AJ. The effect of integrin-specific bioactive coatings on tissue healing and implant osseointegration. *Biomaterials*. 2008 Jul;29(19):2849-57.

[71] Martino MM, Mochizuki M, Rothenfluh DA, Rempel SA, Hubbell JA, Barker TH. Controlling integrin specificity and stem cell differentiation in 2D and 3D environments through regulation of fibronectin domain stability. *Biomaterials*. 2009 Feb;30(6):1089-97.

[72] Greiling D, Clark RA. Fibronectin provides a conduit for fibroblast transmigration from collagenous stroma into fibrin clot provisional matrix. *J Cell Sci.* 1997 Apr;110 (Pt 7):861-70.

[73] Cota E, Clarke J. Folding of beta-sandwich proteins: three-state transition of a fibronectin type III module. *Protein Sci.* 2000 Jan;9(1):112-20.

[74] Kwon YM, Baudys M, Knutson K, Kim SW. In situ study of insulin aggregation induced by water-organic solvent interface. *Pharmaceutical Research.* 2001 Dec;18(12):1754-9.

[75] Qiu Z, Kwon AH, Kamiyama Y. Effects of plasma fibronectin on the healing of full-thickness skin wounds in streptozotocin-induced diabetic rats. *The Journal of surgical research.* 2007 Mar;138(1):64-70.

[76] Livant DL, Brabec RK, Kurachi K, Allen DL, Wu Y, Haaseth R, Andrews P, Ethier SP, Markwart S. The PHSRN sequence induces extracellular matrix invasion and accelerates wound healing in obese diabetic mice. *The Journal of clinical investigation.* 2000 Jun;105(11):1537-45.

[77] Kimura K, Hattori A, Usui Y, Kitazawa K, Naganuma M, Kawamoto K, Teranishi S, Nomizu M, Nishida T. Stimulation of corneal epithelial migration by a synthetic peptide (PHSRN) corresponding to the second cell-binding site of fibronectin. *Investigative ophthalmology & visual science.* 2007 Mar;48(3):1110-8.

[78] Livant DL, Brabec RK, Pienta KJ, Allen DL, Kurachi K, Markwart S, Upadhyaya A. Anti-invasive, antitumorigenic, and antimetastatic activities of the PHSCN sequence in prostate carcinoma. *Cancer research.* 2000 Jan 15;60(2):309-20.

[79] Stoeltzing O, Liu W, Reinmuth N, Fan F, Parry GC, Parikh AA, McCarty MF, Bucana CD, Mazar AP, Ellis LM. Inhibition of integrin alpha5beta1 function with a small peptide (ATN-161) plus continuous 5-FU infusion reduces colorectal liver metastases and improves survival in mice. *International journal of cancer.* 2003 Apr 20;104(4):496-503.

[80] Susuki Y, Hojo K, Okazaki I, Kamata H, Sasaki M, Maeda M, Nomizu M, Yamamoto Y, Nakagawa S, Mayumi T, Kawasaki K. Preparation and biological activities of a bivalent poly(ethylene glycol) hybrid containing an active site and its synergistic site of fibronectin. *Chemical & pharmaceutical bulletin.* 2002 Sep;50(9):1229-32.

[81] Mardilovich A, Craig JA, McCammon MQ, Garg A, Kokkoli E. Design of a novel fibronectin-mimetic peptide-amphiphile for functionalized biomaterials. *Langmuir.* 2006 Mar 28;22(7):3259-64.

[82] Demirgoz D, Garg A, Kokkoli E. PR_b-targeted PEGylated liposomes for prostate cancer therapy. *Langmuir.* 2008 Dec 2;24(23):13518-24.

[83] Garg A, Tisdale AW, Haidari E, Kokkoli E. Targeting colon cancer cells using PEGylated liposomes modified with a fibronectin-mimetic peptide. *International journal of pharmaceutics.* 2009 Jan 21;366(1-2):201-10.

In: Ribosomal Proteins and Protein Engineering... ISBN 978-1-60741-005-8
Editors: V. Ortendhal and H. Salchow, pp.143-174

Chapter 10

HALOPHILIC ADAPTATIONS OF PROTEINS*

Albert Bolhuis*†, *Daniel Kwan and Judith R. Thomas
Department of Biological Sciences, University of Warwick,
Coventry CV4 7AL, UK

ABSTRACT

Halophilic organisms thrive in a wide variety of environments, ranging from mildly saline to saturated brine. Extremely saline conditions are very harsh to proteins and can result in problems such as loss of flexibility, which leads to loss of catalytic activity, or aggregation and precipitation due to salting-out effects. Halophilic proteins have therefore acquired specific adaptations to prevent such problems, and that is clearly reflected in their amino acid composition. As a consequence, halophilic proteins do not function properly in low salt conditions, and many even unfold and denature without sufficient amounts of salt. One of the most obvious adaptations found in halophilic proteins is an excess of acidic residues, while other adaptations include a low lysine content and an increase in small hydrophobic amino acids, the latter of which is at the expense of large hydrophobic residues. Different functions have been suggested for the role of the acidic residues. They are, most likely, important for binding of essential water molecules, but may play also a critical role in binding of salt or prevention of aggregation. The other changes in amino acid composition in halophilic proteins are probably important for the reduction of hydrophobicity, in order to prevent aggregation and maintain flexibility in high salt concentrations. Structural features that may be important for halophilic adaptation are also found, such as an increased number of salt-bridges. That is, however, not used by all halophilic proteins, indicating that different strategies have evolved to cope with very saline conditions.

---

* A version of this chapter was also published in Protein Adaptation in Extremophiles, edited by K. S. Siddiqui and T. Thomas published by Nova Science Publishers, Inc. It was submitted for appropriate modifications in an effort to encourage wider dissemination of research.

† Corresponding author. E-mail: a.bolhuis@bath.ac.uk. Current address: Department of Pharmacy and Pharmacology, University of Bath, Bath BA2 7AY, UK

INTRODUCTION

Evolution of life on Earth has filled almost every possible niche imaginable. Environments that were once considered too harsh to sustain life have been revealed to contain a large variety of organisms. However, survival in the most adverse conditions does require extensive adaptations in order to enable life at the extremes of, for instance, temperature, pH, or salinity. Particularly challenging habitats are those containing high concentrations of salt, which are severely adverse to organisms and their proteins and cellular processes. This chapter focuses in particular on strategies used for halophilic adaptation of proteins. Briefly discussed are also how halophilic organisms are able to grow in highly saline conditions and how some of the cellular processes in these organisms have adapted.

There is a huge diversity of organisms living in various saline environments, and these can be either halophilic or halotolerant. Halophilic (salt-loving) organisms require salt for growth, whereas halotolerant organisms prefer to grow in low salt environments but can tolerate fairly high concentrations of salt. Depending on their salt requirement, halophilic organisms can be classified in different groups. Mild halophiles grow optimally at 1-6% salt, moderate halophiles prefer 7-15% salt and, finally, extreme halophiles require at least 15% salt. The largest ecosystem for halotolerant and halophilic organisms are the Earth's oceans, which contain around 3.5% total salt, with most of the salt being sodium chloride. Hypersaline environments can be found in many countries, and these consist of two main types [1]. The first type has arisen from seawater and has roughly the same salts in similar proportions to the oceans; these are termed thalassohaline. The total salt concentration of these environments can reach saturation, which is approximately 10 times the concentration of sea water. Examples of such hypersaline environments are solar evaporation ponds for the production of sea salt (found in many coastal areas) and the Great Salt Lake in Utah. The second type of hypersaline environments, denoted athalassohaline, are those which arose from non-marine sources. The ion ratios in these environments differ significantly from that of sea water, and examples include the Dead Sea, which contains very high concentrations of magnesium and calcium ions, and soda lakes (found for instance in Africa and China), which have a very high pH value (>10), are rich in carbonate, and are extremely low in magnesium and calcium. Truly remarkable habitats for halophiles (and possibly the most extreme) are deep hypersaline anoxic basins found, for example, on the Mediterranean Sea floor. These basins hardly mix with the overlaying seawater due to the high concentration of salt present. Some of these, such as the Discovery basin in the eastern Mediterranean contain more than 5 M Mg^{2+} while they are relatively low in Na^+ or K^+ [2]. Such a concentration of $MgCl_2$ should not sustain life due to the extremely low water activity [3], yet a recent study showed that the Discovery basin does contain a metabolically active and diverse microbial community containing both bacteria and archaea [4].

Living in high concentrations of salt requires several adaptations. One important problem is the maintenance of the water balance. In a saline environment non-halophilic organisms would quickly loose water through osmosis, and halotolerant and halophilic organisms have to avoid this by keeping their cytoplasm at least iso-osmotic with the environment. Two main mechanisms have evolved to cope with saline environments [5, 6]. Most organisms only use one of the two strategies, although some, such as the halophilic methanogenic archaea, use both [7]. The first is denoted the "salt-out" or "compatible-solute" strategy. Organisms using this strategy balance their osmotic pressure through the accumulation of organic compatible solutes such as betaine, ectoine, glycerol, or trehalose [8]. In many bacteria the main compatible solute is betaine; various moderately halophilic strains were shown to contain, on average, 1 M betaine intracellularly when grown in 20% salt, while lower concentrations of betaine were found when these organisms were grown at lower salt concentrations [9]. Another striking example is the halotolerant algae *Dunaliella salina*. This organism can grow in media containing 0.1 to 5.5 M NaCl using glycerol as the compatible solute. The concentration of intracellular glycerol was shown to be proportional to the extracellular NaCl concentration, with the concentration of glycerol reaching 7-8 M when *D. salina* was grown in 4 M NaCl [10]. An important advantage of the compatible-solute strategy is that organisms can grow in a wide range of salt concentrations. Moreover, proteins not exposed to the extracellular medium are not exposed to very saline conditions and do not require specific adaptations. Naturally, extracytoplasmic proteins (e.g. membrane, cell-wall, and secretory proteins) in organisms using the compatible-solute strategy will be exposed to high levels of salt, so these proteins still need to be functional in saline conditions. A notable disadvantage of the strategy is that production of compatible solutes is energetically very costly; calculations indicated that the amount of energy required for the production of compatible solutes may be much higher than the energy requirement for some other cellular processes, such as the synthesis of proteins and nucleic acids [6].

The second strategy for coping with high concentrations of salt is the "salt-in" strategy, in which the osmotic balance is maintained through the intracellular accumulation of salt. This strategy is used only by a limited number of halophilic bacteria and all halophilic archaea of the order *Halobacteriales*. Notably, whereas the environment usually contains molar concentrations of sodium chloride as the main salt, organisms using the salt-in strategy accumulate equimolar concentrations of potassium chloride intracellularly [11]. Most organisms, including those which are non-halophilic, prefer potassium as the main intracellular cation, presumably because potassium is much less toxic than sodium [12]. The salt-in strategy is energetically cheaper than the salt-out strategy [6], but it does demand that both intra- and extra-cellular proteins are functional in highly saline conditions. This is manifested in extensive evolutionary adaptations in extreme halophilic organisms. As a consequence of this, most proteins from organisms using the salt-in strategy are in fact completely dependent on the presence of salt for activity and structural integrity. Moreover, these extreme halophiles are unable to grow at low salt concentrations and their environmental niche is therefore quite limited. For example, most halophilic archaea will lyse in less than 1 M salt [13].

THE EFFECTS OF SALTS ON NON-HALOPHILIC PROTEINS

Before going into more detail on how halophilic proteins remain functional in highly saline conditions, it is useful to consider first what happens to non-halophilic proteins in concentrated salt solutions. An important effect of high salt concentrations was already described at the end of the 19th century by Franz Hofmeister, who studied the effects of different concentrations of salts on hen egg white proteins [14]. He found that high concentrations of certain salts precipitate these proteins, which was not only strongly dependent on the concentration of the salt but also on the type of salt used. Thus, some salts were better in precipitating egg white proteins than others. This precipitation effect is known as "salting-out". The salting-out properties of the salts depend on the nature of the cation and the anion, and their ability to precipitate proteins is described by the so-called Hofmeister series. The ranking for some of the cations in the Hofmeister series is $NH_4^+ > K^+ > Na^+ > Li^+ > Mg^{2+} > Ca^{2+} > Al^{3+}$, whereas for anions the ranking is $PO_4^{3-} > SO_4^{2-} > F^- > Cl^- > Br^- > I^- > ClO_4^-$. Ions on the left promote salting-out, while those to the right promote the opposite effect, salting-in. Salting-out salts also tend to stabilise proteins, while strongly salting-in salts may denature a protein due to direct binding to proteins [15]. The salts most relevant for halotolerant and halophilic proteins are NaCl and KCl, both of which are mildly salting out. Salting-out of proteins also depends on several other factors, such as the nature of the protein and its concentration, as well as temperature, pH, and pressure.

Several salts increase the surface tension of water, and it is thought that this is an important factor in salting-out [16]. However, some salts such as $MgCl_2$ increase surface tension, while they are in fact salting-in at high concentrations [17], so salting-out cannot be explained only through surface tension. A significant role is also played by the hydrophobic effect, which is thought to be the main driving force of protein folding and a key contributor to protein stability [18]. Non-polar amino acids are poorly soluble in water, and their exposure to water forces the ordering of water molecules around these residues. That has a negative effect on the entropy of the solvent, and shielding as many hydrophobic residues as possible in the core of a protein, away from solvent, is therefore thermodynamically favourable. Both proteins and salts interact with water molecules, and at high salt concentrations proteins will have to compete with salt for binding of water. In these conditions the number of free water molecules becomes limiting and the protein dehydrates. This causes an increase in the strength of hydrophobic interactions and this, in turn, promotes aggregation of proteins. The increase in strength of hydrophobic interactions in high salt also explains the stabilising effect of salting-out salts on proteins, but it is at the expense of flexibility within a protein. This is exemplified by a study on dihydrofolate reductase (DHFR) of *Escherichia coli* [19]. Increasing the salt concentration (either NaCl or KCl) reduced the catalytic activity of *E. coli* DHFR drastically without significant effects on solubility, substrate binding, or structure. The authors concluded that at high ionic strength the protein became more stable, but at the same time also more rigid. Catalytic activity of enzymes depends on conformational flexibility, and as a consequence, many enzymes will have strongly reduced activity at high ionic strength.

In conclusion, saline conditions pose a number of major problems that halophilic proteins have to overcome. They have to be able to compete with salt ions for binding of water molecules and must prevent aggregation and precipitation caused by salting-out effects. Furthermore, they must maintain sufficient amount of flexibility in order to remain catalytically active.

AMINO ACID COMPOSITION OF HALOPHILIC PROTEINS

There are various notable changes in proteins from halophilic organisms when compared to non-halophilic counterparts. A feature that was recognised early, and which is also the most obvious, is the excess of negative charge in halophilic proteins [11]. This is clearly reflected in the low pI values of halophilic proteins. The proteome of the first extreme halophile that was fully sequenced, *Halobacterium salinarum* (sp. NRC-1), has an average (theoretical) pI of 5.1 (Table 1). This acidic character was confirmed by 2D-gel electrophoresis, which showed that most cytoplasmic proteins in *H. salinarum* are resolved in the range of pH 3.5-5.5 [20]. The latter study also showed that the theoretical pI values of halophilic proteins corresponded very well to the experimental data. This is in contrast to the molecular mass of halophilic proteins, which cannot be determined accurately by gel electrophoresis. Also proteomes of other halophiles for which the genome sequence is available have a low average pI value (Table 1), which include the halophilic archaea *Haloarcula marismortui* [21], *Natromonas pharaonis* [22], and *Haloquadratum walsbyi* [3], as well as the halophilic bacterium *Salinibacter ruber* [23]. For comparison, proteomes of non-halophilic organisms such as *E. coli* or *Bacillus subtilis* have an average pI of approximately 7 (Table 1). The low pI value of halophilic proteins is mostly due to increased numbers of the acidic residues glutamate and aspartate [11]. Also other differences in amino acid composition between halophilic and non-halophilic proteins have been reported, such as a decreased number of lysines, which may also contribute to a low pI value, a decrease in large hydrophobic residues, and an increase in serine, threonine, and small hydrophobic residues [11, 24]. Some of the aforementioned studies only examined a limited number of proteins, but the availability of several fully sequenced genomes of halophilic organisms makes it possible to compare the composition of entire proteomes of various organisms. Presented in Table 1 is the analysis of the proteomes of five extremely halophilic prokaryotes (*H. salinarum*, *H. marismortui*, *H. walsbyi*, *N. pharaonis*, and *S. ruber*) and three (non-halophilic) prokaryotes *E. coli*, *B. subtilis*, and *Pseudomonas aeruginosa*. On a note of caution, these data only show the average amino acid content of all proteins in these organisms; differences between, for instance, the relative content of residues in the interior or on the surface of proteins, which is important for halophilic adaptation (see below), are not taken into account. Furthermore, we focus in this chapter mostly on the extreme halophiles, as the effects on for instance amino acid composition are much more pronounced in these organisms than in more moderate halophiles. The latter group of organisms generally use the compatible solute strategy, but halophilic adaptation in these can be observed when analysing extracytoplasmic proteins [25].

As is clear from Table 1, the extreme halophiles have indeed an excess of negative charge compared to the non-halophilic proteomes. The acidity of the halophilic proteomes is not only attributable to a high number of acidic residues (aspartate and glutamate), but also to the exceptionally low percentage of lysine residues. Arginine, which, like lysine, is also a basic residue with a high pKa value, is not significantly different in the proteomes analysed (data not shown). The data in table 1 do show that halophilic proteomes have an increased number of small hydrophobic residues (alanine, glycine, and valine) and a decreased number of large hydrophobic residues (isoleucine, leucine, phenylalanine, and methionine). An increase of serine and threonine residues in halophilic proteomes had also been suggested [11], but that appears not to be an obvious property, with a possible exception in that of *H. walsbyi*. Notably, most of the extreme halophiles have a high genomic GC content (60-70%; see Table 1). Various studies have shown a correlation between GC content and amino acid frequencies (see Ref 26), and so there is the distinct possibility that some of the differences in amino acid composition of non-halophiles versus halophiles is merely due to the nucleotide bias. It is for this reason that *P. aeruginosa* was included in Table 1, as the genome of this organism has a GC content of 67%. *P. aeruginosa* does have, compared to *E. coli* and *B. subtilis*, a relatively low lysine content, as well as a higher number of small hydrophobic residues. Nevertheless, the average charge difference and pI in the proteome of *P. aeruginosa* is still similar to that of the non-halophilic proteomes, suggesting that the increased number of aspartate and glutamate residues is the defining characteristic of halophilic proteins. One study indicated that only the number of aspartate residues deviated significantly from what might be expected of GC-rich genomes; it was suggested that the change in the nucleotide composition in halophilic genomes was merely serving an increase in aspartate residues, and that other deviations in amino acid composition (in lysine, alanine, etc.) were merely a "dragging effect" [27]. However, this is contradicted by the remarkable low GC content of *H. walsbyi* (48%; less than that of *E. coli*, 51%). Similar to other extreme halophiles, the proteome of *H. walsbyi* is strongly negatively charged, has a low lysine content, and is relatively low in large hydrophobic residues. Less obvious is an increase in small hydrophobic residues, but the organism may have compensated for that in a different manner as it is the only halophile with a clear increase in serine and threonine residues. As outlined in more detail below, we believe that all the differences highlighted in Table 1 are, to a greater or lesser degree, important for halophilic adaptation. The high GC content of most halophiles may therefore have been driven by the selective pressure of halophilic adaptation rather than the GC content driving the bias in amino acid composition.

As shown by the statistics above, halophilic proteomes are, on average, less hydrophobic than their non-halophilic counterparts. It is important to note, however, that the proteomes analysed do not only include soluble proteins, but also membrane proteins. In prokaryotes the number of proteins predicted to be membrane located is around 25%, and these proteins are of course, by their nature, much more hydrophobic than soluble proteins. Large soluble domains attached to membrane anchors are likely to display a similar composition to soluble proteins, but integral membrane proteins that mainly reside within the membrane may skew the results from the proteome analysis. As an example, we analysed the composition of the SecY, which is an essential and conserved protein involved in protein translocation. SecY is an integral membrane protein containing ten membrane-spanning domains that are spread around the entire region of the polypeptide. In Table 2 we show average values for five non-halophilic and five halophilic organisms. As obvious from the charge difference and the pI, the halophilic SecY proteins have an acidic character. This is again reflected in an increase in acidic residues and a decrease in lysines. The total number of acidic residues is lower than in the entire proteome, as most acidic residues will only be found in the loops connecting the membrane-spanning domains. The percentage of lysine residues are comparable to that of the entire proteome, which probably reflects the important role that positively charged residues have in determining the topology of membrane proteins [28]. Of the small or strongly hydrophobic residues only glycine appears to differ significantly between halophilic and non-halophilic organisms. Not included in Table 2 is the composition of the halophilic bacterium *S. ruber*, which seems to be atypical. SecY in this organism is much more similar to non-halophilic bacteria with a theoretical pI of 9.2. Notably, the non-halophilic strains analysed were all bacteria, whereas the halophilic strains analysed were all archaea. The chemical composition of the archaeal cytoplasmic membrane is completely different from that of bacteria [29], and for that reason we included in Table 2 also five non-halophilic archaea. The charge difference, average pI, and amino acid composition of SecY proteins from these organisms is much more similar to that of non-halophilic bacteria, indicating that the chemical composition of the membrane does not seem to have a major effect on the composition of its membrane proteins. Note that 3 out of 6 non-halophilic archaea used in Table 2 are thermophiles. Thermophily also affect amino acid composition (see elsewhere in this book), and a better comparison would have been with mesophilic non-halophilic archaea only. However, there are not many other sequences of such archaea available, and some of those that are known, including some of the *Methanosarcina* spp., are of halotolerant strains that can grow in up to 1 M NaCl [30]. Analysis of SecY proteins from these proteins indicates that these have more halophilic-like characteristics, with for instance a theoretical pI of 5.8 for the *Methanosarcina mazei* SecY protein. Comparisons of membrane proteins seem therefore not as straightforward as for soluble proteins and must to be taken with a note of caution.

Nevertheless, halophilic membrane proteins also appear to have an acidic character with an increase in acidic residues and a decrease in lysine residues. Acidity for SecY, and probably other membrane proteins, is less obvious than for the entire proteome and, furthermore, the only notable increase in non-polar residues is found with glycine. It is therefore conceivable that the bias in amino acid composition in soluble proteins is in fact even more pronounced than is indicated by the analysis of the entire proteomes of halophiles.

Characterisation of Halophilic Proteins

As seen from the analysis of proteomes below, the most striking features of halophilic proteins are a large excess of acidic residues and a low lysine content. Moreover, there is a lower hydrophobicity due to an increase in small hydrophobic residues and a decrease in large hydrophobic residues. The key question is of course what the relevance is of these differences for halophilic adaptation. This is discussed in the next few sections using a number of well-studied halophilic proteins as examples, with a particular focus on proteins for which the crystal structures are available.

Halophilic Malate Dehydrogenase

One of the best studied halophilic proteins is malate dehydrogenase from the extremely halophilic archaeon *H. marismortui*. This protein, denoted HmMDH, catalyses the reversible conversion of oxaloacetate into malate, a reaction that requires NADH. The enzyme is a homotetramer in solution and requires high salt concentrations for activity [31]. Similar to many other halophilic proteins, the protein becomes unstable at low salt concentrations [32], and studies showed that this is related to the dissociation of the subunits and unfolding of the monomers [33]. The protein is stable between 2 and 4 M NaCl, but at lower concentrations

Table 1. Amino acid composition in proteomes of halophilic organisms compared to non-halophilic organisms[a]

Organism	GC content DNA (%)	pI	Charge difference	DE (%)	K (%)	AGV (%)	ILFM (%)	ST (%)
E. coli	50.7	6.9 (±1.8)	-3.3 (±10.2)	10.9 (±3.7)	4.7 (±2.4)	23.7 (±5.0)	23.5 (±5.2)	11.1 (±2.9)
B. subtilis	43.5	6.8 (±1.9)	-3.8 (±12.8)	12.5 (±4.5)	7.4 (±3.0)	20.8 (±5.2)	24.6 (±5.8)	11.5 (±2.9)
P. aeruginosa	66.6	6.9 (±1.8)	-3.1 (±10.6)	11.4 (±3.4)	3.0 (±2.2)	26.8 (±4.6)	22.3 (±4.6)	9.6 (±2.7)
H. salinarum	65.9	5.1 (±1.7)	-21.4 (±22.9)	15.9 (±5.8)	1.9 (±1.7)	29.8 (±7.6)	17.2 (±4.8)	12.3 (±3.6)
H. marismortui	60.7	5.0 (±1.5)	-23.5 (±24.3)	16.5 (±5.8)	2.1 (±1.6)	27.0 (±6.1)	18.5 (±5.2)	12.8 (±3.5)
H. walsbyi	47.9	5.1 (±1.4)	-21.1 (±25.7)	15.5 (±5.3)	2.5 (±1.8)	24.1 (±5.5)	19.7 (±5.2)	14.4 (±3.6)
N. pharaonis	63.1	4.8 (±1.3)	-27.1 (±30.1)	17.4 (±5.9)	2.0 (±1.8)	28.6 (±6.6)	18.1 (±5.0)	11.6 (±2.8)
S. ruber	66.1	5.9 (±2.0)	-15.2 (±21.7)	14.1 (±4.6)	2.2 (±1.8)	26.8 (±5.1)	18.9 (±4.3)	11.9 (±2.8)

[a]Protein sequences of all organisms were obtained from the EBI genome pages (http://www.ebi.ac.uk/genomes/). Amino acid composition and theoretical pI of each protein was calculated using web based tools (http://proteome.ibi.unicamp.br/tools/). The data shown are averages per protein, with the standard deviation between brackets. The charge difference per protein was calculated by adding the number of lysines and arginines, and subtracting the number of aspartate and glutamate residues

Table 2. Charge difference, pI, and amino acid composition of non-halophilic, halophilic, and thermophilic organisms of the integral membrane protein SecY

Organisms	Charge difference	pI	Asp, Glu (%)	Lys (%)	Gly (%)
non-halophilic bacteria[a]	+15.4 (±1.4)	9.8 (±0.8)	4.5 (±0.8)	4.3 (±0.4)	8.7 (±1.0)
Halophilic archaea[b]	-3.0 (±1.4)	5.8 (±0.3)	6.3 (±1.3)	1.9 (±0.2)	13.2 (±0.4)
Other archaea [c]	+7.2 (±2.3)	8.8 (±0.9)	5.8 (±0.6)	3.4 (±1.3)	9.2 (±0.8)

[a] *B. subtilis, E. coli, Haemophilus influenzae, Lactococcus lactis*, and *Streptococcus pneumoniae*
[b]*H. salinarum, H. walsbyi, Haloferax volcanii, H. marismortui, N. pharaonis*
[c]*Cenarchaeum symbiosum, Ferroplasma acidarmanus, Methanospirillum hungatei, Methanococcus maripaludis, Pyrococcus furiosis, Thermoplasma volcanicum*

the stability decreases rapidly, albeit that inactivation is a fairly slow process. Initial studies indicated that the activity of HmMDH is maximal only at 1.25 M NaCl and declines with higher salt concentration. This was surprising as the enzyme is unstable below 2 M NaCl [32]. However, later studies showed that the observed effect was due to a strong dependency of the K_m value for oxaloacetate on salt, which increased approximately 6-fold when going from 1 M NaCl to 4 M NaCl [34]. Activity at high salt concentrations was therefore not measured at saturating substrate concentration in high salt. Reanalysing HmMDH activity in saturating oxaloacetate concentrations showed that the enzyme activity increases up to 1 M salt, after which activity remains constant.

A comprehensive study on the effects of several salts on the stability of HmMDH showed that anions with a high charge density are the most efficient in stabilising the protein; the effectiveness of anions follows the order in the Hofmeister series [35]. In contrast, cations of high charge density were shown to be efficient in stabilising the folded form only at low salt concentrations and denature the protein at high concentrations. HmMDH could be stabilised by $MgCl_2$, but only in the range between 0.3 and 1.3 M [36]. In initial studies using techniques such as small angle neutron-scattering, analytical ultracentrifugation, and light scattering measurements, HmMDH (in high concentrations of NaCl or KCl) was shown to bind approximately 0.4 g of water per gram of protein, and 0.1 g of salt, corresponding to about 3000 molecules of water and 200 molecules of salt per tetramer [31]. In more recent studies these figures were somewhat revised, indicating that HmMDH in NaCl binds around 2000 water molecules and 50 salt molecules [37]. That study also showed that in $MgCl_2$, HmMDH binds more water (4100) and salt molecules (85), while in ammonium sulphate (which is strongly salting-out), no salts bind to HmMDH. A number of other halophilic proteins have also been shown to bind substantial amounts of salt, including *H. marismortui* elongation factor Tu [38] and *Haloarcula vallismortis* glyceraldehyde-3-phosphate dehydrogenase [39]. Binding of salt has also been observed for non-halophilic proteins and is possibly only quantitatively different from salt binding of halophilic proteins [37].

In addition to salt, other factors can also stabilise HmMDH. Firstly, the enzyme is stabilised at low salt concentrations by its cofactor, NADH, indicating that binding of NADH adds to enthalpic stabilisation [40, 41]. Secondly, the enzyme is stabilised by heavy water, D_2O [41]. Many non-halophilic proteins are also stabilised by D_2O, which increases rigidity of most proteins [42]. Deuterium bonds are stronger than hydrogen bonds, and the stabilising effect from D_2O may be due to either an increase in hydrophobic interactions (i.e. a salting-out like effect), or stronger intrapeptide deuterium bonds; the study by Cioni & Strambini [42] indicated that for most proteins the stronger hydrophobic effect of D_2O plays the dominant role.

Like other halophilic proteins, HmMDH has a large excess of acidic over basic residues. The crystal structure of the protein, which was first solved at 3.2 Å, showed that most of the acidic residues appear on the surface of the tetramer, while homologues from non-halophilic organisms lack this feature [43]. The increased number of acidic residues on the surface of halophilic proteins compared to their non-halophilic counterparts is evident in most proteins from extreme halophiles, and is exemplified in Figure 1 with the comparison of HmMDH and its non-halophilic homologue, dog-fish lactate dehydrogenase (DfLDH). Like most halophilic proteins, HmMDH has a high content of acidic residues when compared to DfLDH (Fig. 1A and 1B; acidic residues in red). Also the relatively low number of lysine residues in HmMDH is noticeable (Fig. 1A and 1B; lysine residues in blue).

The HmMDH protein was shown to be a dimer of dimers, and the main contacts between the two dimers are electrostatic in nature with two salt-bridge clusters. This appears to be in complete contrast to DfLDH, in which the contacts between the dimers consist primarily of hydrophobic interactions [44]. Other salt bridges were found as well and it was noted that HmMDH contained significantly more salt bridges than its mesophilic and thermophilic counterparts. This finding was surprising, as one might expect salt bridges to be destabilised in a high salt environment. However, a higher resolution structure of a mutant HmMDH (at 2.6 Å) suggested that the salt bridges were protected by specific ion binding sites, with sodium ions at the monomer-monomer interfaces, and chloride ions at the dimer-dimer interface. The location of the chloride ions is indicated in Figure 1B. It was proposed that the negative free energy of binding of these ions counterbalances the screening of charged groups involved in salt-bridges by solvent ions [45].

Zaccai *et al.* [36] proposed a halophilic stabilisation model, which postulates that acidic residues on the surface of a halophilic protein cooperatively bind a network of hydrated salt ions, with local salt concentrations higher than that of the solvent. Thus, according to this model hydrated ions make up an important part of the structure of HmMDH. At low salt concentrations there is insufficient binding of ions to maintain a stable protein-salt-water complex, explaining the instability at low salt. Repulsive electrostatic interactions between the acidic residues, which are normally screened at high salt concentrations, contribute to instability at low salt concentrations as well. At too high concentrations of salt, the concentration of ions in the solvent will approach the localised ion concentration on the protein's surface, and a consequence of this is that the protein can no longer compete with the solvent salt for binding of water, also leading to instability of the protein. This point may be reached more readily at relatively low Mg^{2+} concentrations, because Mg^{2+} ions are more strongly hydrated than Na^+ or K^+ ions. This could explain the instability of HmMDH above 1.3 M $MgCl_2$. Notably, $MgCl_2$ is a salting-in salt, and this may also account for the instability of the protein at high salt concentrations. The latter is corroborated by the observation that for divalent cations there is a balance between hydration and salt-binding [17]; at relatively low concentrations $MgCl_2$ is salting-out (due to increase in surface tension), while at higher concentrations it becomes salting-in (due to direct binding of Mg^{2+} to proteins). Interestingly, in potassium phosphate, in which HmMDH is also active, the protein does not bind significant amounts of salt and behaves like a non-halophilic protein. It was concluded that in potassium phosphate the main contributor of stability is the hydrophobic effect and entropy terms (i.e. a classical salting-out mechanism), whereas in physiological relevant salts (NaCl, KCl) stability mainly depends on the enthalpic contribution of binding to hydrated ions.

Strikingly, a mutation in HmMDH in which a glutamate was replaced by an arginine (E267R) made the protein apparently more halophilic; the mutant protein was less stable than the wild-type protein, and this could be compensated for by increasing the salt concentration (Ref. 24; note that in this paper, in which a different numbering was used, the mutation is denoted as E243R). It was suggested that the precise structural organisation of acidic residues is more important than an increased acidic content. Notably, E267 is located at the dimer-dimer interface (Richard *et al.*, 2000), and an alternative explanation is that the E267R mutation simply destabilises HmMDH, and that this lack in stability can be counterbalanced by the stronger salting-out effects of higher salt concentrations. A similar effect was seen with halolysin R4, a secreted serine protease from *Haloferax mediterranei* [46]. The mutation of two cysteine to serine residues in this protein resulted in lower overall activity that had a salt optimum at approximately 2.6 M NaCl, compared to an optimum of 1.7 M NaCl for the wild-type enzyme. It was not shown whether the cysteines were involved in disulfide bonds or perhaps the binding of Ca^{2+} ions, but these mutations resulted, most likely, in a less stable enzyme. Proteins become structurally more rigid in higher salt concentration due to increased strength of hydrophobic interactions, which explains the shift in salt optimum for enzymes that contain destabilising mutations.

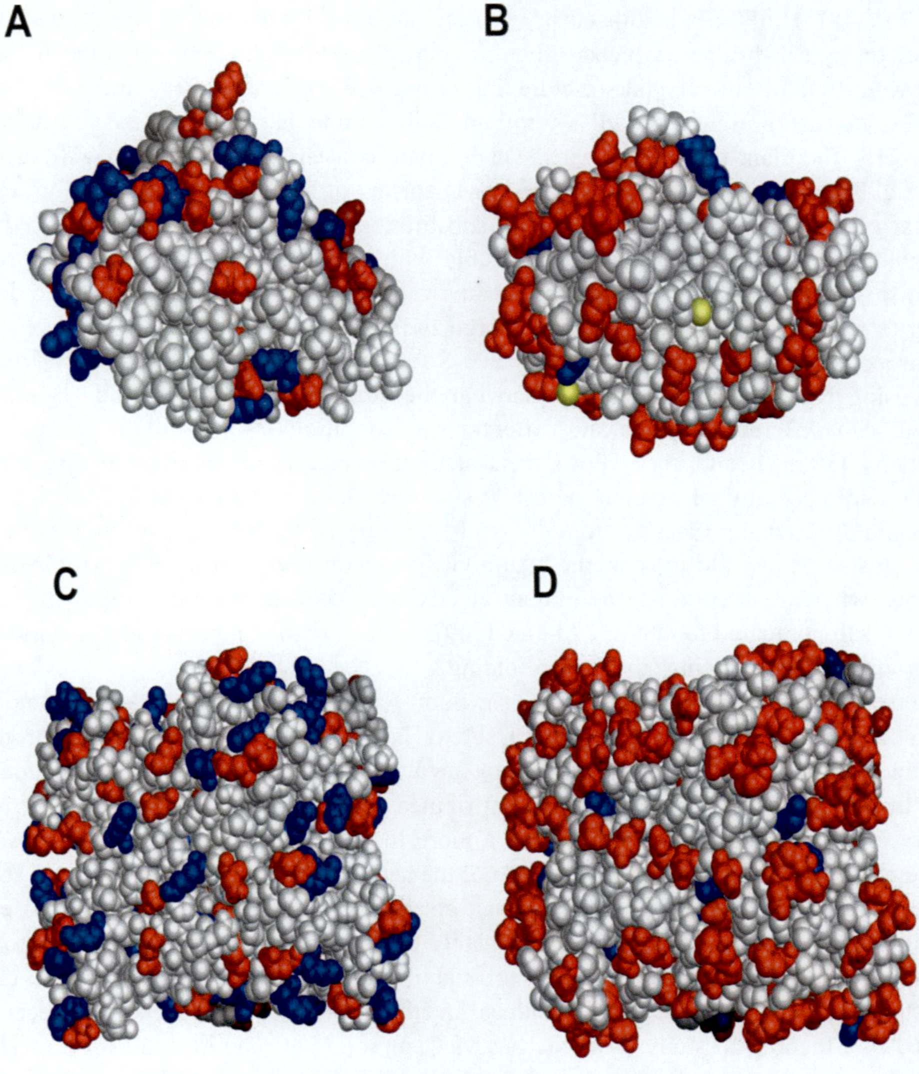

Figure 1. Comparison of the structures of non-halophilic and halophilic proteins. Space-filling models are shown of malate dehydrogenases (A and B), and nucleoside disphosphate kinases (C and D). A, monomer of dog-fish LDH (Brookhaven Protein Database number 6LDH); B, monomer of *H. marismortui* MDH (Brookhaven Protein Database number 1D3A); C, human NDK (Brookhaven Protein Database number 1NUE); D, *H. salinarum* NDK (Brookhaven Protein Database number 2AZ1) Glutamate and aspartate residues are in red, lysine residues are in blue, and chloride ions are in yellow. The structures were visualised with Chimera [114].

Halophilic 2Fe-2S Ferredoxin

Another halophilic protein for which the crystal structure is available is that of the halophilic 2Fe-2S ferredoxin from *H. marismortui* (HmFdx; Ref. 47). The structure of this monomeric protein, which contains two iron-sulphur clusters, was solved at 1.9 Å resolution. Similar to HmMDH and other halophilic proteins, it has a highly acidic surface. Unlike HmMDH however, the protein remains active at fairly low concentrations of salt (0.4 M NaCl; Ref. 48). HmFdx is stabilised by its iron-sulphur clusters and does not need further stabilisation by salt-bridges, which is also in contrast to HmMDH. Several water molecules (>200) were found in the crystal structure, but only a few K^+ ions. A very interesting feature of HmFdx is the presence of an N-terminal domain that is absent in its non-halophilic counterparts, the plant-type ferredoxins. This domain consists of 30 residues, 44% of which are acidic. Frolow *et al.* [47] proposed two mechanisms of halophilic adaptation for HmFdx. The first mechanism is that of "halophilic substitution", which means that HmFdx has as many acidic residues on its surface as possible without affecting its structure. The second mechanism is "halophilic addition", which is the acquirement of an additional acidic domain in order to provide more negative surface charge and thus more hydration.

In a recent publication the structure of 2Fe-S ferredoxin from *H. salinarum* (HsFdx) was solved using NMR [49]. Despite the fact that the structures of HmFdx and HsFdx were obtained with different methods and different salt conditions (4 M sodium phosphate for HmFdx and 450 mM NaCl for HsFdx), the structures were very similar. HsFdx also contains a highly acidic N-terminal domain, and it was shown that a mutant form of HsFdx lacking this domain differed significantly from the wild-type enzyme in the reconstitution of the iron-sulphur cluster. The wild-type enzyme could only be reconstituted efficiently with in at least 2 M NaCl whereas, in contrast, the mutant HsFdx was reconstituted more efficiently in the absence of salt compared to that in 2 M NaCl [49], indicating that the acidic domain in HsFdx is important both for protein stability and folding.

Bandyopadhyay *et al.* [50] used intrinsic fluorescence and circular dichroism studies to analyse the salt-dependent stabilisation of HsFdx. They showed that in low salt the protein is less stable and is more readily denatured by urea. Unlike HmMDH, the protein does not unfold in the absence of salt, which was attributed to the presence of salt bridges in the structure. The protein does, however, have a more loosely folded structure, as fluorescence measurements showed that either one or both of the tryptophans in the protein become solvent exposed in low salt. Strikingly, the low salt form of the protein could be stabilised by lowering the pH to 4.5. At this pH approximately 50% of the acidic residues are protonated and neutral, which suggests that the loss of stability at neutral pH (in low salt) is due to electrostatic repulsion of these acidic residues. Interestingly, such pH dependent effects were predicted in a theoretical study by Elcock & McCammon [51]. The low salt form of HsFdx could also be stabilised by the addition of polylysine, which proved to be effective already at 2 μM. It was suggested that stabilisation by polylysine was due to salt bridges formed between polylysine and surface exposed carboxylate groups, as well as enhancing hydrophobic interactions [50]. However, polylysine could not substitute salt fully, and the authors suggested that salting-out effects are also important for structural stability.

Halophilic Nucleoside Diphosphate Kinase

The crystal structure of nucleoside disphosphate kinase from *H. salinarum* (HsNDK) has been solved with and without its substrate at 2.2 and 2.35 Å, respectively [52]. Similar to other halophilic proteins it has a high content of acidic residues, and this is clearly noticeable when comparing the structures of HsNDK with human NDK (Fig. 1C and 1D; acidic residues are indicated in red). HsNDK is also low in lysine residues, and this is also an obvious difference when comparing the structure of HsNDK with that of human NDK (Fig. 1C and 1D; lysine residues indicated in blue). Some studies have suggested that halophilic proteins have an increased arginine to lysine ratio because arginines are able to bind more water [53]. However, if that were the sole reason for a low lysine content one would expect a significant increase in arginine residues, but that is not really noticeable when analysing halophilic and non-halophilic proteomes (see above in the section "Amino acid composition of halophilic proteins").

A reduction of surface-exposed lysine residues was also observed in a homology-based modelling study on the glutamate dehydrogenase from *H. salinarum* [54], and that indicated that loss of lysine is, surprisingly, a dominant factor in the reduction of exposed hydrophobic surface. Lysine is of course normally considered hydrophilic, but analysis of accessible surface of residues in a collection of nearly 600 proteins showed that of all the residues, lysine is in fact the one with the largest hydrophobic accessible surface in folded proteins [55]. Comparing a model of halophilic glutamate dehydrogenase with the structures of a mesophilic (*Clostridium symbiosum*) or thermophilic (*Pyrococcus furiosis*) glutamate dehydrogenase indicated a reduction in solvent-exposed hydrophobic surface of 3600 and 6100 $Å^2$, respectively. This reduction was much more significant than that for the reduction of surface exposed hydrophobic residues. Similar calculations were also made for halophilic malate dehydrogenase and its non-halophilic counterpart dogfish malate dehydrogenase, showing that in HmMDH the decrease in lysine residues accounted for a reduction in solvent-exposed hydrophobic surface of 5100 $Å^2$, compared to a reduction of only 900 $Å^2$ attributable to strongly hydrophobic residues. Thus, the lower lysine content of halophilic proteins appears to be a key factor in reducing surface exposed hydrophobicity, which is probably important for preventing aggregation in highly saline conditions.

In contrast to most halophilic proteins, HsNDK is stable and active at low salt concentrations [56]. Similar behaviour was found for a nucleoside diphosphate kinase from a haloalkaliphilic archaeon *Natrialba magadii* (NmNDK; Ref. 57). Remarkably, in the presence of NaCl this protein remained active in buffer containing up to 8 M urea. We made similar observations for a secreted alpha-amylase from the haloarchaeon *Haloarcula hispanica* (AmyH; Ref. 58). This enzyme has maximum activity at 4-5 M NaCl, but it does not unfold in low salt conditions. The enzyme also remains fully folded in the presence of molar concentrations of NaCl and 6 M urea. This was determined by intrinsic fluorescence emission spectra, which are useful in assessing structural changes in a protein. In particular the fluorescence of tryptophan residues is sensitive to changes in the solvent, and increased exposure of tryptophan residues to water results in a red-shift of the maximum wavelength emission of a protein and a change in the fluorescence intensity. When excited at 280 or 295 nm a fully unfolded protein usually displays a higher emission maximum wavelength than that of a fully folded protein, which is exemplified in Figure 2. As shown in Figure 2a, in the

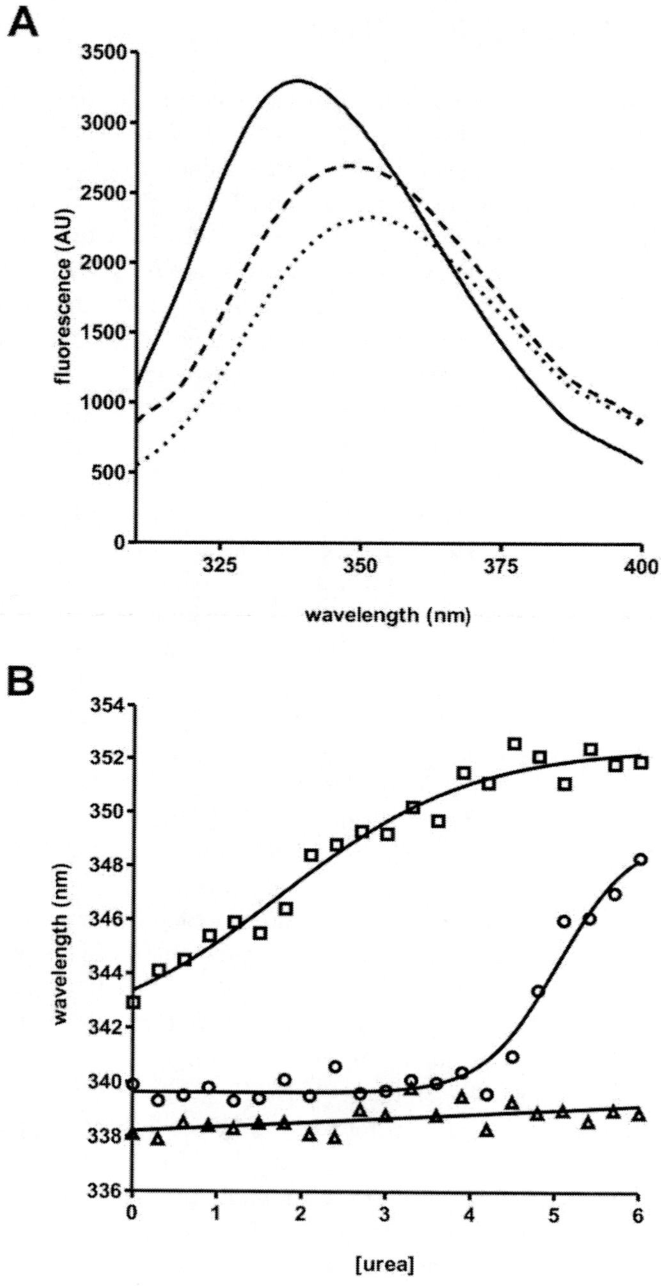

Figure 2. Salt-dependent stability of alpha-amylase from *Haloarcula hispanica* (AmyH). A, fluorescence emission spectra of AmyH in buffer with 6 M urea and 0 (dotted line), 2 (dashed line), or 4 M NaCl (solid line). B, urea denaturation curves in 0 (squares), 2 (circles), and 4 M NaCl (triangles), plotted as the wavelength of maximum emission versus urea concentration.

presence of 4 M NaCl and 6 M urea, the maximum wavelength of emission of AmyH is 338 nm, which is the same as for the protein in the absence of urea (not shown). When the salt concentration is lowered, the maximum emission wavelength shifts, reaching 352 nm in the absence of salt, which is typical for an unfolded protein. In the absence of both salt and urea, AmyH is partially active (around 30% of the activity under optimal conditions, which is 4-5 M NaCl). When AmyH is dialysed against a low salt buffer and then activity determined at high salt, nearly 100% of activity is recovered (Hutcheon *et al.*, 2005). We also observed that in a buffer without salt, the maximum emission wavelength of the protein is intermediate between fully folded and unfolded. These results indicate that in low salt the main structure of the protein is still intact, but that the protein becomes more loosely folded. Thus, the stability of the protein is clearly dependent on the salt concentration, and explains the observation that less urea is required to denature the protein when the salt concentration is lowered (Figure 2b).

Halophilic Dihydrofolate Reductase

H. volcanii contains two related dihydrofolate reductases, HvDHFR1 and HvDHFR2. Despite the fact that both proteins are fairly similar (39% identity), they behave quite differently with respect to activity in different salt concentrations. Both proteins are moderately acidic (Asp + Glu: 18.4% in HvDHFR1, 20.3% in HvDHFR2, and 15.7% in *E. coli* DHFR). Similar to other halophilic proteins, HvDHFR1 and 2 are less hydrophobic than the *E. coli* DHFR, and contain less lysine residues. Both enzymes are unstable and unfold below 0.5 M salt. Activity of HvDHFR1 increases with salt concentration and has an optimum at 3-4 M KCl. In contrast, HvDHFR2 appears much less halophilic as it shows its maximal activity around 0.6-1 M KCl [19]. This result was surprising, as HvDHFR2 is more acidic and less hydrophobic than HvDHFR1, but this could be explained, at least in part, by a reduction in affinity for the cofactor NADPH at high salt concentration. Activity of HvDHFR1 decreases at lower salt concentrations mainly due to a decreased affinity for its substrate [59].

The structure of the HvDHFR1 was solved at 2.6 Å [53]. The moderate increase in acidic residues in this protein is mainly localised in the C-terminus, which is reminiscent of the highly acidic N-terminal domain of the halophilic 2Fe-2S ferredoxin proteins discussed above. The protein contains only two lysines residues, and both of these are associated with the substrate-binding site. A comparative modelling study by Bohm & Jaenicke [60] suggested the presence of a number of clusters of acidic residues on the surface of HvDHFR. The crystal structure of HvDHFR1 showed, however, that two main clusters were modelled incorrectly [53]. There are some clusters of negative charge on the surface of HvDHFR1, but these do not interact closely. Thus, electrostatic repulsion at such clusters is not very strong, which could explain the observation that the protein remains stable in relatively low salt concentrations.

Other Soluble Proteins

A number of other halophilic proteins have been crystallised. The lab of Dieter Oesterhelt used an inverse structural genomics approach in which they systematically screened the *H. salinarum* proteome for crystallisable proteins. Two proteins that were identified in this manner include a halophilic dodecin, a dodecameric flavin binding-protein with unknown function [61], and DspA, a ferritin [62]. The structures of both proteins showed a large excess of acidic residues. Similar to HmMDH, dodecin has an increased number of complex salt-bridges on the subunit interfaces, and nearly all of the observed salt-bridges were not conserved in non-halophilic dodecins. DspA was shown to have an elongated N-terminal tail enriched in negatively charged residues, which is again reminiscent of the HmFdx protein discussed above.

Very recently the crystal structure of a secretory alpha-amylase (AmyA) from the halophilic and thermophilic bacterium *Halothermothrix orenii* was published [63]. *H. orenii* is a moderate halophile, and grows optimally in medium with around 10% NaCl. Its optimum growth temperature is 60 °C [64]. AmyA requires around 1 M NaCl for optimal activity [65], but it retains substantial activity without salt (45%) and in 4 M NaCl (90%). Optimum activity of this protein is at 65 °C. The crystal structure showed that AmyA does not have a significant excess of acidic residues on its surface, nor does it appear to have other characteristics of halophilic proteins, such as a lower content of lysine residues or a reduced hydrophobicity, suggesting a novel mechanism of stabilisation. It is noteworthy pointing out that *H. orenii* and its proteins need to be able to cope with a wide range of salinities. The environments where the organism grows, Tunesian salt lakes in the Sahara desert, have large seasonal fluctuations in their salt concentration, varying from salt crusts covering soil soaked in briny water in summer to saline lakes in rainy periods in winter [66]. Because of these variations it might be advantageous for AmyA and other *H. orenii* proteins to use a different type of halophilic adaptation. This is in line with results from a random sequence analysis of genomic DNA of *H. orenii* [67], which indicated that most, if not all, *H. orenii* proteins do not have an excess of acidic residues. It is also possible that the both hot and salty environment (poly-extremophilic) demands different mechanisms of adaptation. Biophysical studies showed that AmyA has a remarkable mechanism of coping with low salt concentrations [63]. In the absence of salt the protein forms very large aggregates (4-5 million Da) that remain active and retain secondary structure. Addition of only 5 mM NaCl converted the oligomers into monomers. Notably, the concentration of salt in the Tunesian salt lakes where *H. orenii* lives is unlikely ever to fall below 5 mM, and it is therefore not clear whether this phenomenon is biologically relevant. Nevertheless, the aggregates were shown to be much more thermostable than the monomers, indicating that inter-subunit interactions stabilise the protein, something which is not uncommon in oligomeric proteins. It was speculated that aggregation of AmyA was promoted by high affinity ion binding sites on the surface that in the presence of salt are occupied by ions; binding of ions to these sites could then prevent aggregation.

Membrane Proteins

Few integral membrane proteins from halophiles have been studied extensively. The aforementioned SecY, together with its partner protein SecE, from the haloarchaeon *H. volcanii* interacts at the membrane with ribosomes that synthesise secretory proteins. This interaction is salt-dependent and requires at least 2.5 M KCl [68]. This is in complete contrast to non-halophilic organisms, in which high salt is actually used to remove ribosomes from membranes. It has not been shown, however, which membrane component of the transport machinery (SecY, SecE, or other subunits of the Sec complex) mediates the salt-dependent binding of ribosomes.

One of the few halophilic membrane proteins that has been characterised thoroughly is bacteriorhodopsin (BR), a light-driven proton pump from *H. salinarum*. BR was one of the first membrane proteins for which an atomic resolution structure became available [69]. It is a retinal protein that contains seven membrane-spanning domains, and it is synthesised as a precursor with a 13-residue leader peptide that is removed after membrane insertion [70]. In the cytoplasmic membrane, trimers of BR pack into patches of natural-forming 2-dimensional crystals (called the purple membrane; Ref. 71). The amino acid composition of BR appears to have the same characteristics as other halophilic proteins. For instance, the precursor of BR has a theoretical pI of 4.6, which is low compared to a pI of 8.3 for its closest related non-halophilic homologue, the opsin-like protein CarO from the fungus *Gibberella fujikuroi* [72]. Also, the overall charge for BR is -7, whereas that for CarO is +2. However, unlike most soluble halophilic proteins (see above), BR is stable at low salt and has its optimal activity in 0.5 M NaCl [73]. Nevertheless, when purple membrane is exposed to the detergent Triton X-100 (which results in monomers of BR), both high salt (> 2 M) and native lipids are important for the re-formation of a well-ordered 2D array and normal photocycle behaviour [74, 75].

Halophilic Proteins in Low-Water Environments

Hypersaline environments, such as salt lakes, may be subject to long periods of drought, and organisms living in these conditions are regularly exposed to desiccation. One would expect therefore that halophiles are capable of coping with desiccation. This was indeed shown for *H. salinarum*: after being exposed to desiccation for up to 20 days either at normal pressure or in vacuum, a significant proportion (25%) of cells survived [76]. Cells were apparently protected from desiccation stress through the formation of salt crystals. Strikingly, halophilic and halotolerant organisms have even been isolated from ancient rock salt up to 250 million years old, in which they appear to survive in fluid inclusions in the salt crystals [77]. Whether the organisms from these salt deposits are truly ancient remains controversial (see Ref. 78, and references therein), but it does seem clear that halophilic organisms can survive for long periods in the presence of very little amounts of water.

One study has also examined the effect of desiccation on a halophilic protein [79]. In that study the thermostability of *H. marismortui* malate dehydrogenase was measured after drying under vacuum and compared to the stability of a non-halophilic homologue, pig heart malate dehydrogenase. It showed that drying drastically destabilised the non-halophilic protein, either in the presence or absence of 4 M NaCl. In contrast, the halophilic enzyme was completely resistant to drying, and had in fact a higher thermal stability: in dry salt crystals more than 90% of enzyme activity was recovered after 1 hour at 60 °C, while only 15% of activity remained after incubation in salt solution.

A likely reason for the resistance of halophilic proteins to dry conditions is that they are in fact adapted to an environment that has a low water activity *per se*. Water activity (a_w) is basically a measure of the available amount of free water molecules, and it is usually expressed as the ratio between the measured vapour pressure and the vapour pressure of pure water. It is on a scale from 0 to 1, with pure water having an a_w of 1. Saturated sodium chloride (~6 M) has, at room temperature, a water activity of 0.75. This water activity is similar to that of 60 % (v/v) dimethyl formamide [80], or when comparing with something more familiar, it is in the same range as for instance honey or dried fruits. The water activity of saturated potassium chloride (~4.8 M) is slightly higher than that of sodium chloride at 0.84. It is generally presumed that no growth is possible below a water activity of 0.6 [81], although life has been observed in deep hypersaline anoxic basins (see above), some of which contain up to 5 M magnesium chloride; saturated $MgCl_2$ (~5.8 M) has an a_w of 0.33. Because of the low water activity of the saline environments in which extreme halophiles live, one can therefore wonder whether halophilic proteins are also stable and active in mixtures of water and organic solvents, which also have a low water activity.

In general, various co-solutes such as sucrose or glycerol have been used to stabilise proteins, and the stabilising effects of these co-solvents has been attributed to preferential exclusion of the co-solvent from the protein through an increase in surface tension [15]. This exclusion causes the presence of an ordered shell of water molecules around the protein. It is thermodynamically favourable to minimise this; thus, as a natively folded protein is more compact than the protein in an unfolded conformation, the native conformation is stabilised by the addition of polyols. This is, in effect, similar to salting-out salts, which are also preferentially excluded from protein surfaces. If such co-solvents stabilise the folded state of a protein in general, one might expect that lack of stability of halophilic proteins in low salt buffers might be compensated for by organic solvents. For the halophilic catalase from *H. salinarum (cutirubrum)* this was indeed shown to be the case [82]. Activity assays in different salt concentrations showed a bell shaped curve with an optimum around 2 M NaCl or KCl. Similarly shaped curves were found in molar concentrations of dimethyl sulfoxide (DMSO), ethylene glycol, and glycerol, showing that halophilic proteins do not necessarily require salt for their activity and that salt can be replaced by organic solutes. The activity of a mesophilic catalase (from beef liver) was reduced upon addition of organic solutes, perhaps caused by a more rigid structure of the enzyme.

The aforementioned halophilic nucleoside diphosphate kinase from *H. salinarum* (HsNDK) was shown to require salt for refolding [83]. Both heat-denatured and urea-denatured HsNDK did not refold in a buffer without salt. Strikingly, the protein could be efficiently refolded in the absence of salt when 4 M trimethylamine N-oxide (TMAO) was added. TMAO is, like glycerol and sucrose, preferentially excluded from the protein, and it was concluded that preferential hydration of HsDNK in high concentrations of TMAO can drive the folding reaction to a state with smaller solvent-exposed surface. Thus, TMAO can replace salts in the refolding reaction, showing that folding of halophilic proteins is not strictly dependent on salt.

Ryu *et al.* [84] analysed the catalytic activity of a secreted serine protease from the archaeon *H. salinarum (halobium)*. Similar to many other halophilic proteins, this protease was very unstable in low salt conditions and in fact inactivated irreversibly. Activity was also affected by low salt, but it was shown that the reduction in activity in low salt was mainly caused by the strong increase in the K_m value (a measure of the strength of the enzyme-substrate complex), while the k_{cat} (turnover number) was largely unaffected. The enzyme was also active in mixtures of water and dimethylformamide (DMF). The effect of DMF on the catalytic activity was similar to low salt concentrations, with drastically increased K_m values due to diminished substrate binding and similar k_{cat} values in 33% DMF. Strikingly, in solutions containing 33% DMF and saturating NaCl concentrations (~2.8 M), the enzyme was capable of synthesising peptides, showing a reverse catalytic activity in these conditions. Stability of the halophilic protease in organic solvent-water mixtures was shown to be dependent on the salting-out like properties of the solvent. In particular organic solvents such as DMSO and ethylene glycol had a stabilising effect, while a salting-in like organic solvent such as tetrahydrofuran had a destabilising effect [85]. Using inactivation assays and fluorescence studies it was concluded that organic solvents can replace salts to increase the stability of the hydrophobic core of the protease, which is weaker than in its non-halophilic counterparts. Similarly, for a halophilic amylase from the archaeon *Haloarcula* sp. S-1 it was also shown that salts could be replaced, in part, by organic solvents; this enzyme was more active at low salt concentrations in the presence of chloroform, while at high concentrations of salt the activity did not change when incubated in the presence of chloroform [86].

Another interesting area in the field of biocatalysis in organic solvents is that of so-called reverse micelles. These are small water droplets (1-10 nm) in a nonpolar organic solvent, with a surfactant on the interphase between water and organic solvents [87]. The surfactant is amphiphilic, with the polar heads in contact with the water phase, and the aliphatic tails sticking into the organic solvent. Reverse micelles are of interest because they may mimic the intracellular environment much more closely than that of the dilute aqueous environments biochemists usually favour. A number of halophilic proteins have been tested in reversed micelles, including p-nitrophenylphosphate phosphatase (pNPPase) and malate dehydrogenase (both from *H. salinarum*). The enzyme pNPPase is typically halophilic, requiring at least 0.6 M NaCl for activity and stability [88]. In the absence of salt the protein has only 15% of its activity and denatures rapidly, and even at concentrations where the protein is already close to its maximum activity (1 M NaCl) it is not very stable. In a system with an anionic surfactant commonly used for reverse micelles (i.e. dioctyl sodium sulphosuccinate (AOT) in cyclohexane (solvent) and 1-butanol (co-surfactant)), pNPPase remained active and stable in 50 mM NaCl [89]. Activity depended strongly on the molar ratio of water to surfactant (w_o), showing a bell-shaped curve with an optimum w_o value of approximately 8. A cationic surfactant, hexadecyltrimethylammonium bromide (CTAB) showed similar results, albeit that the optimal w_o value was somewhat higher (~11). Similarly, halophilic malate dehydrogenase was stabilised in reverse micelles using CTAB as surfactant and cyclohexane as solvent. Piera-Velazquez *et al.* [90] showed that the half life of HmMDH at 25 °C was approximately 1 day in aqueous buffer containing 1 M NaCl, while this could be drastically increased in reverse micelles (12-50 days depending on conditions such as the w_o value). The reason for the importance of the w_o value lies in the fact that the water : surfactant ratio determines the size of the micelle. At low w_o values the micelle is small, whereas the micelle size increases with higher w_o values. Importantly, the size of the micelle determines the ratio of structured water to free water. With very small micelles nearly all of the water will be structured at the surfactant-water interface. Increasing the size of the micelle will increase the amount of free water molecules, causing an increase in activity. However, above a certain w_o ratio the activity decreases as the system resembles more of an aqueous environment [91]. Thus, instead of salt ions structuring water molecules, it is the polar heads of the surfactant structuring the water.

Biocatalysis in organic media is by no means unique to halophilic proteins (for a review, see Ref. 92). However, the studies mentioned above do clearly show that halophilic proteins are also able to function in non-saline conditions, suggesting that it is the low water activity of concentrated salt solutions rather than the salt itself that is important for halophilic proteins.

Common Themes and Different Strategies of Halophilic Protein Adaptation

As outlined in the sections above, there are a number of common themes in the halophilic adaptation, with certain features found in nearly all proteins. The amino acid composition of halophilic proteins is characterised by a high content of acidic residues on the surface of proteins, a low lysine content, and a relatively low hydrophobicity of the core of the protein. The latter two features are probably the easiest to explain. In molar salt concentrations proteins may suffer from an increase in hydrophobic interactions, leading to aggregation caused by interactions of hydrophobic patches on a protein's surface and a decrease in flexibility in the core of a protein. Aggregation must of course be prevented, and an important factor in this is the reduction in exposed hydrophobic surface, which is achieved mainly through a reduction in lysine residues. Flexibility in a protein is also important for activity, and the increase of small hydrophobic residues in favour of large hydrophobic residues is likely to play an important role in this. A consequence of the lower hydrophobicity of halophilic proteins is that it may be part of the reason for the decrease in stability at low salt concentrations.

The increased number of acidic residues on the surface of halophilic proteins has been interpreted in different ways and the role of these acidic residues is therefore less clear. Studies on HmMDH and a few other proteins (see above) showed that these bind significant numbers of salt molecules, and it was suggested that the acidic residues are important for binding a large network of hydrated salt ions [31]. However, while high resolution crystal structures showed the presence of several water molecules, they failed to show large numbers of ions. This discrepancy could be caused by a disorder of bound solvent molecules, which makes it difficult to see them in a crystal structure. Also, sodium ions are difficult to distinguish from water as they have the same number of electrons. Nevertheless, there is ample evidence for binding of salt ions to some halophilic proteins, indicating that these seem to harness the high ionic strength of their environment rather than being protected from it [45]. Notably, as the binding of salt ions requires molar concentrations of salt, protein-salt interactions can only be weak [15] although, at least in some cases, the interactions are specific.

HmMDH is sensitive to cations and anions, both of which play an important role in stabilisation of the tetrameric protein. This is in contrast to the behaviour of isocitrate dehydrogenase from *H. volcanii* (HvICDH), a dimeric protein that was found not to be sensitive to the type of anion [93]. The sensitivity of HvICDH to cations is similar to that of HmMDH, with for instance the enzyme stabilised by low concentrations and deactivated by high concentrations of $MgCl_2$. The studies on HvICDH suggested that it might represent a novel type of halophilic adaptation, being an oligomeric protein lacking intersubunit anion-binding sites. There is, however, no structural information on HvICDH, which will be needed to understand the biophysical characteristics of this protein in more detail.

The binding of salt to halophilic proteins is not required *per se*, as some halophilic proteins are also active and stable in conditions where no or little salt binds, such as in ammonium sulphate or in organic solvents. Carboxylate groups of glutamate and aspartate are strongly hydrated [94], and it seems therefore likely that the acidic residues are important for binding essential water molecules. Also non-halophilic proteins bind water, but halophilic proteins have to compete with the salt for binding of water, and the acidic residues probably play an important role in this.

Repulsive electrostatic interactions between the acidic residues probably contribute to a large extent to the destabilisation of halophilic proteins, which is in addition to the lower hydrophobicity of halophilic proteins. Elcock & McCammon [51] suggested that the main role for acidic residues is to prevent aggregation of proteins through these repulsive interactions. This explanation was ruled out for the halophilic dodecin, as the core of that protein has a similar negative charge as the outer surface [61]. Mevarech *et al.* [95] suggested that the acidic residues have two important roles, being (a) binding of essential water molecules, and (b) providing flexibility within a protein through electrostatic repulsion.

In many halophilic proteins the acidic residues are distributed fairly equally in the polypeptide. However, in some proteins the acidic residues appear to be clustered in short regions (see above). Short stretches of very acidic regions have been found in several halophilic proteins; regions of 30 amino acids with more than 50% Asp + Glu were found in around 240 proteins in *H. salinarum* [49]. It was suggested that this may allude to a mechanism for rapid adaptation of non-halophilic proteins. Thus, a protein encoded by a gene acquired from a non-halophilic organisms could first adapt rapidly through insertion (or perhaps replacement) of a sequence encoding a short stretch rich in acidic residues. Proteins that have had the chance to evolve for a longer period in a halophilic environment may end up with a more equal distribution of acidic residues.

A question that remains is how some halophilic proteins such as *H. hispanica* AmyH, HsNDK, and HmFdx remain folded in the absence of salt, while most other halophilic proteins unfold under these conditions. It is striking to note that both HsNDK [52] and HmFdx (see above) have short regions that are highly acidic. In HsNDK, the surplus of acidic over basic residues is only 7% in 122 residues of the N-terminus, but 35% in the 39 C-terminal residues. Similarly, HmFdx has an acidic/basic surplus of 41% in the N-terminus (37 residues), but only 14% in the C-terminus (92 residues). Such an asymmetric distribution is not found in *H. hispanica* AmyH, but this protein has an overall acidic to basic surplus of only 10%, which is low compared with that of its homologue from the haloalkaliphilic *Natronococcus amylolyticus*. The latter, which has a acidic to basic surplus of nearly 19%, is unstable in low salt conditions [96]. Thus, it seems possible that in some halophilic proteins repulsive interactions are relatively weak, because of an acidic content that is either relatively low overall, or localised in only a small part of the protein. This could explain why some halophilic proteins remain stable, but it does not explain why a protein such as AmyH has its highest activity in 4-5 M NaCl and remains folded in 6 M urea. It is obvious that detailed biophysical studies are required to understand the halophilic nature of these proteins.

In summary, the increase in acidic residues in halophilic proteins has been suggested to serve a number of purposes. These are (a) binding of water molecules, (b) binding of salt ions, (c) preventing aggregation through repulsive interaction and (d), maintaining flexibility within a protein. Notably, the latter two can also be accomplished through other means, i.e. low hydrophobicity of the surface of proteins, chiefly through a low number of lysines, and a relatively low hydrophobicity in the core of the protein, through replacement of large hydrophobic residues with smaller ones. Some of the interpretations for the role of the acidic residues appear to be mutually exclusive, such as binding of salt and the prevention of aggregation through repulsive interactions. However, effects such as instability at low salt can be explained through both, and it may therefore be difficult to address this experimentally [51].

Another form of halophilic adaptation in proteins is an increase in salt-bridges as found for *H. salinarum* dodecin and HmMDH, with (as found for HmMDH) chloride or sodium ion binding at specific sites involved in stabilisation of these salt-bridges. This is, however, not a strategy used by all halophilic proteins, as for instance halophilic 2Fe-2S ferredoxin does not contain additional salt bridges when compared to its non-halophilic counterparts.

Other Halophilic Adaptations in the Cell

In addition to adaptation of proteins, halotolerant and halophilic organisms also have adapted their cellular processes. The natural environment of many halotolerant and halophilic organisms are salt lakes and, depending on for instance rain fall, the concentration of salt in the lake may vary significantly. One obvious cellular process that needs to be regulated accurately is the response to osmotic stress. For an organism using the compatible solute strategy, this must involve the ability to increase or decrease the cytoplasmic concentration of the compatible solute; increasing the concentration of compatible solutes can occur through synthesis and/or uptake of compatible solutes, while decreasing this will involve either excretion or metabolic degradation. For the halophilic algae *Dunaliella parva* it was shown that when cells were resuspended in a significantly lower or higher salt concentration, osmotic equilibrium was reached within 90 minutes [97]. When cells were grown at 1.5 M NaCl, and then resuspended in 0.6 M NaCl, no extracellular glycerol was detected, indicating metabolic degradation of glycerol. At lower salt concentrations glycerol leaks out or is excreted. When the salt concentration is increased, glycerol is synthesised from starch. An important trigger for osmotic changes in *Dunaliella* spp. is the change in cell volume, and plasma membrane sterols were shown to play an important role in sensing osmotic changes [98].

Halophilic organisms using the salt-in strategy utilise various ion and proton pumps to maintain the osmotic balance. The concentration of potassium inside the cell far exceeds that of the outside. For instance for halophilic archaea, the intracellular potassium concentration is usually around 4 M, while the concentration in the growth medium is in the range of 10-50 mM [6]. In contrast, the intracellular sodium concentration is about 4-fold lower than in the medium. Halophilic archaea contain various pumps for the transport of protons, cations, and anions (for a review, see Ref. 6). *H. salinarum* contains several active K^+ transporters, including an ATP-driven K^+ transport system and a low-affinity K^+ transporter that is driven by the membrane potential. Furthermore, there is active Na^+ efflux, most likely through a Na^+/H^+ antiporter. The intracellular chloride concentration (the main anion) is similar to the extracellular concentration. The chloride concentration is maintained using chloride pumps, which for *H. salinarum* includes the light-driven chloride pump halorhodopsin [99] and a Na^+ Cl^- symporter [100].

In addition to proteins, membranes must also be resistant to very saline conditions in halophilic organisms in order to maintain their integrity and function. It was shown that liposomes derived from lipids of the archaea *H. salinarum* and *Halorubrum vacuolatum* (a halophilic alkaliphile) are stable in a wide range of salt concentrations up to 4 M NaCl or KCl, whereas *E. coli* liposomes aggregate at or above 0.5 M NaCl. Proton and sodium permeability for the halophilic liposomes was low, even at very high salt concentrations [101]. In general, archaeal lipids are characterised by ether bonds and branched isoprenoid chains rather than ester bonds and fatty acyl chains as found in bacteria and eucarya. Archaeal lipids are chemically much more stable than those of bacteria and eucarya, which is one of the reasons that archaea are capable of surviving extreme conditions [29]. Membranes of halophilic archaea are abundant in archaetidylglycerol methylphosphate (PGP-Me), a diacidic phospholipid that can make up 50-80% of the polar lipids, while non-halophilic archaea do not contain PGP-Me. It was recently shown that liposomes prepared from non-halophilic archaea were leaky and aggregated at high salt concentration, but that adding PGP-Me to these liposomes made them much more resistant and stable in high salt [102]. PGP-Me is doubly charged, so one could speculate that electrostatic repulsion is important for its stabilising role. However, the headgroup of PGP-Me is also large, and Tenchov *et al.* [102] showed that it is probably steric repulsion that prevents close approach and aggregation. Less is known about halophilic adaptation of lipids from extremely halophilic bacteria. The extremely halophilic bacterium *S. ruber* was recently shown to contain a novel type of sulfonolipid, representing about 10% of the total cellular lipid. It was suggested that the novel lipid might provide a highly negatively charged surface (indicating a role for electrostatic repulsion), or have an effect on the membrane fluidity [103].

One striking example of a cellular process adapted to halophilic conditions is protein translocation. In prokaryotes, most proteins utilise one of two pathways for export, these being the Sec pathway and the Twin-arginine translocation (Tat) pathway [104, 105]. The export route used depends on the folded state of the protein: proteins that can fold after translocation utilise the Sec pathway, whereas proteins that fold before translocation are exported *via* the Tat pathway. It is arguably much easier to export a protein in an unfolded conformation, as the size of the pore does not need to vary when an unfolded protein is threaded through. The Tat system is also energetically very costly [106], and it is therefore not a big surprise that in most prokaryotes more than 90% of proteins are Sec dependent [107]. The *raison d'etre* for the Tat system is that some proteins either have to fold before translocation, for example because they require the incorporation of a complex cofactor, or that they merely fold too quickly [108]. All Sec and Tat substrates can be recognised by a transient amino-terminal signal sequence that is required for targeting and initiation of translocation. During, or shortly after translocation this signal peptide is removed by a specific signal peptidase. Tat substrates are recognised by the presence of a specific motif in the signal peptide in which two adjacent arginines play a critical role [109] - hence the name Twin-arginine translocation pathway. When the genome of the first halophilic archaea became available, it was noted that most proteins containing a putative signal peptide appeared to contain a twin-arginine motif [107, 110], suggesting that the Tat system plays a dominant role in halophilic archaea. It was recently shown that the components of the Tat system are in fact essential in the halophilic archaea *H. volcanii* and *H. salinarum* [111, 112], which is in contrast to all other prokaryotes thus far tested. The question is of course why the Tat system plays such an important role in halophilic archaea, even for proteins that do not bind cofactors. The most likely reason seems to lie in the extremely high cytoplasmic salt concentration. The hydrophobic effect is one of the most important factors in protein folding (see above). As hydrophobic interactions are stronger at high ionic strength, it is conceivable that proteins fold very rapidly in the cytoplasm of organisms using the salt-in strategy. This would mean that most secretory proteins fold before ever reaching the cytoplasmic membrane. Folded proteins are incompatible with the Sec pathway; forcing the export of a folded protein through the Sec translocase may be lethal to a cell [113]. Export of most halophilic proteins may therefore only be possible through the Tat pathway, and the dominant role of this pathway in halophilic archaea is therefore clearly a halophilic adaptation. Surprisingly, the halophilic bacterium *S. ruber* does not appear to utilise the Tat pathway as the major route for export [111], indicating that this organism uses other means of dealing with the "folding-problem" in high salt. It is worth pointing out that the Sec pathways of bacteria and archaea probably differ fundamentally in their mechanism [104]. Halophilic archaea may lack the tools to keep secretory proteins in an unfolded conformation, forcing them to use the Tat pathway, whereas such tools may be present in halophilic bacteria, allowing the use of the Sec pathway [111].

CONCLUSION

Proteins from halophilic organisms have adopted unique features to enable them to function in extremely saline conditions, which have resulted in a substantial effect on the amino acid composition of halophilic proteins. As already suggested by Lanyi [11], halophilic proteins are probably not qualitatively different from non-halophilic proteins, but have merely increased certain interactions (e.g. binding of water and salt), and decreased others (e.g. hydrophobicity) to remain both active and stable in highly saline conditions. It should therefore be possible, in theory, to convert a non-halophilic protein into a halophilic one using our knowledge on halophilic adaptation, experiments which, as far as we know, have not yet been performed. It is important to note that the behaviour of only very few halophilic proteins have been analysed thoroughly in a wide range of salts. It seems therefore clear that detailed biophysical studies on other halophilic proteins are required, as well as more high resolution crystal structures, to increase our understanding of halophilic adaptation. This will not only increase our knowledge on how proteins are able to function in highly saline environments, but it will also provide more general insights into, for instance, interactions of non-halophilic proteins with their solvent environment.

An area of research that has not really been addressed in detail is the folding of halophilic proteins. The observation that halophilic archaea mostly use the Tat pathway for secretion (which exports folded proteins) rather than the Sec pathway (which exports unfolded proteins) suggests that halophilic proteins may fold more quickly than their counterparts in non-halophilic organisms, but no information is available on the rate of folding of halophilic proteins. The main factor in protein folding is the hydrophobic effect and, because hydrophobic interactions are stronger in high salt concentrations, it is indeed conceivable that the rate of protein folding is very fast in these conditions. On the other hand, the lower hydrophobicity and/or repulsive interactions of acidic residues of halophilic proteins may provide a counterbalance to this. Thus, future studies on halophilic adaptation should not only address the biophysical and structural properties of halophilic proteins, but also the effect of salt on cellular processes such as protein folding and translocation.

ACKNOWLEDGMENTS

A.B. is a Royal Society University Research Fellow. D.K. and J.R.T. are supported by the Biotechnology and Biological Sciences Research Council.

REFERENCES

[1] Litchfield, C.D., & Gillevet, P.M. (2002) Microbial diversity and complexity in hypersaline environments: a preliminary assessment. *J Ind Microbiol Biotechnol* 28: 48-55.

[2] Wallmann, K., Suess, E., Westbrook, G.H., Winckler, G., & Cita, M.B. (1997) Salty brines on the Mediterranean sea floor. *Nature* 387: 31-31.

[3] Bolhuis, H., Palm, P., Wende, A., Falb, M., Rampp, M., Rodriguez-Valera, F., Pfeiffer, F., & Oesterhelt, D. (2006) The genome of the square archaeon *Haloquadratum walsbyi* : life at the limits of water activity. *BMC Genomics* 7: 169.

[4] van der Wielen, P.W., Bolhuis, H., Borin, S., Daffonchio, D., Corselli, C., Giuliano, L., D'Auria, G., de Lange, G.J., Huebner, A., Varnavas, S.P., Thomson, J., Tamburini, C., Marty, D., McGenity, T.J., & Timmis, K.N. (2005) The enigma of prokaryotic life in deep hypersaline anoxic basins. *Science* 307: 121-123.

[5] Dennis, P.P., & Shimmin, L.C. (1997) Evolutionary divergence and salinity-mediated selection in halophilic archaea. *Microbiol Mol Biol Rev* 61: 90-104.

[6] Oren, A. (1999) Bioenergetic aspects of halophilism. *Microbiol Mol Biol Rev* 63: 334-348.

[7] Lai, M.C., Sowers, K.R., Robertson, D.E., Roberts, M.F., & Gunsalus, R.P. (1991) Distribution of compatible solutes in the halophilic methanogenic archaebacteria. *J Bacteriol* 173: 5352-5358.

[8] Roberts, M.F. (2005) Organic compatible solutes of halotolerant and halophilic microorganisms. *Saline Systems* 1: 5.

[9] Imhoff, J.F., & Rodriguez-Valera, F. (1984) Betaine is the main compatible solute of halophilic eubacteria. *J Bacteriol* 160: 478-479.

[10] Avron, M. (1986) The osmotic components of halotolerant algae. *Trends Biochem Sci* 11: 5-6.

[11] Lanyi, J.K. (1974) Salt-dependent properties of proteins from extremely halophilic bacteria. *Bacteriol Rev* 38: 272-290.

[12] Serrano, R. (1996) Salt tolerance in plants and microorganisms: toxicity targets and defense responses. *Int Rev Cytol* 165: 1-52.

[13] Kushner, D.J. (1964) Lysis and dissolution of cells and envelopes of an extremely halophilic bacterium. *J Bacteriol* 87: 1147-1156.

[14] Hofmeister, F. (1888) Zur Lehre von der Wirkung der Salze. *Arch Exp Pathol Pharmakol* 24: 247-260.

[15] Timasheff, S.N. (1993) The control of protein stability and association by weak interactions with water: how do solvents affect these processes? *Annu Rev Biophys Biomol Struct* 22: 67-97.

[16] Baldwin, R.L. (1996) How Hofmeister ion interactions affect protein stability. *Biophys J* 71: 2056-2063.

[17] Arakawa, T., & Timasheff, S.N. (1984) Mechanism of protein salting in and salting out by divalent cation salts: balance between hydration and salt binding. *Biochemistry* 23: 5912-5923.

[18] Matsumura, M., Becktel, W.J., & Matthews, B.W. (1988) Hydrophobic stabilization in T4 lysozyme determined directly by multiple substitutions of Ile 3. *Nature* 334: 406-410.

[19] Wright, D.B., Banks, D.D., Lohman, J.R., Hilsenbeck, J.L., & Gloss, L.M. (2002) The effect of salts on the activity and stability of *Escherichia coli* and *Haloferax volcanii* dihydrofolate reductases. *J Mol Biol* 323: 327-344.

[20] Tebbe, A., Klein, C., Bisle, B., Siedler, F., Scheffer, B., Garcia-Rizo, C., Wolfertz, J., Hickmann, V., Pfeiffer, F., & Oesterhelt, D. (2005) Analysis of the cytosolic proteome of *Halobacterium salinarum* and its implication for genome annotation. *Proteomics* 5: 168-179.

[21] Baliga, N.S., Bonneau, R., Facciotti, M.T., Pan, M., Glusman, G., Deutsch, E.W., Shannon, P., Chiu, Y., Weng, R.S., Gan, R.R., Hung, P., Date, S.V., Marcotte, E., Hood, L., & Ng, W.V. (2004) Genome sequence of *Haloarcula marismortui*: a halophilic archaeon from the Dead Sea. *Genome Res* 14: 2221-2234.

[22] Falb, M., Pfeiffer, F., Palm, P., Rodewald, K., Hickmann, V., Tittor, J., & Oesterhelt, D. (2005) Living with two extremes: conclusions from the genome sequence of *Natronomonas pharaonis*. *Genome Res* 15: 1336-1343.

[23] Mongodin, E.F., Nelson, K.E., Daugherty, S., Deboy, R.T., Wister, J., Khouri, H., Weidman, J., Walsh, D.A., Papke, R.T., Sanchez Perez, G., Sharma, A.K., Nesbo, C.L., MacLeod, D., Bapteste, E., Doolittle, W.F., Charlebois, R.L., Legault, B., & Rodriguez-Valera, F. (2005) The genome of *Salinibacter ruber*: convergence and gene exchange among hyperhalophilic bacteria and archaea. *Proc Natl Acad Sci USA* 102: 18147-18152.

[24] Madern, D., Pfister, C., & Zaccai, G. (1995) Mutation at a single acidic amino acid enhances the halophilic behaviour of malate dehydrogenase from *Haloarcula marismortui* in physiological salts. *Eur J Biochem* 230: 1088-1095.

[25] Coronado, M.J., Vargas, C., Mellado, E., Tegos, G., Drainas, C., Nieto, J.J., & Ventosa, A. (2000) The alpha-amylase gene *amyH* of the moderate halophile *Halomonas meridiana*: cloning and molecular characterization. *Microbiology* 146: 861-868.

[26] Banerjee, T., Gupta, S.K., & Ghosh, T.C. (2005) Role of mutational bias and natural selection on genome-wide nucleotide bias in prokaryotic organisms. *Biosystems* 81: 11-18.

[27] Fukuchi, S., Yoshimune, K., Wakayama, M., Moriguchi, M., & Nishikawa, K. (2003) Unique amino acid composition of proteins in halophilic bacteria. *J Mol Biol* 327: 347-357.

[28] von Heijne, G.V. (1986) The distribution of positively charged residues in bacterial inner membrane proteins correlates with the trans-membrane topology. *EMBO J* 5: 3021-3027.

[29] van de Vossenberg, J.L., Driessen, A.J., & Konings, W.N. (1998) The essence of being extremophilic: the role of the unique archaeal membrane lipids. *Extremophiles* 2: 163-170.

[30] Sowers, K.R., Boone, J.E., and Gunsalus, R.P. (1993) Disaggregation of Methanosarcina spp. and Growth as Single Cells at Elevated Osmolarity. *Appl Environ Microbiol* 59: 3832-3839.

[31] Bonnete, F., Ebel, C., & Zaccai, G. (1993) Biophysical study of halophilic malate dehydrogenase in solution: revised subunit structure and solvent interactions of native and recombinant enzyme. *J Chem Soc Faraday Trans* 89: 2659-2666.

[32] Mevarech, M., & Neumann, E. (1977) Malate dehydrogenase isolated from extremely halophilic bacteria of the Dead Sea. 2. Effect of salt on the catalytic activity and structure. *Biochemistry* 16: 3786-3792.

[33] Zaccai, G., Bunick, G.J., & Eisenberg, H. (1986) Denaturation of a halophilic enzyme monitored by small-angle neutron scattering. *J Mol Biol* 192: 155-157.

[34] Hecht, K., Wrba, A., & Jaenicke, R. (1989) Catalytic properties of thermophilic lactate dehydrogenase and halophilic malate dehydrogenase at high temperature and low water activity. *Eur J Biochem* 183: 69-74.

[35] Ebel, C., Faou, P., Kernel, B., & Zaccai, G. (1999) Relative role of anions and cations in the stabilization of halophilic malate dehydrogenase. *Biochemistry* 38: 9039-9047.

[36] Zaccai, G., Cendrin, F., Haik, Y., Borochov, N., & Eisenberg, H. (1989) Stabilization of halophilic malate dehydrogenase. *J Mol Biol* 208: 491-500.

[37] Ebel, C., Costenaro, L., Pascu, M., Faou, P., Kernel, B., Proust-De Martin, F., & Zaccai, G. (2002) Solvent interactions of halophilic malate dehydrogenase. *Biochemistry* 41: 13234-13244.

[38] Ebel, C., Guinet, F., Langowski, J., Urbanke, C., Gagnon, J., & Zaccai, G. (1992) Solution studies of elongation factor Tu from the extreme halophile *Halobacterium marismortui*. *J Mol Biol* 223: 361-371.

[39] Ebel, C., Altekar, W., Langowski, J., Urbanke, C., Forest, E., & Zaccai, G. (1995) Solution structure of glyceraldehyde-3-phosphate dehydrogenase from *Haloarcula vallismortis*. *Biophys Chem* 54: 219-227.

[40] Pundak, S., Aloni, H., & Eisenberg, H. (1981) Structure and activity of malate dehydrogenase from the extreme halophilic bacteria of the Dead Sea. 2. Inactivation, dissociation and unfolding at NaCl concentrations below 2 M. Salt, salt concentration and temperature dependence of enzyme stability. *Eur J Biochem* 118: 471-477.

[41] Madern, D., & Zaccai, G. (1997) Stabilisation of halophilic malate dehydrogenase from *Haloarcula marismortui* by divalent cations - effects of temperature, water isotope, cofactor and pH. *Eur J Biochem* 249: 607-611.

[42] Cioni, P., & Strambini, G.B. (2002) Effect of heavy water on protein flexibility. *Biophys J* 82: 3246-3253.

[43] Dym, O., Mevarech, M., & Sussman, J.L. (1995) Structural features that stabilize halophilic malate dehydrogenase from an archaebacterium. *Science* 267: 1344-1346.

[44] Abad-Zapatero, C., Griffith, J.P., Sussman, J.L., & Rossmann, M.G. (1987) Refined crystal structure of dogfish M4 apo-lactate dehydrogenase. *J Mol Biol* 198: 445-467.

[45] Richard, S.B., Madern, D., Garcin, E., & Zaccai, G. (2000) Halophilic adaptation: novel solvent protein interactions observed in the 2.9 and 2.6 Å resolution structures of the wild type and a mutant of malate dehydrogenase from *Haloarcula marismortui. Biochemistry* 39: 992-1000.

[46] Kamekura, M., Seno, Y., & Dyall-Smith, M. (1996) Halolysin R4, a serine proteinase from the halophilic archaeon *Haloferax mediterranei*; gene cloning, expression and structural studies. *Biochim Biophys Acta* 1294: 159-167.

[47] Frolow, F., Harel, M., Sussman, J.L., Mevarech, M., & Shoham, M. (1996) Insights into protein adaptation to a saturated salt environment from the crystal structure of a halophilic 2Fe-2S ferredoxin. *Nat Struct Biol* 3: 452-457.

[48] Werber, M.M., & Mevarech, M. (1978) Purification and characterization of a highly acidic 2Fe-ferredoxin from *Halobacterium* of the Dead Sea. *Arch Biochem Biophys* 187: 447-456.

[49] Marg, B.L., Schweimer, K., Sticht, H., & Oesterhelt, D. (2005) A two-alpha-helix extra domain mediates the halophilic character of a plant-type ferredoxin from halophilic archaea. *Biochemistry* 44: 29-39.

[50] Bandyopadhyay, A.K., Krishnamoorthy, G., & Sonawat, H.M. (2001) Structural stabilization of [2Fe-2S] ferredoxin from *Halobacterium salinarum*. *Biochemistry* 40: 1284-1292.

[51] Elcock, A.H., & McCammon, J.A. (1998) Electrostatic contributions to the stability of halophilic proteins. *J Mol Biol* 280: 731-748.

[52] Besir, H., Zeth, K., Bracher, A., Heider, U., Ishibashi, M., Tokunaga, M., & Oesterhelt, D. (2005) Structure of a halophilic nucleoside diphosphate kinase from *Halobacterium salinarum*. *FEBS Lett* 579: 6595-6600.

[53] Pieper, U., Kapadia, G., Mevarech, M., & Herzberg, O. (1998) Structural features of halophilicity derived from the crystal structure of dihydrofolate reductase from the Dead Sea halophilic archaeon, *Haloferax volcanii*. *Structure* 6: 75-88.

[54] Britton, K.L., Stillman, T.J., Yip, K.S., Forterre, P., Engel, P.C., & Rice, D.W. (1998) Insights into the molecular basis of salt tolerance from the study of glutamate dehydrogenase from *Halobacterium salinarum*. *J Biol Chem* 273: 9023-9030.

[55] Lins, L., Thomas, A., & Brasseur, R. (2003) Analysis of accessible surface of residues in proteins. *Protein Sci* 12: 1406-1417.

[56] Ishibashi, M., Tokunaga, H., Hiratsuka, K., Yonezawa, Y., Tsurumaru, H., Arakawa, T., & Tokunaga, M. (2001) NaCl-activated nucleoside diphosphate kinase from extremely halophilic archaeon, *Halobacterium salinarum*, maintains native conformation without salt. *FEBS Lett* 493: 134-138.

[57] Polosina, Y.Y., Zamyatkin, D.F., Kostyukova, A.S., Filimonov, V.V., & Fedorov, O.V. (2002) Stability of *Natrialba magadii* NDP kinase: comparisons with other halophilic proteins. *Extremophiles* 6: 135-142.

[58] Hutcheon, G.W., Vasisht, N., & Bolhuis, A. (2005) Characterisation of a highly stable alpha-amylase from the halophilic archaeon *Haloarcula hispanica*. *Extremophiles* 9: 487-495.

[59] Blecher, O., Goldman, S., & Mevarech, M. (1993) High expression in *Escherichia coli* of the gene coding for dihydrofolate reductase of the extremely halophilic archaebacterium *Haloferax volcanii*. Reconstitution of the active enzyme and mutation studies. *Eur J Biochem* 216: 199-203.

[60] Bohm, G., & Jaenicke, R. (1994) A structure-based model for the halophilic adaptation of dihydrofolate reductase from *Halobacterium volcanii*. *Protein Eng* 7: 213-220.

[61] Bieger, B., Essen, L.O., & Oesterhelt, D. (2003) Crystal structure of halophilic dodecin: a novel, dodecameric flavin binding protein from *Halobacterium salinarum*. *Structure* 11: 375-385.

[62] Zeth, K., Offermann, S., Essen, L.O., & Oesterhelt, D. (2004) Iron-oxo clusters biomineralizing on protein surfaces: structural analysis of *Halobacterium salinarum* DpsA in its low- and high-iron states. *Proc Natl Acad Sci USA* 101: 13780-13785.

[63] Sivakumar, N., Li, N., Tang, J.W., Patel, B.K., & Swaminathan, K. (2006) Crystal structure of AmyA lacks acidic surface and provide insights into protein stability at poly-extreme condition. *FEBS Lett* 580: 2646-2652.

[64] Cayol, J.L., Ollivier, B., Patel, B.K., Prensier, G., Guezennec, J., & Garcia, J.L. (1994) Isolation and characterization of *Halothermothrix orenii* gen. nov., sp. nov., a halophilic, thermophilic, fermentative, strictly anaerobic bacterium. *Int J Syst Bacteriol* 44: 534-540.

[65] Mijts, B.N., & Patel, B.K. (2002) Cloning, sequencing and expression of an alpha-amylase gene, *amyA*, from the thermophilic halophile *Halothermothrix orenii* and purification and biochemical characterization of the recombinant enzyme. *Microbiol* 148: 2343-2349.

[66] Kbir-Ariguib, N., Chehimi, D.B.H., & Zayani, L. (2001) Treatment of Tunesian salt lakes using solubility phase diagrams. *Pure Appl Chem* 73: 761-770.

[67] Mijts, B.N., & Patel, B.K. (2001) Random sequence analysis of genomic DNA of an anaerobic, thermophilic, halophilic bacterium, *Halothermothrix orenii*. *Extremophiles* 5: 61-69.

[68] Ring, G., & Eichler, J. (2004) Membrane binding of ribosomes occurs at SecYE-based sites in the archaea *Haloferax volcanii*. *J Mol Biol* 336: 997-1010.

[69] Luecke, H., Schobert, B., Richter, H.T., Cartailler, J.P., & Lanyi, J.K. (1999) Structure of bacteriorhodopsin at 1.55 A resolution. *J Mol Biol* 291: 899-911.

[70] Dunn, R., McCoy, J., Simsek, M., Majumdar, A., Chang, S.H., Rajbhandary, U.L., & Khorana, H.G. (1981) The bacteriorhodopsin gene. *Proc Natl Acad Sci U S A* 78: 6744-6748.

[71] Haupts, U., Tittor, J., & Oesterhelt, D. (1999) Closing in on bacteriorhodopsin: progress in understanding the molecule. *Annu Rev Biophys Biomol Struct* 28: 367-399.

[72] Prado, M.M., Prado-Cabrero, A., Fernandez-Martin, R., & Avalos, J. (2004) A gene of the opsin family in the carotenoid gene cluster of Fusarium fujikuroi. *Curr Genet* 46: 47-58.

[73] Lind, C., Hojeberg, B., & Khorana, H.G. (1981) Reconstitution of delipidated bacteriorhodopsin with endogenous polar lipids. *J Biol Chem* 256: 8298-8305.

[74] Sternberg, B., L'Hostis, C., Whiteway, C.A., & Watts, A. (1992) The essential role of specific Halobacterium halobium polar lipids in 2D-array formation of bacteriorhodopsin. *Biochim Biophys Acta* 1108: 21-30.

[75] Mukhopadhyay, A.K., Dracheva, S., Bose, S., & Hendler, R.W. (1996) Control of the integral membrane proton pump, bacteriorhodopsin, by purple membrane lipids of Halobacterium halobium. *Biochemistry* 35: 9245-9252.

[76] Kottemann, M., Kish, A., Iloanusi, C., Bjork, S., & DiRuggiero, J. (2005) Physiological responses of the halophilic archaeon *Halobacterium* sp. strain NRC1 to desiccation and gamma irradiation. *Extremophiles* 9: 219-227.

[77] Vreeland, R.H., Rosenzweig, W.D., & Powers, D.W. (2000) Isolation of a 250 million-year-old halotolerant bacterium from a primary salt crystal. *Nature* 407: 897-900.

[78] Vreeland, R.H., Rosenzweig, W.D., Lowenstein, T., Satterfield, C., & Ventosa, A. (2006) Fatty acid and DNA analyses of Permian bacteria isolated from ancient salt crystals reveal differences with their modern relatives. *Extremophiles* 10: 71-78.

[79] Tehei, M., Franzetti, B., Maurel, M.C., Vergne, J., Hountondji, C., & Zaccai, G. (2002) The search for traces of life: the protective effect of salt on biological macromolecules. *Extremophiles* 6: 427-430.

[80] Sellek, G.A., & Chaudhuri, J.B. (1999) Biocatalysis in organic media using enzymes from extremophiles. *Enzyme Microb Technol.* 25: 471-482.

[81] Grant, W.D. (2004) Life at low water activity. *Philos Trans R Soc Lond B Biol Sci* 359: 1249-1266.

[82] Lanyi, J.K., & Stevenson, J. (1969) Effect of salts and organic solvents on the activity of *Halobacterium cutirubrum* catalase. *J Bacteriol* 98: 611-616.

[83] Ishibashi, M., Sakashita, K., Tokunaga, H., Arakawa, T., & Tokunaga, M. (2003) Activation of halophilic nucleoside diphosphate kinase by a non-ionic osmolyte, trimethylamine N-oxide. *J Protein Chem* 22: 345-351.

[84] Ryu, K., Kim, J., & Dordick, J.S. (1994) Catalytic properties and potential of an extracellular protease from an extreme halophile. *Enzyme Microb Technol* 16: 266-275.

[85] Kim, J., & Dordick, J.S. (1997) Unusual salt and solvent dependence of a protease from an extreme halophile. *Biotechnol Bioeng* 55: 471-479.

[86] Fukushima, T., Mizuki, T., Echigo, A., Inoue, A., & Usami, R. (2005) Organic solvent tolerance of halophilic alpha-amylase from a haloarchaeon, *Haloarcula* sp. strain S-1. *Extremophiles* 9: 85-90.

[87] Melo, E.P., Aires-Barros, M.R., & Cabral, J.M. (2001) Reverse micelles and protein biotechnology. *Biotechnol Annu Rev* 7: 87-129.

[88] Bonet, M.J., LLorca, F.I., & Cadenas, E. (1991) Purification and some properties of an atypical alkaline p-nitrophenylphosphatase activity from *Halobacterium halobium*. *Int J Biochem* 23: 1445-1451.

[89] Marhuenda-Egea, F.C., Piera-Velazquez, S., Cadenas, C., & Cadenas, E. (2002) An extreme halophilic enzyme active at low salt in reversed micelles. *J Biotechnol* 93: 159-164.

[90] Piera-Velazquez, S., Marhuenda-Egea, F., & Cadenas, E. (2002) Increased stability of malate dehydrogenase from *Halobacterium salinarum* at low salt concentration in reverse micelles. *Extremophiles* 6: 407-412.

[91] Marhuenda-Egea, F.C., & Bonete, M.J. (2002) Extreme halophilic enzymes in organic solvents. *Curr Opin Biotechnol* 13: 385-389.

[92] Khmelnitsky, Y.L., & Rich, J.O. (1999) Biocatalysis in nonaqueous solvents. *Curr Opin Chem Biol* 3: 47-53.

[93] Madern, D., Camacho, M., Rodriguez-Arnedo, A., Bonete, M.J., & Zaccai, G. (2004) Salt-dependent studies of NADP-dependent isocitrate dehydrogenase from the halophilic archaeon *Haloferax volcanii*. *Extremophiles* 8: 377-384.

[94] Kuntz, I.D. (1971) Hydration of macromolecules. IV. Polypeptide conformation in frozen solutions. *J Am Chem Soc* 93: 516-518.

[95] Mevarech, M., Frolow, F., & Gloss, L.M. (2000) Halophilic enzymes: proteins with a grain of salt. *Biophys Chem* 86: 155-164.

[96] Kobayashi, T., Kanai, H., Hayashi, T., Akiba, T., Akaboshi, R., & Horikoshi, K. (1992) Haloalkaliphilic maltotriose-forming alpha-amylase from the archaebacterium *Natronococcus* sp. strain Ah-36. *J Bacteriol* 174: 3439-3444.

[97] Ben-Amotz, A., & Avron, M. (1973) The role of glycerol in the osmotic regulation of the halophilic alga *Dunaliella parva*. *Plant Physiol* 51: 875-878.

[98] Zelazny, A.M., Shaish, A., & Pick, U. (1995) Plasma membrane sterols are essential for sensing osmotic changes in the halotolerant alga *Dunaliella*. *Plant Physiol* 109: 1395-1403.

[99] Schobert, B., & Lanyi, J.K. (1982) Halorhodopsin is a light-driven chloride pump. *J Biol Chem* 257: 10306-10313.

[100] Duschl, A., & Wagner, G. (1986) Primary and secondary chloride transport in *Halobacterium halobium*. *J Bacteriol* 168: 548-552.

[101] van de Vossenberg, J.L., Driessen, A.J., Grant, W.D., & Konings, W.N. (1999) Lipid membranes from halophilic and alkali-halophilic archaea have a low H+ and Na+ permeability at high salt concentration. *Extremophiles* 3: 253-257.

[102] Tenchov, B., Vescio, E.M., Sprott, G.D., Zeidel, M.L., & Mathai, J.C. (2006) Salt tolerance of archaeal extremely halophilic lipid membranes. *J Biol Chem* 281: 10016-10023.

[103] Corcelli, A., Lattanzio, V.M., Mascolo, G., Babudri, F., Oren, A., & Kates, M. (2004) Novel sulfonolipid in the extremely halophilic bacterium Salinibacter ruber. *Appl Environ Microbiol* 70: 6678-6685.

[104] Bolhuis, A. (2004) The archaeal Sec-dependent protein translocation pathway. *Philos Trans R Soc Lond B Biol Sci* 359: 919-927.

[105] Robinson, C., & Bolhuis, A. (2004) Tat-dependent protein targeting in prokaryotes and chloroplasts. *Biochim Biophys Acta* 1694: 135-147.

[106] Alder, N.N., & Theg, S.M. (2003) Energetics of protein transport across biological membranes. a study of the thylakoid delta pH-dependent/cpTat pathway. *Cell* 112: 231-242.

[107] Bolhuis, A. (2002) Protein transport in the halophilic archaeon *Halobacterium* sp. NRC-1: a major role for the twin-arginine translocation pathway? *Microbiology* 148: 3335-3346.

[108] Robinson, C., & Bolhuis, A. (2001) Protein targeting by the twin-arginine translocation pathway. *Nat Rev Mol Cell Biol* 2: 350-356.

[109] Chaddock, A.M., Mant, A., Karnauchov, I., Brink, S., Herrmann, R.G., Klosgen, R.B., & Robinson, C. (1995) A new type of signal peptide: central role of a twin-arginine motif in transfer signals for the delta pH-dependent thylakoidal protein translocase. *EMBO J* 14: 2715-2722.

[110] Rose, R.W., Brüser, T., Kissinger, J.C., & Pohlschröder, M. (2002) Adaptation of protein secretion to extremely high-salt conditions by extensive use of the twin-arginine translocation pathway. *Mol Microbiol* 45: 943-950.

[111] Dilks, K., Gimenez, M.I., & Pohlschroder, M. (2005) Genetic and biochemical analysis of the Twin-arginine translocation pathway in halophilic archaea. *J Bacteriol* 187: 8104-8113.
[112] Thomas, J.R., & Bolhuis, A. (2006) The *tatC* gene cluster is essential for viability in halophilic archaea. *FEMS Microbiol Lett* 256: 44-49.
[113] Bassford, P.J., Jr., Silhavy, T.J., & Beckwith, J.R. (1979) Use of gene fusion to study secretion of maltose-binding protein into *Escherichia coli* periplasm. *J Bacteriol* 139: 19-31.
[114] Pettersen, E.F., Goddard, T.D., Huang, C.C., Couch, G.S., Greenblatt, D.M., Meng, E.C., & Ferrin, T.E. (2004) UCSF Chimera - a visualization system for exploratory research and analysis. *J Comput Chem* 25: 1605-1612.

In: Ribosomal Proteins and Protein Engineering… ISBN 978-1-60741-005-8
Editors: V. Ortendhal and H. Salchow, pp. 175-210

Chapter 11

ALKALINE ADAPTATION OF PROTEINS*

Tsuyoshi Shirai[1,†], Tohru Kobayashi[2], Susumu Ito[3] and Koki Horikoshi[2]

[1] Department of Bioscience, Nagahama Institute of Bio-science and Technology, and JST-BIRD, 1266 Tamura, Nagahama 526-0829, Japan

[2] Japan Agency for Marine-Earth Science and Technology (JAMSTEC), 2-15 Natsushima, Yokosuka 237-0061, Japan

[3] Creative Research Initiative (CRIS), Hokkaido University, North-21, West-10, Sapporo 001-0021, Japan

ABSTRACT

Alkaliphilic microorganisms (alkaliphiles) produce a variety of proteins, mostly enzymes, that are stable and optimally active at alkaline pH. The alkaline enzymes represent most of the currently known proteins adapted to high pH. Alkaline enzymes are widely used for industrial or commercial purposes, for example, as additives to laundry detergents. They have been improved in the course of evolution to be resistant to low proton density conditions that would cause protein denaturation or dissociation. In order to elucidate the mechanism of high-pH adaptation, the alkaline enzymes have been characterized and their structures have been compared with homologous non-alkaliphilic proteins. These comparative studies, together with several experimental and theoretical studies, have suggested three major strategies for alkaline adaptation: p*K*a modulation, Asp+Glu gain, and Asp+Lys loss–Glu+Arg gain. The history of representative alkaline enzymes, proposed strategies for high-pH adaptation, and protein engineering studies performed in order to adapt proteins to alkaline conditions will be reviewed in this chapter.

---

* A version of this chapter was also published in Protein Adaptation in Extremophiles, edited by K. S. Siddiqui and T. Thomas published by Nova Science Publishers, Inc. It was submitted for appropriate modifications in an effort to encourage wider dissemination of research.

† Corresponding author. E-mail: t_shirai@nagahama-i-bio.ac.jp.

INTRODUCTION

Alkaliphiles are defined as microorganisms that grow optimally at pH 9 or higher [1, 2]. The alkaliphiles include various species of bacteria, archaea, cyanobacteria, yeast, and filamentous fungi. The major clusters of alkaliphiles are found in the bacterial genera *Bacillus*, *Micrococcus*, and *Pseudomonas* [2]. Although alkaliphiles were used in the ancient indigo fermentation industry, the biological study of the organisms did not begin until the 1960s. Since then, many alkaliphiles have been isolated from alkaline environments such as soda lakes or soda deserts. However, it is also known that near-neutral pH environments can serve as natural habitats for alkaliphiles [2].

Most alkaliphilic proteins characterized to date are extracellular enzymes, which are secreted by cells for nutrient-scavenging purposes [2]. These enzymes show optimal activity at pH between 8 and 13, and are called alkaline or high-alkaline enzymes. In early studies, the alkaliphiles were found to produce a number of extracellular hydrolases such as protease, amylase, pullulanase, and xylanase [3-6].

The alkaliphiles are currently attracting attention mainly for the exploitation of their protein products [7-15]. Some industrial enzymes must work under alkaline conditions. For example, the alkaline enzymes have a long history as detergent additives [7-14], which have been commercialized for more than 20 years. Alkaline proteases, lipases, cellulases, and amylases are used in laundry detergents to digest stains or condition fabrics. The ability to function in alkaline solutions containing ingredients such as surfactants, chelating reagents, and oxidative bleaches is required of detergent enzymes. Therefore, most of the currently used enzymes have been obtained from alkaliphiles [11-13].

The current knowledge of protein adaptation to high pH has been mainly obtained from analyses of alkaline enzymes. Although alkaline adaptation is one of the most important properties for utilizing proteins in industry, the underlying molecular mechanisms of the adaptation have not been studied in great details as well as their adaptation to the other extreme conditions such as high temperature. In this chapter, the current understanding of the molecular properties and adaptation mechanisms of high-pH–adapted proteins will be reviewed, focusing mainly on the previous studies on alkaline enzymes.

ALKALINE ENZYMES AND THEIR STRUCTURES

Some of alkaline enzymes, mainly hydrolases, have been studied in detail. Table I summarizes the known three-dimensional structures of alkaline enzymes, which are the current representatives of high-pH–adapted proteins. They are the high-alkaline proteases, alkaline amylases, alkaline cellulases, alkaline xylanase, alkaline galactanase, high-alkaline pectate lyase, and alkaline phosphoserine aminotransferase. The following sections will discuss these enzymes in greater detail.

Table I. Three-dimensional structures of alkaline enzymes

Name	Origin	Protein Family	Protein Database Brookhaven (PDB) codes	Ref.*[1]
Protease				
Subtilisin BL	*Bacillus lentus*	Subtilisin	1st3	[16]
Savinase	*Bacillus lentus*	Subtilisin	1tk2, 1ndq, 1ndu, 1svn	[17]
Subtilisin PB92	*Bacillus alcalophilus* PB92	Subtilisin	1iav, 1ah2	[18, 19]
M-protease	*Bacillus* sp. KSM-K16	Subtilisin	1mpt, 1wsd	[20, 21]
α-Amylase				
□□□	*B. halmapalus*	GH13	1w9x, 2gjp, 2gjr	[22, 23]
□my□□□	*Bacillus* sp. KSM-K38	GH13	1ud2, 1ud3, 1ud4, 1ud5, 1ud6, 1ud8	[24]
AmyK	*Bacillus* sp. KSM-1378	GH13	2die	[25]
G6-amylase	*Bacillus* sp. 707	GH13	1wp6, 1wpc, 2d3l, 2d3n	[26, 27]
CGTase	*Bacillus* sp. 1011	GH13	1ukq, 1uks, 1ukt, 1v3k, 1v3l, 1v3m, 1v3j, 1d7f, 1ded, 1i75, 1pam	[28-31]
Cellulase				
Cel5A	*Bacillus agaradherans*	GH5	1h11, 1h2j, 1ocq, 1w3k, 1w3l, 1a3h, 1h5v, 1hf6, 3a3h, 5a3h, 6a3h, 7a3h	[32-35]
CelK	*Bacillus* sp. KSM-635	GH5	1g01, 1g0c	[36]
Cellulase 103	*Bacillus* sp. CBS 670.93	GH5	1lf1	[37]
CBM28	*Bacillus* sp. 1139	CBM28	1uww	[38]
Xylanase	*Bacillus* sp. NG-27	GH10	2f8q, 2fgl	[39]
Galactanase	*Humicola insolens*	GH53	1hjq	[40]
Pectate lyase	*Bacillus* sp. KSM-P15	PL3	1ee6	[41]
Phosphoserine aminotransferase	*Bacillus alcalophilus*	PLP enzyme	1w23	[42]
Phosphoserine aminotransferase	*Bacillus circulans*	PLP enzyme	1w3u	[42]

[1] Only selected references are shown.

High-Alkaline Proteases

Proteases are classified into serine, aspartate, cysteine, and metallo proteases, based on their catalytic site residues or co-factors. Serine proteases are a major class of the proteases that are characterized by conserved Ser, Asp, and His residues known as the catalytic triad [43]. This class includes the subtilase superfamily, which is further classified into families A to F [44]. Subtilisins belong to family A and are mainly found in *Bacillus* strains. Subtilisin was the first enzyme isolated from an alkaliphilic, and the majority of the known alkaliphile proteases are classified into this family [1, 12, 44]. The major activity of subtilisin is the hydrolysis of peptide bonds (EC 3.4.21.14). Subtilisins often show optimal activity at pH 9-10, even if they are derived from non-alkaliphilic organisms [43].

In 1971, the alkaliphilic *Bacillus* sp. strain 221 was found to secrete a protease that showed an optimal pH of 11.5 [3]. The alkaliphilic subtilisin genes of elastase YaB (*Bacillus* sp. YaB) and subtilisin PB92 (*Bacillus alcalophilus* PB92) were cloned in 1989 and 1991, respectively [45, 46]. Since then, a number of alkaline proteases of this family were isolated and characterized from alkaliphilic *Bacillus* and other species. The subtilisins from alkaliphiles show a rather high optimal pH for enzymatic activity compared to those of non-alkaliphiles. For example, Savinase, subtilisin PB92, and elastase YaB have pH optima between 10 and 12 [17, 45, 46]. Subtilisin AprM, subtilisin AH101, and M-protease demonstrate even higher optimal pH, with values of greater than 12 [47-50]. Therefore, these subtilisins are often called high-alkaline proteases, in contrast to the alkaline proteases of non-alkaliphiles. High-alkaline proteases were also isolated from, for example, *Bacillus* sp. JB-99, *Bacillus clausii* GMBAE 42, *Stenotrophomonas maltophilia* S-1, and *Thermoactinomyces* sp. strains E79 and HS682 [51-55]. Interestingly, the last one was identified as an intracellular high-alkaline protease [52].

Both the alkaline and high-alkaline proteases possess a large potential for industrial applications [8-13]. Alkaline and high-alkaline proteases comprise 30-40% of all worldwide enzyme products for commercial uses. They are used in laundry and dishwashing detergents to digest protein stains like blood, milk, or gravy sauces, by utilizing their high-pH resistant characters. They are also used for the cleaning of membranes, surgical instruments, and contact lenses [12]. Many of the high-alkaline proteases are known by their trade names, such as Savinase (Novo Nordics), Maxacal (subtilisin PB92, Genencor), KAP (*Bacillus* sp. KSM-K16, Kao), and Esperase (subtilisin BL, Novo Nordics) [13].

The crystal structure of the high-alkaline protease of subtilisin BL from *B. alcalophilus* was determined in 1990 to a resolution of 2.4 Å [56, 57]. Since then, the structures of subtilisin BL, Savinase, subtilisin PB92, and M-protease have also been reported [16-21]. Figure 1a shows the three-dimensional structure of a high-alkaline protease, M-protease, determined to a 1.5 Å resolution (PDB code 1wsd) [21]. Subtilisins typically have a molecular weight of ~30 kDa, and fold into a conserved structure characterized by layers of the central α-helix, parallel β-sheet, and outer helices [43, 44]. This fold is named the subtilisin-like fold [58]. The conserved catalytic sites Asp32, His64, and Ser221 are close to the N-terminal of the central α-helix. The Asp-His pair and Ser are thought to be the general acid-base and nucleophilic catalysts, respectively [43, 44].

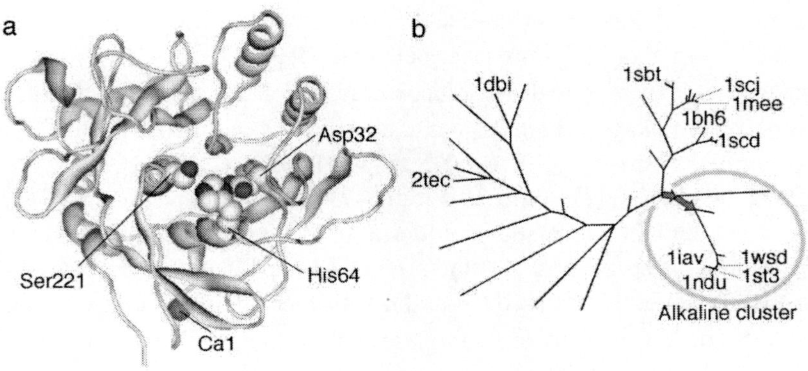

Figure 1. **a.** Structure of high-alkaline protease, M-protease (1msdA). The catalytic site residues (triad) and Ca^{2+} ion (Ca1) are shown in space-filling models. The figures were prepared by using MOE (Ryouka Systems Inc.). **b.** Molecular phylogeny of the high-alkaline and related proteases. Only those with known three-dimensional structures are shown with the PDB codes. The arrow indicates the alkaline adaptation branch used for analysis of evolutionary process (see "Alkaline-adaptation strategies of proteins"). Two consecutive arrows mean the two branches were used for analysis because a single branch was not reliable enough.

Based on comparisons of high-alkaline and alkaline protease structures, several potential mechanisms for high-pH adaptation have been proposed. The crystal structure analysis of subtilisin PB92 showed a loss of Asp and Lys residues and a gain of Arg residues when compared to its non-alkaliphilc counterparts. This would result in an alkaline shift of the isoelectric point (pI) of the protein and could be an important factor in the alkaline adaptation strategy [18]. The PB92 and Savinase structure analyses further showed that the number of Tyr residues, which would be negatively charged at the optimal pH of these enzymes, was decreased in high-alkaline proteases [17, 18]. A similar strategy of charge modulation caused by a decrease in Asp+Glu content and the replacement of Lys with Arg residues, was also identified in the comparison of the amino acid contents of the high-alkaline proteases AprM, PB92, and elastase YaB with those of alkaline proteases [59]. Structure and phylogeny analyses of M-protease also supported this hypothesis [21]. Because most Arg residues have a higher p*K*a than Lys, replacements of Lys with Arg may help to retain positive charges on the protein molecule under high-alkaline conditions. These observations suggest that a shift towards higher pI values is an important adaptation process in high pH environments.

Alkaline Amylase (α-amylase, Pullulanase, G6-amylase and Cyclomaltodextrin Glucanotransferase)

The α-amylase from alkaliphilic *Bacillus* sp. A-40-2 was discovered in 1971 [4]. α-Amylase (1,4-α-D-glucan glucanohydrolase, EC 3.2.1.1) hydrolyzes α1-4 glucoside linkages of amylose [14]. Alkaline α-amylases from other alkaliphiles, such as *Bacillus* sp. strains TS-23, GM8901, KMS-1378, and KMS-K38, *Bacillus coagulans* CUMC 512, *Bacillus licheniformis* CUMC 305, *Anaerobranca gottschalkii*, and *Bacillus halmapalus* [60-65] have also been characterized. Glucohydrolases are classified into more than 100 families according to their amino acid sequences [66, 67]. The α-amylases are categorized into six different families, among which GH13, GH14, and GH15 are the most numerous. The catalytic domains of the known alkaline α-amylases belong to GH13.

Pullulanase (α-dextrin endo-1,6-α-glucosidase, EC 3.2.1.41) is a variant of α-amylase that hydrolyzes α1-6 linkages of amylopectin, dextrin, and pullulan. The alkaliphiles, namely, *Bacillus* sp. strains KSM-1378, XAL601, KSM-1876, S-1, and 202-1 were reported to produce extracellular alkaline pullulanases [5, 68-73]. The pullulanases from *Bacillus* sp. S-1 and 202-1 showed optimal enzymatic activities at pH values between 8.5 and 9.0 [5, 68]. The enzymes from *Bacillus* sp. KSM-1378 and XAL601 (AapT) had molecular weights over 200 kDa, and showed dual activities as α1-6 and α1-4 glucosidases [69, 70]. Subsequently they were found to be fusion proteins of α-amylase and pullulanase domains. The pH optima of the KSM-1378 enzymes for α-amylase and pullulanase activities were 8.5 and 9.0, respectively [70]. The alkaline pullulanases and pullulanase domains of the fusion proteins also belong to the GH13 family.

G6-amylase (maltohexaose-producing amylase, EC 3.2.1.98) hydrolyzes α1-4 linkages of amylose, as α-amylase does, but it predominantly produces maltohexaose as a product. An alkaline G6-amylase was isolated from the alkaliphilic *Bacillus* sp. 707, and this enzyme was also classified as a member of the GH13 family [74].

CGTase (cyclodextrin glycosyltransferase, EC 2.4.1.19) produces cyclodextrin from starch by coupling α1-4 hydrolysis and transglycosylation reactions. A *Bacillus* species was found to produce CGTase with an optimal pH at 8.0 in 1977, and similar enzymes from *Bacillus agaradhaerens* LS-3C and *Bacillus* sp. 1011 have also been characterized [75-77]. The alkaline CGTases are distant homologues of α-amylase, but also belong to the GH13 family.

α-Amylases and related enzymes are widely used, for example in the food industry for starch liquefaction, brewing, and flour improvement [14, 78, 79]. G6-amylases are especially useful in starch processing because they specifically produce maltoligosaccharides. CGTases are utilized in the food, chemical, and pharmaceutical industries to produce cyclodextrin, which can be used as a molecular capsule in drug delivery system. For example, the alkaline CGTase of *Bacillus* sp. 38-2 was applied for the large-scale production of cyclodextrin, and a ~200 times lower production cost was achieved [80]. The α-amylases are also used in laundry and dishwashing detergents to digest starch stains [78]. The AmyK from *Bacillus* sp. KSM-1378 shows significantly higher activity against starch, amylose, and glycogen at pH 9.0 than non-alkaliphilic enzymes and it has also been improved by protein engineering to be used as a detergent additive [81-84].

The crystal structures of alkaline α-amylases have been determined for BHA (*B. halmapalus*), AmyK38 (*Bacillus* sp. KMS-K38), and AmyK (*Bacillus* sp. KMS-1378) [22-25]. Figure 2a shows the structure of AmyK (PDB core 2die) as an example [25]. The catalytic domain of GH13 α-amylase is a $(\beta/\alpha)_8$-barrel structure, which has the catalytic residues Asp236, Glu266, and Asp333 at the C-terminal end of the barrel β-sheet [14]. The residues Glu266, Asp333 and Asp236 are thought to be general acid-base and nucleophilic catalysts, respectively. Most α-amylases bind calcium ions, which are required for structural integrity (Ca1-3 in Figure 2a). These structural features are common among the three known alkaline α-amylases. However, AmyK38 represents an interesting exception as it has lost its three calcium-binding sites and hence has become highly resistant to chelating agents [24]. The three-dimensional structure of alkaline G6-amylase from *Bacillus* sp. 707 has been determined [26, 27], and it shared the structural features with most other α-amylases.

The crystal structure of the alkaline CGTase from *Bacillus* sp. 1011 is also available [28-31]. (PDB code 1i75, Figure 2c). The catalytic domain of this enzyme forms a $(\beta/\alpha)_8$-barrel that has the catalytic residues Asp229, Glu257, and Asp328, and two calcium-binding sites (Ca1 and Ca2 in Figure 2c).

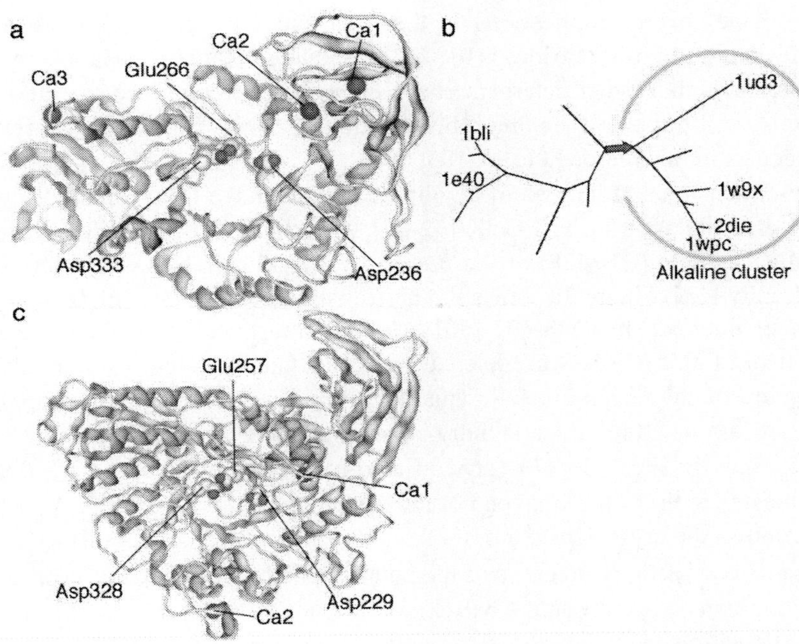

Figure 2. **a.** Structure of alkaline α-amylase AmyK (2dieA). The catalytic site residues and Ca^{2+} ions (Ca1-3) are shown in space-filling models. **b.** Molecular phylogeny of the alkaline and relative amylases (drawn as explained in the legend of Figure 1b). **c.** Structure of alkaline CGTase (1i75A). The catalytic site residues and Ca^{2+} ions (Ca1 and Ca2) are shown in space-filling models.

Alkaline Cellulase

Cellulose is the most abundant carbohydrate molecule on earth. Generally, cellulose is degraded by the synergistic action of three kinds of enzymes as follows. Endoglucanases randomly hydrolyze internal β-glycosidic bonds in carboxymethylcellulose, acid/alkaline-swollen cellulose, or cellulose in amorphous regions. Cellobiohydrolases remove cellobiose units from the non-reducing end of crystalline regions of cellulose. β-Glycosidases degrade cellobiose and cellotriose into glucose [85]. Most alkaliphilic cellulases are endoglucanases (β1-4 endo-glucanase, EC 3.2.1.4). The catalytic domains of cellulases are classified into 15 different families, of which GH5, GH9, GH12, and GH45 are the major families [66, 67].

The alkaliphilic *Bacillus* sp. strains 1139 and N-4 were found to produce extracellular cellulases, and the enzyme of N-4 showed an optimal pH for its activity around pH 9 [86, 87].

The basic pH preference of the alkaline enzyme is remarkable, because the cellulases of non-akaliphiles generally show optimum activity at pH 4–6. Cloning the two cellulase genes showed their similarity to family 5A (GH5) cellulases [88, 89]. Subsequently, alkaline cellulases have been characterized from, for example, *Streptomyces lividans*, *Bacillus* sp. strains KSM-635, KSM-64, KSM-19, KSM-520, KSM-S237, KSM-N252, and HSH-810, *Humicola insolens, Bacillus pumilus, Bacillus circulans* KSM-N257, and *Bacillus sphaericus* JS1 [90-100]. The majority of the alkaline cellulases so far characterized belong to the GH5 family with an exception being the one from *B. circulans* KSM-N257, which belongs to GH8 [98].

Cellulases have several applications in the food and fuel industries because they can convert cellulose into oligosaccharides [101, 102]. Alkaline cellulases have also been used in laundry detergents. Unlike other detergent enzymes, which are employed to digest stains, the cellulases release soil trapped in an amorphous region of cotton fibers by their reaction with cellulose molecules in this region [11, 91, 103, 104].

The three-dimensional structures of alkaline cellulases have been reported for Cel5A of *B. agaradherans*, cellulase 103, and CelK [32-37]. Additionally, the structure of the isolated cellulose-binding module (CBM28) of the *Bacillus* sp. 1139 cellulase was determined by X-ray crystallography [38]. Figure 3a shows the crystal structure of CelK (PDB code 1g01), the GH5 cellulase of *Bacillus* sp. KMS-635 [36]. Although most cellulases are associated with a CBM, the CBM of this CelK was deleted via gene manipulation without affecting the alkaliphilic nature of the enzyme [104]. This engineered cellulase shows an optimal pH of 8.3, and is used as an additive of laundry detergents. GH5 cellulase has a $(\beta/\alpha)_8$-barrel structure. The catalytic sites, Glu373 (general acid-base catalyst) and Glu485 (nucleophilic catalyst), are located at the C-terminal end of the barrel β-sheet.

Figure 3c shows the crystal structure of alkaline cellulase N257 (unpublished result). This enzyme has an $(\alpha/\alpha)_6$-barrel structure, and the putative catalytic sites, Glu93 and Asp154, are located at the bottom of the cleft on the barrel.

Alkaline adaptation mechanisms of the cellulases were investigated by comparing the structures of alkaline and non-alkaline enzymes. Based on the sequence comparison of alkaline cellulase NK1 from *Bacillus* sp. N-4 with its close non-alkaline relative, BSC from *B. subtilis*, two residues were hypothesized to be responsible for the alkaliphilic nature of NK1. The two sites in NK1 were substituted by the corresponding amino acid of BSC by site-directed mutagenesis (S287N/A296S). As a result, the pH-activity profile of the mutant NK1 resembled that of BSC, proving the relevance of the two residues to the alkaline adaptation [105].

A comparison of the amino acid sequences of the alkaline cellullase from *Bacillus* sp. KSM-635 and its relatives revealed 9 (2 Asp, 2 Ala, Ser, Thr, Val, Glu, and Pro) and 18 (5 Lys, 4 Asp, 4 Asn, 3 Ser, Thr, and Gly) residue sites that were specifically conserved in alkaline and non-alkaline enzymes, respectively, which implies that the substitutions in these sites were responsible for the alkaline adaptation [103]. The evolutionary process of these enzymes was also analyzed by using the calculated ancestral sequences [36], and it was proposed that a decrease in Lys and Asp contents and an increase in Arg and His contents occurred during the alkaline adaptation process. As a consequence, the number of Lys-Asp ion pairs in the current alkaline cellulase is smaller than that in non-alkaline enzymes. However, the total number of ion pairs did not decrease, because the loss of Lys-Asp pairs was compensated by the gain of Arg-Asp/Glu pairs in the alkaline enzymes.

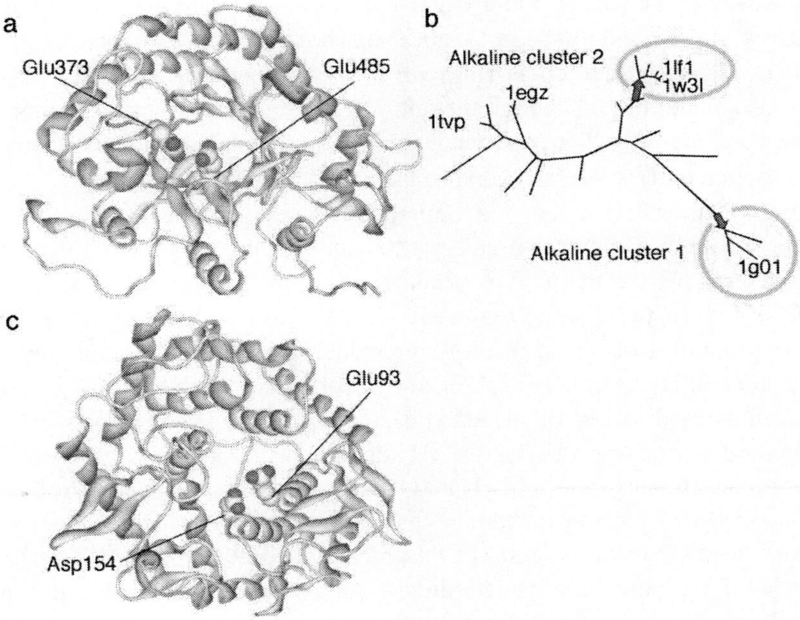

Figure 3. **a.** Structure of alkaline cellulase CelK (1g01A). The catalytic site residues are shown in space-filling models. **b.** Molecular phylogeny of the alkaline and related cellulases (drawn as explained in the legend of Figure 1b). **c.** Structure of alkaline cellulase N257. The putative catalytic site residues are shown in space-filling models.

Alkaline Xylanase

Xylanase (1,4 β-D-xylan xylanohydrolase, EC 3.2.1.8) hydrolyzes the β1-4 linkage of xylopyranoses in xylane (hemicellulose), which is, along with cellulose, a major biomass on earth [15, 106]. Xylanases are classified into 13 families, and most of them belong to GH10 or GH11 [66, 67]. Xylanases are used in the paper industry for bleaching wood pulp. The alkaliphilic and thermostable xylanases constitute an ecologically sound resource for this purpose because they can replace the toxic chlorine-containing chemicals traditionally used [15, 107, 108].

The first xylanase from alkaliphilic *Bacillus* sp. C-59-2, which was found in 1973, showed a broad pH optimum between 6 and 8 [6]. Alkaline xylanases with optimal pH in the basic range were reported for *Bacillus halodulans* C-125 (GH10), *Bacillus* sp. strains 41M-1 (xylanase J, GH11), N137 (xylanase A, GH10), NG-27 (GH10), *Geobacillus stearothermophilus* No. 236 (Xyn A, GH11) [109-113]. Alkaline xylanases were also found in *Bacillus* sp. strains W1, AR-009, and NCL87-6-10, *Micrococcus* sp. AR-135, *Streptomyces* sp. QG-11-3, *Cephalosporium* sp. strains NCL87-11-9 and RYM-202, *Aspergillus nidulans* KK-99, and *Staphylococcus* sp. SG-13, though their family identities are not known [114-122].

Figure 4a shows the crystal structure of GH11 xylanase Xyl11 from the alkaliphilic strain *B. agaradherans* (PDB code 1qh6), which exhibits an optimal pH of 5.5 for enzymatic activity [123]. The GH11 xylanase has a β-sandwich fold in a jellyroll topology. The catalytic sites Glu94 (nucleophilic catalyst) and Glu184 (general acid/base catalyst) are located in the cleft formed by the wound β-sheets. Figure 4b shows the three-dimensional structure of GH10 xylanase, XT-6 (PDB code 1hiz) of *B. stearothermophilus* T-6 [124, 125]. This bacterium is also alkaliphilic, although the optimal pH of XT-6 xylanase is reported to be 6.5. The putative catalytic sites Glu160 (nucleophilic catalyst) and Glu266 (general acid/base catalyst) are located at the C-terminal end of the barrel β-sheet.

It has been demonstrated that the optimal pH value of GH11 xylanases is generally dependent on the amino acids proximate to the catalytic center [126, 127]. In the enzymes with a near-neutral pH optimum, Asn occupies the 45th site (in the numbering system of 1qh6, Asn45 in Figure 4a) close to the active site, while the equivalent site in enzymes with an acidic pH optimum is occupied by an Asp residue. A point mutation that replaced the Asp site with Asn resulted in a basic shift of the optimal pH by about 3 units [128]. It demonstrated that modulation of a catalytic site's pKa by the presence or absence of a proximal negative charge was an effective pH adaptation strategy.

Another mode of adaptation of GH11 xylanases was also suggested from an analysis of the dipeptide contents of the enzymes [129]. The optimal pH values of twenty-three GH11 xylanases were predicted as a function of the fractional contents of dipeptides that was fitted with respect to the experimentally determined optimal pH values. The coefficients in the fitted function were interpreted as the relevance of each dipeptide for the pH optima. The coefficients were negative (disfavored) for YS and positive (favored) for SY, GR, MR, and KR, suggesting Arg residues play a role in the alkaline adaptation.

It was also proposed from a comparison of GH10 xylanases that the ratio of Asp+Glu to Arg+Lys contents on the molecular surface was higher for the alkaline enzyme BSX from *Bacillus* sp. NG-27 [39]. The excess of negative charge on the molecular surface is thought to protect the protein core from hydroxide ions that exist in high concentration under alkaline conditions.

Alkaline Galactanase

β-1,4-Galactanase (EC 3.2.1.89) hydrolyzes β-1,4 glycoside bonds in galactan and arabinogalactan, which are the polysaccharide branches of pectin [130]. The β-1,4-galactanases are classified into GH53 and GH5 [66, 67].

The alkaliphilic fungus *Humicolla insolens* produces a GH53 β-1,4-galactanse (HIGAL) which shows an optimal activity at pH 8.5 [40]. The structure of HIGAL (PDB code 1hjq) is shown in Figure 5. It forms a $(\beta/\alpha)_8$-barrel, and the putative catalytic sites, Glu135 (general acid/base catalyst) and Glu245 (nucleophilic catalyst), are at the C-terminal end of the barrel β-sheet.

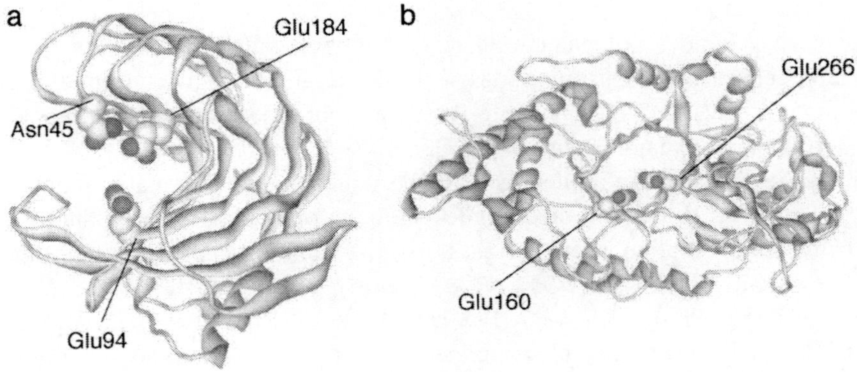

Figure 4. **a.** Structure of GH11 xylanase Xyl11 (1qh6A). The catalytic site and the proximal Asn45 residues are shown in space-filling models. **b.** Structure of GH10 xylanase XT-6 (1hizA). The catalytic site residues are shown in space-filling models.

A comparison of the amino acid content of HIGAL with that of the non-alkaliphilic homologue of *Aspergillus aculeatus* (AAGAL) revealed that the concentration of negatively charged residues around the catalytic center was lower for HIGAL. The difference in charge distribution was thought to be a part of the alkaline adaptation strategy of HIGAL. This hypothesis was experimentally confirmed by creating the point mutation D180N in AAGAL, which led to an upward shift in the pH optimum for the non-alkaline enzyme [40].

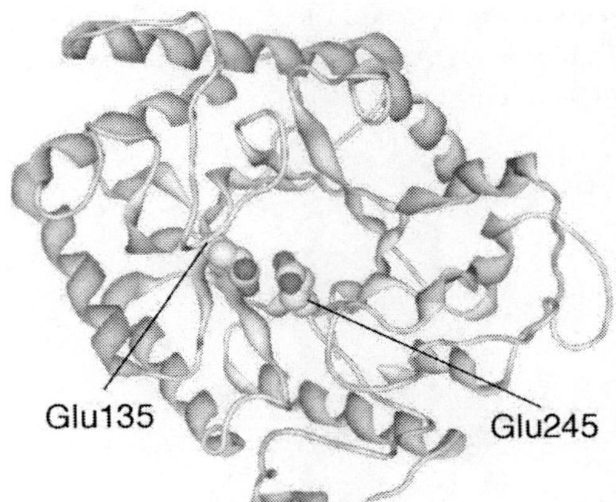

Figure 5. Structure of alkaline galactanase HIGAL (1hjqA). The catalytic site residues are shown in space-filling models.

High-Alkaline Pectate Lyase

Pectate lyase degrades polygalacturonate or low-methylated pectin by a β-elimination reaction and generates oligo-galacturonides with 4,5 unsaturated galacturonide at their non-reducing ends (EC 4.2.2.2) [131]. Pectate lyases are classified into six families with PL1, PL3, and PL10 being the major groups [66, 67].

Pectate lyases of non-alkaliphiles usually show an optimal enzymatic activity between pH 8 and 9. However, those from alkaliphiles show an optimal pH higher than 10, and are therefore called high-alkaline pectate lyases. High-alkaline pectate lyases of *Bacillus* sp. strains RK9, GIR-277, NT-2, NT-33, NT-82 (family unknown), KSM-P15 (Pel15, PL3; Pel15H, PL9; Pel15E, PL10), KSM-P7 (Pel-7, PL1), KSM-P103 (Pel-103, PL1), and P-4-N (Pel-4A and B, PL1) have been characterized [132-141]. The optimal pH for two high-alkaline enzymes, Pel15 and Pel15H from *Bacillus* sp. KSM-P15, are a remarkable 10.5 and 11.5, respectively [135, 139].

Some plant pathogenic bacteria produce pectate lyases that digest cell walls and cause soft-rot disease [131]. Because of their ability to digest plant bodies, high-alkaline and alkaline pectate lyases are often used in the paper industry. The alkaline pectate lyase from *Bacillus* sp. GIR 277 was used for retting of mitsumata (Oriental paper bush) bast [133]. The use of *Bacillus* sp. GIR 621-7 for digestion of vegetable fibers in waste treatment has also been reported [142].

The three-dimensional structure of high-alkaline pectate lyase has been reported for Pel15 from *Bacillus* sp. KSM-P15 (Figure 6, PDB code 1ee6) [41]. Pel15 has a simple β-helix fold and the putative catalytic sites Lys107, Lys129, and Arg132 are located on one of the three parallel β-sheets. Pectate lyases require calcium ions for the formation of their structure and for their catalytic activity. Pel15 has a conserved calcium-binding site (Ca1 in Figure 6), and an additional binding site (Ca2 in Figure 6) was found in the crystal grown under an alkaline condition. The second site appeared to stabilize the protein structure under high-pH conditions. An increased number of Arg residues in Pel15 was also noted and could play a role in the adaptation of this enzyme.

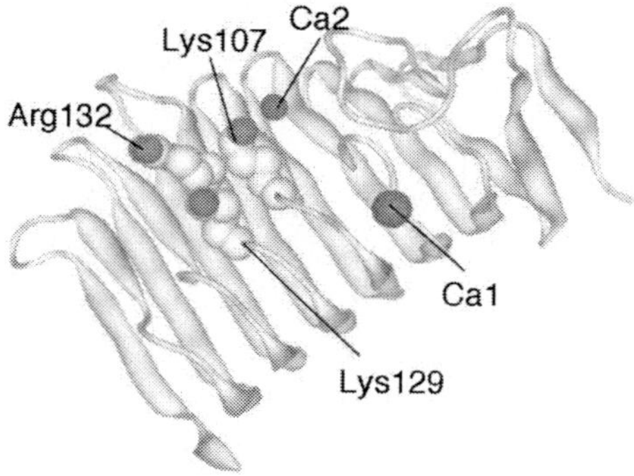

Figure 6. Structure of high-alkaline pectate lyase Pel15 (1ee6A). The catalytic site residues and Ca^{2+} ions (Ca1 and Ca2; Ca2 is a water molecule in this structure) are shown in space-filling models.

Alkaline Phosphoserine Aminotransferase

Phosphoserine aminotransferase (EC 2.6.1.52) catalyzes the conversion of 3-phoshohydroxypyruvate into L-phosphoserine [143]. X-ray crystallographic analyses of two alkaline phosphoserine aminotransferases from *B. alcalophilus* (BALC PSAT) and *B. circulans* subsp. *alkalophilus* (BCIR PSAT) at atomic resolution have been reported [42]. Although these enzymes are cytoplasmic proteins, BALC PSAT showed an optimal pH for enzymatic activity at pH 9.

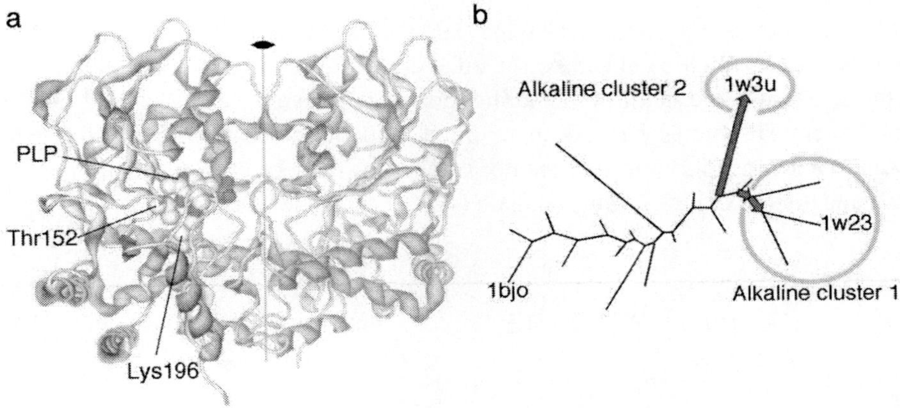

Figure 7. **a.** Structure of alkaline phosphoserine aminotransfearse BALC PSAT dimer (1w23A and B). The catalytic site residues, cofactor (PLP), and Thr152 residue are shown in space-filling models. **b.** Molecular phylogeny of the alkaline and related phosphoserine aminotransfearses (drawn as explained in the legend of Figure 1b).

The structure of BALC PSAT (PDB code 1w23) is shown in Figure 7a. This enzyme is a homodimer, and each subunit contains the cofactor pyridocal-5'-phosphate (PLP in Figure 7a) that is covalently bound to the catalytic site residue Lys196.

Comparisons of the alkaline enzyme structures with that of the non-alkalphilic homologue from *Escherichia coli* revealed increased numbers of H-bonds and negatively charged amino acids on the molecular surface, enhanced hydrophobic interactions at the dimer interface, decreased molecular surface area, and fewer ion pairs on the molecular surface [42]. Interestingly, disfavor of amino acids with a shorter side chain was also observed in the alkaline enzymes. In the catalytic center, a hydrogen bond between PLP and Thr152 was only observed in the alkaline enzymes (Thr152 in Figure 7a), and this additional hydrogen bond was thought to raise the p*K*a of the catalytic site, and shift the pH optimum upward.

Alkaline-Adaptation Strategies of Proteins

One of the ultimate goals of alkaline enzyme studies is the elucidation of molecular mechanisms that enable these proteins to function optimally under alkaline conditions. An insight into the high-pH adaptation mechanism would benefit the design of advanced industrial enzymes, as well as to understand the evolutionary strategies by which organisms find their ways into open niches under extreme conditions.

In high-pH solutions, non-adapted proteins often denature either reversibly or irreversibly [144-147]. Alkaline conditions also cause dissociation or aggregation of proteins in some cases [148, 149]. The denaturing effect of alkaline conditions are thought to be caused by the repulsion between excess negative charges, isolated charges in the interior of the molecule, or by the attenuation of hydrogen bonds due to low proton density. These effects might have the most critical impact on the catalytic site residues. Alkaline conditions are also known to cause deamination in Asp and Asn residues, which may result in isomerization and cleavage of peptides [150-152].

Most characterized alkaline enzymes have non-alkaliphilic homologues and hence the most feasible approach to elucidate the adaptation mechanism is a comparison between the structures of alkaline and non-alkaline enzymes. Several experimental and theoretical physico-chemical studies have been reported, and the current theories of protein high-pH adaptation appear to fall into three major categories, namely, p*K*a modulation, Asp+Glu gain (+DE), and Asp+Lys loss – Glu+Arg gain (-DK+ER) strategies (Figure 8).

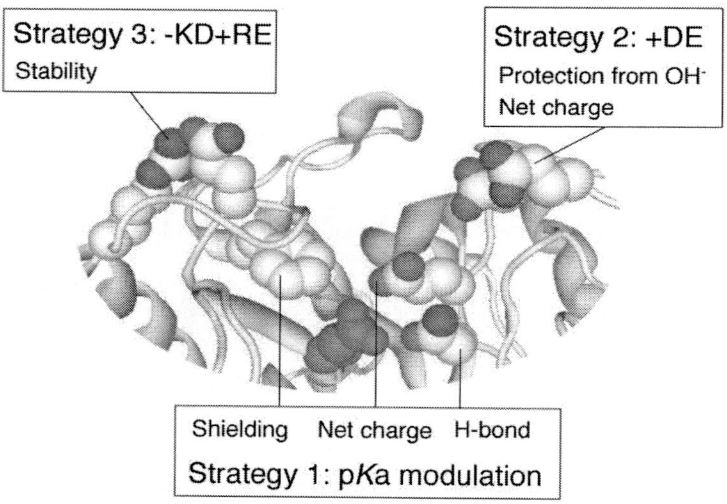

Figure 8. Schematics of the three major strategies of alkaline adaptation.

PKA MODULATION STRATEGY

The pH-dependent activity of an enzyme is typically presented as a bell-shaped curve, which consists of an acidic limb, a peak (optimal pH), and a basic limb. The shape of the profile is most likely a direct consequence of the p*K*a values for the catalytic residues. Therefore, a plausible strategy for the adaptation of enzyme function to alkaline conditions is modulation of the p*K*a of a catalytic residue toward a higher pH [153]. This can be achieved by either modifying the hydrogen bonds of catalytic residues, shielding the catalytic residues from the solvent, or changing the net charge of the protein molecule.

Modification of hydrogen bond of the catalytic residues: The formation of Hydrogen bond depends on the protonation status of acceptor and donor groups. Therefore, the hydrogen bond of an ionizable amino acid group can modulate the p*K*a value of the group. This concept has been extensively studied in the GH11 xylanases. As already mentioned, these xylanases with a relatively high pH optimum have an Asn proximal to the catalytic Glu (nucleophilic catalyst), while an Asp occupies the same site in the homologous enzymes with lower pH optima (Figure 4a) [126, 127]. The proximal Asp residue was found to have a quite low p*K*a value and formed a short hydrogen bond with the catalytic Glu in the high-resolution crystal structure [154]. This Asp (COOH) to Glu (COO^-) hydrogen bond favors the deprotonated state of the catalytic Glu, and therefore lowers its p*K*a. Replacement of the Asp residue with Asn should disrupt this hydrogen bond, and hence should raise the p*K*a of the catalytic Glu towards the optimal pH of the alkaline enzyme. Experimantal confirmation was achieved by the site-directed mutagenesis in GH11 xylanase. D37N, D37S, and D37E mutations of non-alkaline xylanase C from *Aspergillus kawachii* raised the optimal pH by 1.5-3.0 units [128]. In contrast, the N35D mutation of xylanase BCX from *B. circulans* lowered the pH optimum from 5.7 to 4.6 [154]. Similarly, the mutations N71D and N71S of *Bacillus pumilus* A-30 xylanase brought down the pH optimum from 6.7 to 6.3 with a substantial loss of activity at alkaline pH [155]. Upwards modulations of the optimal pH by disrupting the hydrogen bond of catalytic residue were also reported for the D180N mutation of *Aspergillus aculeatus* β1,4-galactanase, the S441G/A/C/H/D mutations of *Aspergillus awamori* glucoamylase, and the Y80F mutation of the *B. circulans* GH10 xylanase BCX [40, 156, 157]. These examples demonstrate that the p*K*a modulation strategy was adopted in the natural proteins.

However, hydrogen bond formation with a catalytic site did not always result in downward modulation of the optimal pH. The disruption of a hydrogen bond at the catalytic site with the H327N mutation in *Bacillus* sp. 1011 CGTase resulted in a low-pH shift of the basic limb of the activity curve, without affecting the optimum pH [77]. Also, the A224H mutation of *Trichoderma reesei* cellulase Cel7A caused an upward shift of optimal pH by the formation of a hydrogen bond [158]. For these cases, it was anticipated that the hydrogen bonds actually favored protonation of the catalytic site residues.

Shielding catalytic residues from solvent: Sufficient access to the solvent molecule is crucial for the deprotonation of ionizable groups. Under alkaline conditions, retaining protons at a catalytic site becomes a critical issue because of the low proton density. Theoretically, amino acid residues with a large side chain could help to prevent solvent molecules from accessing the positively charged group at a catalytic site.

For example, the mutants W169Y/F/L/H/V/A of *Achromobactor* protease I showed a downward shift of the p*K*a for His210 and the acidic limb of the pH activity profile [159]. His210 interacts with the catalytic residue Asp113, and the positive charge of protonated His210 is important for the enzymatic activity. Trp169 is close to the His210-Asp113 pair, and it shields the pair from the solvent. It was hypothesized that the smaller side chain at the Trp169 site allowed solvent access, and subsequent deprotonation of His210 occurred at lower pH values than in the native protein. This example demonstrates that solvent shielding can be used to raise the p*K*a of catalytic site residues.

Changing net charge: Either local or global low electrostatic potential (negative charges) can attract protons and increase the p*K*a of ionizable groups. Thus, introduction of the negatively charged amino acids, Asp and Glu, into a protein molecule will result in a shift toward a higher p*K*a of catalytic site residues, while introduction of positively charged Lys and Arg will produce the opposite effect. This concept was first proposed for and tested on *Bacillus amyloliquefaciens* subtilisin [153]. As predicted, the p*K*a of the catalytic site His64 was shifted downward by the single mutations D99S and E156S, and even further decreased when the same residues were mutated into Lys (D99K and E156K). The effects of the individual mutations were cumulative; the double mutants D99S/E156S and D99K/E156K demonstrated a rather low catalytic site p*K*a compared to the corresponding single mutants. These findings imply that introduction of Asp/Glu and removal of Lys/Arg would result in an alkaline adaptation of enzymes.

Because these mutations are not necessarily directed to the catalytic center, this adaptive strategy is highly feasible. Successful applications of this strategy to alkaline adaptation were reported, for example, for the S411D mutant of the *A. awamori* glucomaylase, the R112N mutant of the *B. circulans* xylanase, the K168N/N164D mutants of the *Neocallimastix patriciarum* xylanase, the D142N mutant of the *Manduca sexta* chitinase, and the N326D mutant of the chimera α-amylase Ba [156, 157, 160-162]. However, cases were also reported where the direction of the p*K*a/optimal pH shift was contrary to the prediction [162-164]. The extensive mutational study of α-amylase revealed that not only the net charge of the molecule but also other factors, such as the protein dynamics, significantly affected the p*K*a values of catalytic site residues, and made prediction difficult [162].

These strategies for modulating p*K*a have been extensively tested, and have led in many cases to successful protein engineering. However, there still remains a degree of uncertainty in the prediction, especially when mutations are directed toward catalytic centers. Secondary effects of the mutations proximal to the catalytic sites may also impact on the substrate affinity (K_m), the stability of the reaction intermediate, and the catalytic site conformation.

Asp+Glu Gain (+DE) Strategy

This strategy was initially derived from the comparative study of alkaline and non-alkaline phosphoserine aminotransferases [42]. The ratio of surface-exposed Asp+Glu to Lys+Arg was higher for the alkaline enzymes ({Asp+Glu}/{Lys+Arg} = 44/32 and 43/34 for BALC PSAT from *B. alcalophilus* and BCIR PSAT from *B. circulans,* respectively) compared to the *E. coli* homologue ({Asp+Glu}/{Lys+Arg} = 36/41). Glu was favored over Asp among the negatively charged residues, and the total numbers of charged amino acids were not significantly different between alkaline and non-alkaline enzymes. This Asp+Glu gain (+DE) strategy is also consistent with the p*K*a modulation strategy of catalytic sites (see above), because it changes the net charge of the protein molecule toward negative values.

This trend in amino acid composition was also observed for the GH10 alkaline xylanase BSX from *Bacillus* sp. NG-27 [39]. Comparisons of the models of the alkaline xylanase and those of thermophilic homologues demonstrated that the (Arg+Lys)/(Asp+Glu) ratio of residues having more than 30% solvent accessibility was much lower in the alkaline enzymes. This tendency was not evident for the residues buried inside the protein molecules.

The cell walls of alkaliphilic bacteria are known to contain more acidic polymers such as galacturonic acid, gluconic acid, glutamic/aspartic acids, and phosphoric acid than non-alkaliphilic bacteria [1, 2]. The negative charges of the polymers are thought to help absorb rare sodium (Na^+) and hydronium (H_3O^+) cations and to repel abundant hydroxide anions (OH^-) under high-pH conditions [39]. The increase in negative charges at the molecular surface was thought to be an alkaline adaptation strategy of the proteins by mimicking the character of the cell wall. It was also observed that the extracellular loops of the membrane proteins of alkaliphiles are rich in acidic amino acids [165, 166]. The negative charges on the outside of membrane proteins were thought to help retain protons at the membrane surface.

This +DE strategy resembles the adaptation strategy of proteins to high-salt environments [167, 168; see chapter by Bolhuis et al. in this book], and in fact, the alkaline xylanase BCX has also been described to be adapted to high-salt conditions [39]. The +DE strategy might be an adaptation process for both high-pH and high-salt environments.

Asp+Lys Loss – Glu+Arg Gain (-DK+ER) Strategy

This strategy was derived from the crystal structure analyses of the high-alkaline M-protease, the alkaline cellulase CelK, and the alkaline α-amylase AmyK, together with ancestral sequence evolutionary trace (ASET) analyses of their alkaline adaptation process [21, 25, 36].

The ASET analysis was based on a simple assumption: the evolutionary events that were responsible for the foundation of the alkaline enzymes should have occurred on the branches that separate the alkaline enzyme cluster from others (shown by arrows in the molecular phylogenies in Figures 1b, 2b, 3b, and 7b) [25]. The molecular phylogenies of alkaline and non-alkaline examples of proteases (10/47), cellulases (10/56), α-amylases (8/45), and phosphoserine aminotransferases (5/84) were used for this analysis (in parentheses are the numbers of alkaline/non-alkaline enzymes used). The changes in residue contents and patterns of amino acid replacement were inferred by directly comparing the calculated ancestral sequences before and after the alkaline enzyme's root branches.

Table II (ASET-balance column) shows the amino acid balances of the alkaline-adapted branches. Lys and Asp tended to decrease, and Arg, His, and Glu tended to increase in the alkaline-adaptation process. Some hydrophobic amino acid contents also showed changes. Leu was lost, and the relatively smaller hydrophobic residues Ala and Val appeared to be acquired in the adaptation process, which suggests that changes in hydrophobic interactions were involved in the alkaline adaptation.

The amino acid contents, intra-molecular interactions, and other molecular properties of known three-dimensional structures of alkaline enzymes were compared with those of non-alkaline enzymes. The differences between structures were assessed by Student's t-value [25] (see Tables II and III; a negative sign was added to a value if the average was smaller for the alkaline enzymes). The most significant difference between alkaline and non-alkaline enzymes was the increase in Leu content (see Table II, $p = 0.009$). Decreases in Phe and Ile contents were also significant (see Table II, $p = 0.037$ and 0.039, respectively). These results suggest that hydrophobic interactions were modified during the alkaline adaptation process. However, the Leu acquisition identified by the Student's t-test was contrary to the replacement pattern observed in the ASET analysis (ASET-balance in Table II). In addition, the number of non-polar contacts in the molecules was not significantly different before and after adaptation (non-polar contacts in Table III, $p = 0.550$). Thus, the role of hydrophobic interactions in alkaline adaptation was not confirmed from these analyses.

The second most discriminating feature was the decrease in lysine content of the alkaline enzymes (see Table II, $p = 0.026$). The number of Lys-Asp/Glu ion pairs (K-D/E in Table III, $p = 0.032$) also decreased in the alkaline enzymes, and the number of Arg-Glu ion pairs increased (R-E in Table III, $p = 0.030$). The loss of Lys residues and Lys-Asp/Glu ion pairs, and the acquisition of Arg-Glu ion pairs showed the same tendency for most of the four enzyme families investigated. These observations are also consistent with the results from the ASET analyses. It was therefore concluded that an ion pair remodeling disfavoring Lys and Asp and favoring Arg and Glu (-DK+ER strategy) was relevant for alkaline adaptation. This kind of ion pair remodeling might be important for stabilizing protein structures in alkaline solutions, because Lys loses a positive charge under high-pH conditions.

Table II. Amino acid contents of alkaline and non-alkaline enzymes

Amino acid	*t*-Value[1]	*p*-Value[2]	ASET-balance[3]
V	0.008	0.994	10
A	-1.980	0.060	7
H	1.375	0.182	7
R	0.858	0.400	6
G	1.301	0.206	6
E	1.538	0.138	6
Q	1.040	0.309	5
M	1.873	0.074	5
W	-0.974	0.340	2
F	-2.216	0.037	0
C	n.a.	n.a.	-1
T	-0.590	0.561	-2
Y	-1.358	0.188	-2
I	-2.192	0.039	-3
D	-1.169	0.250	-4
P	0.911	0.372	-5
N	2.131	0.044	-5
S	0.229	0.821	-6
K	-2.387	0.026	-12
L	2.830	0.009	-14
Degree of freedom	23		

[1] Student's *t*-values of the amino acid contents distribution of the alkaline and non-alkaline enzymes. Negative sign indicates the mean is lower for alkaline enzymes. n.a. = not applicable.

[2] Probability of obtaining the observed *t*-value by chance.

[3] Amino acid balance in alkaline adaptation processes obtained by ASET analysis: balance = (number of each amino acid after alkaline adaptation)-(that before alkaline adaptation).

Table III. Molecular features of alkaline and non-alkaline enzymes

Features	*t*-Value[1]	*p*-Value[2]
Interactions		
No. H-bonds	1.411	0.172
No. nonpolar contacts	0.606	0.550
No. ion pairs (all)	-0.537	0.596
R-D/E	1.443	0.163
K-D/E	-2.280	0.032
H-D/E	-0.577	0.151
R-D	1.146	0.264
R-E	2.310	0.030
K-D	-2.065	0.050
K-E	-1.418	0.170
Other features		
MW	0.557	0.583
pI_{calc}	1.379	0.181
ASA[3]	0.018	0.986
Polar ASA[3]	0.995	0.330
Volume	-0.255	0.801
No. cavities	0.644	0.526
Cavity volume	1.393	0.177
Degrees of freedom	23	

[1] Student's *t*-values of the distribution of the values between alkaline and non-alkaline enzymes. A negative sign indicates the mean is lower for alkaline enzymes.
[2] Probability of obtaining the observed *t*-value by chance.
[3] ASA stands for accessible surface area.
Features with $p < 0.05$ are highlighted by underlines.

The fractions of positively (K+R in Table III, $p = 0.114$*) and negatively charged (D+E in Table III,* $p = 0.547$*) amino acids did not show significant differences, as reflected in the total charge balances of the molecules. The estimated isoelectric point (pI) shifted only slightly toward an alkaline pH (*pI_{calc} *in Table III,* $p = 0.181$*). Therefore the –KD+RE strategy did not change significantly the net charge or total number of ion pairs of the protein molecules. Also, the numbers of H-bonds, non-polar interactions, molecular volume, accessible surface area (ASA), and cavity number and volume did not exhibit significant and consensus differences between the alkaline and non-alklaine enzymes. Therefore, a drastic change in the molecular properties of proteins would not be generally required for alkaline adaptation.*

ENGINEERING ALKALINE ADAPTED PROTEINS

Several cases of successful protein engineering have been reported resulting in upward shift of the optimal pH of an enzymatic activity or in improvement of protein stability at alkaline pH (Table IV).

Most of the reported studies are point mutation experiments using both site-directed and random mutagenesis. Though not listed in Table IV, constructions of chimeric proteins of homologous alkaline and non-alkaline enzymes were also reported for cellulase, pectate lyase, and xylanase in order to identify the protein regions responsible for the alkaliphilic nature of the enzymes [105, 138, 179, 180].

More than half of the reported point mutation experiments were designed to manipulate the enzyme activities by directing the mutations to the catalytic center (shown by "yes" in "catalytic center" column of Table IV), and some of them were based on the p*K*a modulation strategy. Adaptations were estimated by an upward shift of the pH optimum (opt. pH in "properties" column) or the basic limb of the activity profile (limb shift in "properties" column) in most cases, though some were evaluated by stabilization of the proteins under alkaline pH conditions (stability in "properties" column). Although caution is required in interpreting these results, because these experiments were performed with different purposes, experimental conditions, and criteria for result assessment, they nevertheless provide some practical information on protein engineering for alkaline adaptation. Of the 63 point mutations (multiple mutations were counted separately by excluding redundant mutation sites) listed in Table IV, thirty-eight mutations involved charged amino acids, and 25 did not. Eleven of the latter 25 mutations were replacements between hydrophobic amino acids. The maximum improvement in pH optimum with point mutations was as high as one pH unit for most cases ("pH shift" column), with the remarkable exceptions of *A. kawachii* xylanase C, where D37N, D37S, and D37E mutants produced an upward shift in pH optimum by 3, 2.5, and 1.5 units, respectively [128]. Apparently, a drastic modulation of pH optimum by point mutation would be quite difficult.

The set of point mutations was evaluated with respect to the alkaline adaptation strategies outlined above: p*K*a modulation by changing net charge (*i.e.*, +DE–KR), +DE, and –KD+ER strategies. Note that this assessment is totally independent of the purposes and intentions of the original works. The numbers of point mutations that were consistent/inconsistent with p*K*a modulation, +DE, and –KD+ER strategies were 7/23, 5/9, and 20/15, respectively, implying that none of these three strategies can fully explain the results.

The three proposed strategies involve modifications of charged amino acids as alkaline adaptation has to compensate for the impact of low proton density. However, the mutational studies demonstrated that the contributions of non-charged amino acids should not be neglected. Although some of the mutations of non-charged amino acids are consistent with the p*K*a modulation strategy by modifying a hydrogen bond or shielding catalytic sites, a certain fraction of them still eluded the proposed strategies. Particularly, the effects of hydrophobic interactions on alkaline adaptation appear to be a major oversight in the current theory.

FUTURE DIRECTIONS

Although the current studies of alkaline adaptation have been based mainly on extracellular enzymes, alkaliphilic organisms require many other proteins for survival in alkaline environments. In particular, membrane proteins must be adapted to a large transmembrane pH gradient. For example, the Na^+ (Li^+)/H^+ and K^+/H^+ antiporters are membrane proteins required by alkaliphiles to maintain the cytoplasmic pH and to transport

Table IV. Alkaline-adaptation protein engineering with site-directed/random mutagenesis.

Enzyme	Origin	Mutation(s) *1	Catalytic. center[2]	Properties[3]	pH shift[4]	Strategy[5]			Ref.
						p*K*a	+DE	-KD+RE	
α-Amylase	*B. licheniformis*	Y290K	no	opt. PH	1	-		-	[164]
	B. amyloliquefaciens	L13P/W192R/S197R/E356D/ N414S	no	opt. pH	1*	- -		++-	[169]
	G. stearothermophilus	N329K	yes	stability	-*	-		-	[170]
Chimera α-amylase Ba2	*B. licheniformis/ B. amyloliquefaciens*	F290E	yes	limb shift	-*	+	+	+	[162]
		F290K			-*	-		-	
		F290A			-*				
		N326D			-*	+	+	-	
		N326L			-*				
Glucoamylase	*A. awamori*	S411G	yes	opt. pH	0.26				[156]
		S411A			0.84				
		S411C			0.87				
		S411H			0.15				
		S411D			0.46	+	+	-	
CGTase	*Bacillus* sp. 1011	F183L	yes	limb shift	1*				[171]
		F259L			0.5*				
Endoglucanase III	*T. reesei*	N231T	no	opt. pH	0.6				[172]
Cellobiohydrolase Cel7A		E223S/A224H/L225A/T226A/ D262G	yes	opt. pH	1*	- +	- -	- +	[158]
Cellobiohydrolase Cel6A		E107Q	yes	stability	-*	-	-	-	[173]
		E107Q/D170N/D366N		stability	-*	- - -	- - -	-++	
Chimera xylanase Stx15	*S. lividans/T. fusca*	A49V	no	limb shift	0.5*				[174]
Xylanase	*N. patriciarum*	E24V/I97S/S128F/N164D/ K168N/G214V/V218A	no	limb shift	0.5*	- + +	+-	+- -	[160]

Table IV. (Continued)

Enzyme	Origin	Mutation(s) [*1]	Catalytic. center[2]	Properties[3]	pH shift[4]	Strategy[5]			Ref.
						p*K*a	+DE	-KD+RE	
	B. circulans	Y80F	yes	opt. pH	0.6				[157]
		R112N			0.5			-	
		E172D			0.4			-	
Xylanase II	*T. reesei*	T26R/Q34R/S40R/N67R/S186R	no	opt. pH	0.5	-----		+++++	[175]
		T26R/Q34R/N67R/N69R/S186R			1	-----		+++++	
		T2C/T28C/K58R/-191D	no	stability	-*	+	+	+ -	[176]
		T2C/T28C/K58R/F93W/N97R/ /H144K/-191D			-*	- - +	+	++ - -	
		T2C/T28C/K58R/N92C/H144C/ -191D			-*	+	+	+ -	
		T2C/T28C/K58R/N92C/H144C/ F180Q/-191D			-*	+	+	+ -	
		T2C/H22K/T28C/K58R/N92C/ F180Q/H144C/-191D			-*	- +	+	+ - -	
Xylanase C	*A. kawachii*	D37N	yes	opt. pH	3	-	-	+	[128]
		D37S			2.5	-	-	+	
		D37E			1.5			+	
Galactanase	*A. aculeatus*	D180N	yes	limb shift	1	-	-	+	[40]
Alcohol dehydrogenase	*B. stearothermophilus*	H43R	yes	opt. pH	1.2	-		+	[177]
Ascorbate oxidase	*Acremonium* sp. HI-25	V193P	no	limb shift	1				[178]
		P190I/V193P			0.5				

*1 Multiple mutation sites are separated by forward slashes.
*2 Yes, mutations were directed to proximity of catalytic center; no, otherwise.
*3 Alkaline adaptation criteria. Opt.pH, pH of optimal enzymatic activity shifted toward alkaline pH; limb shift, optimal pH was not changed, but relative activity at alkaline pH was increased; stability, stability of enzymes was increased in alkaline pH range.
*4 Shift of optimal pH or shift of relative activity curve toward alkaline pH. An asterisk means the value was estimated from drawings in the corresponding literature or exact values could not be guessed. Otherwise, the values are mentioned in the literature.
*5 Plus or minus sign indicates whether mutation is consistent or inconsistent with the proposed strategies.

materials across the membrane [181-183]. Cytochrome *c,* cytochrome *c* oxidase, and F_1F_0-ATPase are essential components of energy production at the cell surface [165, 184, 185]. These proteins are required for the transport of protons under unfavorable conditions, that is, lower proton concentrations at the outside of the cell. The structure-based mechanisms of alkaline adaptation for these membrane and non-enzymatic proteins are still largely an open question.

One of the major difficulties in the studies of protein alkaline adaptation is the number of available samples, which is restricted compared to, for example, high-temperature adapted proteins. Complete genome sequences have been determined for the alkaliphiles *B. halodurans* and *Oceanobacillus iheyensis* HTE831 [186, 187], thus enabling a systematic genome-wide investigation of protein adaptation. A comparison of all gene products of the alkaliphiles and those of close non-alkaliphilic relatives will provide far more abundant data than available before. A difficulty in this approach, however, is the fact that the interior of the alkaliphile cell is usually maintained at near neutral pH [2]. This demands that the target genes for analysis should be selected carefully, because not every protein is alkaline-adapted in these organisms. Even so, a genome-wide study would be most useful in studies of protein adaptation to high pH.

CONCLUSION

The current studies of alkaline adaptation of proteins have focused mainly on the alkaline enzymes, such as proteases, cellulases, amylases, and xylanases, as they have various industrial applications. Comparative studies of the structures of alkaline and non-alkaline enzymes have been the primary approach in elucidating the mechanisms of high-pH adaptation. So far, three major strategies, namely, the p*K*a modulation, the +DE, and the –KD+RE strategies, have been proposed from a large body of studies. The first strategy aims at shifting the p*K*a of the catalytic site residues toward alkaline pH. The second one is a likely strategy to protect proteins from anions. The last one would be employed to stabilize protein structures under alkaline conditions. Protein engineering experiments indicate that these three strategies are applicable to a certain extent. However, undiscovered strategies still likely exist, and improvement in hydrophobic interactions may be one of these undefined strategies. Genome-wide analyses, particularly of membrane proteins, appears promising way to refine the current theories and deepen our understanding in alkaline adaptation of proteins.

REFERENCES

[1] Horikoshi, K., Akiba, T. (1982). Alkalophilic microorganisms: a new microbial world. Springer-Verlag KG, Heidelberg, Germany.

[2] Horikoshi, K. (1999). Alkaliphiles: some applications of their products for biotechnology. *Microbiol. Mol. Biol. Rev.,* 63, 735-750.

[3] Horikoshi, K. (1971). Production of alkaline enzymes by alkalophilic microorganisms. II. Alkaline protease produced by *Bacillus* no. 221. *Agric. Biol. Chem.*, 35, 1407-1414.

[4] Hoirikoshi, K. (1971). Production of alkaline enzymes by alkalophilic microorganisms. II. Alkaline amylase produced by *Bacillus* No. A-40-2. *Agric. Biol. Chem*., 35, 1783-1791.

[5] Nakamura, N., Watanabe, K., Horikoshi, K. (1975). Purification and some properties of alkaline pullulanase from a strain of *Bacillus* no. 202-1, an alkalophilic microorganism. *Biochim. Biophys. Acta*, 397, 188-193.

[6] Horikoshi, K., Atsukawa, Y. (1973). Xylanase produced by alkalophilic *Bacillus* no. C-59-2. *Agric. Biol. Chem*., 37: 2097-2103.

[7] Guntelberg, A.V., Ottesen, M. (1954). Purification of the proteolytic enzyme from *Bacillus subtilis*. *Compt. -rend. Lab. Carlsberg,* 29, 36-48.

[8] Dambmann, C., Holm, P., Jenssen, V., Nielsen, M.H. (1971). How enzymes got into detergents. *Develop. Ind. Microbiol.,* 12, 11-23.

[9] Aunstrup, K., Ottrup, H., Andersen, O., Dambmann, C. (1972). Proteases from alkalophilic *Bacillus* species. *Ferment. Technol. Today*, 4, 299-305.

[10] Riisgaard, S. (1990). The enzyme industry and modern biotechnology. *Genet. Eng. Biotechnol.,* 10, 11-14.

[11] Ito, S., Kobayashi, T., Ara, K., Ozaki, K., Kawai, S., Hatada, Y. (1998). Alkaline detergent enzymes from alkaliphiles: enzymatic properties, genetics, and structures. *Extremophiles*, 2, 185-90.

[12] Gupta, R., Beg, Q.K., Lorenz, P. (2002). Bacterial alkaline proteases: molecular approaches and industrial applications. *Appl. Microbiol. Biotechnol*., 59, 15-32.

[13] Mauer, K.H. (2004). Detergent proteases. *Curr. Opin. Biotechnol*., 15, 330-334.

[14] Nielsen, J.E., Borchert, T.V. (2000). Protein engineering of bacterial α-amylases. *Biochim. Biophys. Acta*, 1543, 253-274.

[15] Beg, Q.K., Kapoor, M., Mahajan, L., Hoondal, G.S. (2001). Microbial xylanases and their industrial applications: a review. *Appl. Microbiol. Biotechnol.*, 56, 326-338.

[16] Goddette, D.W., Paech, C., Yang, S.S., Mielenz, J.R., Bystroff, C., Wilke, M.E., Fletterick, R.J. (1992). The crystal structure of the *Bacillus lentus* alkaline protease, subtilisin BL, at 1.4 A resolution. *J. Mol. Biol*., 228, 580-595.

[17] Betzel, C., Klupsch, S., Padendorf, G., Hastrup, S., Branner, S., Wilson, K.S. (1992). Crystal structure of the alkaline proteinase Savinase^{TM} from *Bacillus lentus* at 1.4 Å resolution. *J. Mol. Biol.*, 223, 427-445.

[18] Van der Laan, J.M., Teplyakov, A.V., Kelders, H., Kalk, K.H., Misset, O., Mulleners, L.J.S.M., Dijkstra, B.W. (1992). Crystal structure of the high-alkaline serine protease PB92 from *Bacillus alcalophilus*. *Protein Eng*., 5, 405-411.

[19] Martin, J.R., Mulder F.A.A., Karimi-Nejad, Y., Van der Zwan, J., Mariani, M., Schipper, D., Boelens, R. (1997). The solution structure of serine protease PB92 from *Bacillus alcalophilus* presents a rigid fold with a flexible substrate-binding site. *Structure*, 5, 521-532.

[20] Yamane, T., Kani, T., Hatanaka, T., Suzuki, A., Ashida, T., Kobayashi, T., Ito, S., Yamashita, O. (1995). Structure of a new alkaline serine protease (M-protease) from *Bacillus* sp. KSM-K16. *Acta Crystallog. Sect. D*, 51, 199-206.

[21] Shirai, T., Suzuki, A., Yamane, T., Ashida, T., Kaobayashi, T., Hitomi, J., Ito, S. (1997). High-resolution crystal structure of M-protease: phylogeny aided analysis of the high-alkaline adaptation mechanism. *Protein Eng*., 10, 627-634.

[22] Davies, G.J., Brzozowski, A.M., Dauter, Z., Rasmussen, M.D., Borchert, T.V., Wilson, K.S. (2005). Structure of *Bacillus halmapalus* family 13 α-amylase, BHA, in complex with an acarbose-derived nonasaccharide at 2.1 Å resolution. *Acta Crystallogr. Sect. D*, 61, 190-193.

[23] Lyhne-Iversen, L., Hobley, T.J., Kaasgaard, S.G., Harris, P. (2006). Structure of *Bacillus halmapalus* α-amylase crystallized with and without the substrate analogue acarbose and maltose. *Acta Crystallogr. Sect. F*, 62, 849-854.

[24] Nonaka, T., Fujihashi, M., Kita, A., Hagihara, H., Ozaki, K., Ito, S., Miki, K. (2003). Crystal structure of calcium-free α-amylase from *Bacillus* sp. strain KSM-K38 (AmyK38) and its sodium ion binding sites. *J. Biol. Chem.*, 278, 24818-24824.

[25] Shirai, T., Igarashi, K., Ozawa, T., Hagihara, H., Kobayashi, T., Ozaki, K., Ito, S. (2007) Ancestral sequence evolutionary trace and crystal structure analyses of alkaline α-amylase from *Bacillus* sp. KSM-1378 to clarify the alkaline adaptation process of proteins. *Proteins*, 66, 600-610.

[26] Kanai, R., Haga, K., Akiba, T., Yamane, K., Harata, K. (2004). Biochemical and crystallographic analyses of maltohexaose-producing amylase from alkalophilic *Bacillus* sp. 707. *Biochemistry*, 43, 14047-14056.

[27] Kanai, R., Haga, K., Akiba, T., Yamane, K., Harata, K. (2006). Role of Trp140 at subsite-6 on the maltohexaose production of maltohexaose-producing amylase from alkalophilic *Bacillus* sp. 707. *Protein Sci.*, 15, 468-477.

[28] Ishii, N., Haga, K., Yamane, K., Harata, K. (2000). Crystal structure of alkalophilic asparagine 233-replaced cyclodextrin glucanotransferase complexed with an inhibitor, acarbose, at 2.0 Å resolution. *J. Biochem.*, 127, 383-391.

[29] Kanai, R., Haga, K., Yamane, K., Harata, K. (2001). Crystal structure of cyclodextrin glucanotransferase from alkalophilic *Bacillus* sp. 1011 complexed with 1-deoxynojirimycin at 2.0 Å resolution. *J. Biochem.*, 129, 593-598.

[30] Haga, K., Kanai, R., Sakamoto, O., Aoyagi, M., Harata, K., Yamane, K. (2003). Effects of essential carbohydrate/aromatic stacking interaction with Tyr100 and Phe259 on substrate binding of cyclodextrin glycosyltransferase from alkalophilic *Bacillus* sp. 1011. *J. Biochem.*, 134, 881-891.

[31] Kanai, R., Haga, K., Akiba, T., Yamane, K., Harata, K. (2004). Role of Phe283 in enzymatic reaction of cyclodextrin glycosyltransferase from alkalophilic *Bacillus* sp. 1011: substrate binding and arrangement of the catalytic site. *Protein Sci.*, 13, 457-465.

[32] Davies, G.J., Dauter, M., Brzozowski, A.M., Bjornvad, M.E., Andersen, K.V., Schulein, M. (1998). Structure of the *Bacillus agaradherans* family 5 endoglucanase at 1.6 Å and its cellobiose complex at 2.0 Å resolution. *Biochemistry,* 37, 1926-1932.

[33] Davies, G.J., Mackenzie, L., Varrot, A., Dauter, M., Brzozowski, A.M., Schulein, M., Withers, S.G. (1998). Snapshots along an enzymatic reaction coordinate: analysis of a retaining β-glycoside hydrolase. *Biochemistry*, 37, 11707-11713.

[34] Varrot, A., Schulein, M., Fruchard, S., Driguez, H., Davies, G.J. (2001). Atomic resolution structure of endoglucanase Cel5A in complex with methyl 4, 4II, 4III, 4Iv-tetrathio-α-cellopentoside highlights the alternative binding modes targeted by substrate mimics. *Acta Crystallogr. Sect. D*, 57, 1739-1742.

[35] Gloster, T.M., Macdonald, J.M., Tarling, C.A., Stick, R.V., Withers, S.W., Davies, G.J. (2004). Structural, thermodynamic, and kinetic analyses of tetrahydrooxazine-derived inhibitors bound to β-glucosidases. *J. Biol. Chem.*, 279, 49236-49242.

[36] Shirai, T., Ishida, H., Noda, J., Yamane, T., Ozaki, K., Hakamada, Y., Ito, S. (2001). Crystal structure of alkaline cellulase K: insight into the alkaline adaptation of an industrial enzyme. *J. Mol. Biol.*, 310, 1079-1087.

[37] Shaw, A., Bott, R., Vonrhein, C., Bricogne, G., Power, S., Day, A.G. (2002). A novel combination of two classic catalytic schemes. *J. Mol. Biol.*, 320, 303-309.

[38] Jamal, S., Nurizzo, D., Boraston, A.B., Davies, G.J. (2004). X-ray crystal structure of a non-crystalline cellulose-specific carbohydrate-binding module: CBM28. *J. Mol. Biol.*, 339, 253-258.

[39] Manikandan, K., Bhardwaj, A., Gupta, N., Lokanath, N.K., Ghosh, A., Reddy, V.S., Ramakumar, S. (2006). Crystal structures of native and xylosaccharide-bound alkali thermostable xylanase from an alkalophilic *Bacillus* sp. NG-27: structural insights into alkalophilicity and implications for adaptation to polyextreme conditions. *Protein Sci.,* 15, 1951-1960.

[40] Nours, J.E., Rtttersgaard, C., Leggio, L.L., Ostergaard, P.R., Borchert, T.V., Christensen, L.L.H., Larsen, S. (2003). Structure of two fungal β-1.4-galactanases: searching for the basis for temperature and pH optimum. *Protein Sci.*, 12, 1195-1204.

[41] Akita, M., Suzuki, A., Kobayashi, T., Ito, S., Yamane, T. (2001). The first structure of pectate lyase belonging to polysaccharide lyase family 3. *Acta Crystallogr. Sect. D,* 57, 1786-1792.

[42] Dubnovitsky, A.J., Kapetaniou, E.G., Papageorgiou, A.C. (2005). Enzyme adaptation to alkaline pH: Atomic resolution (1.08 Å) structure of phosphoserine aminotransferase from *Bacillus alcalophilus*. *Protein Sci.*, 14, 97-110.

[43] Markland, F.S., Smith, E.L. Subtilisins: primary structure, chemical and physical properties. In The Enzymes (Boyer R.D., ed), 3rd ed. New York and London: Academic Press, 1971; 561-608.

[44] Siezen, R.J., Leunissen, J. (1997). Subtilases: the superfamily of subtilisin-like serine proteases. *Protein Sci.*, 6, 501-523.

[45] Kaneko, R., Koyama, N., Tsai, Y.C., Juang, R.Y., Yoda, K., Yamasaki, M. (1989). Molecular cloning of the structural gene for alkaline elastase YaB, a new subtilisin produced by an alkalophilic *Bacillus* strain. *J. Bacteriol.*, 171, 5232-5236.

[46] Van der Laan, J.C., Gerristse, G., Mulleners, L.J.S.M., Van der Hoek, R.A.C., Quax, W.J. (1991). Cloning, characterization, and multiple chromosomal integration of a *Bacillus* alkaline protease gene. *Appl. Environ. Microbiol.*, 57, 901-909.

[47] Fujiwara, N., Masui, A., Imanaka, T. (1993). Purification and properties of the highly thermostable alkaline protease from an alkaliphilic and thermophilic *Bacillus* sp. *J. Biotechnol.*, 30, 245-256.

[48] Takami, H., Kobayashi, T., Aono, R., Horikoshi, K. (1992). Molecular cloning, nucleotide sequence and expression of the structural gene for a thermostable alkaline protease from *Bacillus* sp. no. AH-101. *Appl. Microbiol. Biotechnol.*, 38, 101-108.

[49] Hakamada, Y., Kobayashi, T., Hitomi, J., Kawai, S., Ito, S. (1994). Molecular cloning and nucleotide sequencing of the gene for an alkaline protease from the akalophilic *Bacillus* sp. KSM-K16. *J. Ferment. Bioeng.*, 78, 105-108.

[50] Kobayashi, T., Hakamada, Y., Hitomi, J., Yoshimatsu, T., Koike, K., Kawai, S., Ito, S. (1995). Purification and properties of an alkaline protease from alkalophilic *Bacillus* sp. KSM-K16. *Appl. Microbiol. Biotechnol.*, 43, 473-481.

[51] Lee, J.K., Kim, Y.O., Kim, H.K., Park, Y.S., Oh, T.K. (1996). Purification and characterization of a thermostable alkaline protease from *Thermoactinomyces* sp. E79 and the DNA sequence of the encoding gene. *Biosci. Biotechnol. Biochem.*, 60, 840-846.

[52] Tsuchiya, K., Ikeda, I., Tsuchiya, T., Kimura, T. (1997). Cloning and expression of an intracellular alkaline protease gene from alkalophilic *Thermoactinomyces* sp. HS682. *Biosci. Biotechnol. Biochem.*, 61, 298-303.

[53] Johnvesly, B., Manjunath, B.R., Naik, G.R. (2002). Pigeon pea waste as a novel, inexpensive, substrate for production of a thermostable alkaline protease from thermoalkalophilic *Bacillus* sp. JB-99. *Bioresource Technol.*, 82, 61-64.

[54] Miyaji, T., Otta, Y., Shibata, T., Mitsui, K., Nakagawa, T., Watanabe, T., Niimura, Y., Tomizuka, N. (2005). Purification and characterization of extracellular alkaline serine protease from *Stenotrophomonas maltophilia* strain S-1. *Lett. Appl. Microbiol.*, 41, 253-257.

[55] Kazan, D., Denizci, A.A., Oner, M.N., Erarslan, A. (2005). Purification and characterization of a serine alkaline protease from *Bacillus clausii* GMBAE 42. *J. Ind. Microbiol. Biotechnol.*, 32, 335-344.

[56] Sobek, H., Hecht, H.J., Hofmann, B., Aehle, W., Schomburg, D. (1990). Crystal structure of an alkaline protease from *Bacillus alcalophilus* at 2.4 Å resolution. *FEBS Lett.*, 274, 57-60.

[57] Sobek, H., Hecht, H.J., Aehle, W., Schomburg, D. (1992). X-ray structure determination and comparison of two crystal forms of a variant (Asn115Arg) of the alkaline protease from *Bacillus alcalophilus* refined at 1.85 Å resolution. *J. Mol. Biol.*, 228, 108-117.

[58] Murzin, A.G., Brenner, S.E., Hubbard, T., Chothia, C. (1995). SCOP: a structural classification of proteins database for the investigation of sequences and structures. *J. Mol. Biol.*, 247, 536-540.

[59] Masui, A., Fujiwara, N., Imanaka, T. (1994). Stabilization and rational design of serine protease AprM under highly alkaline and high-temperature conditions. *Appl. Environ. Microbiol.*, 60, 3579-3584.

[60] Medda, S., Chandra, A.K. (1980). New strains of *Bacillus licheniformis* and *Bacillus coagulans* producing thermostable α-amylase active at alkaline pH. *J. Appl. Bacteriol.*, 48, 47-58.

[61] Kim, T.U., Gu, B.G., Jeong, J.Y., Byun, S.M., Shin, Y.C. (1995). Purification and characterization of a maltotetraose-forming alkaline α-amylase from an alkalophilic *Bacillus* strain, GM8901. *Appl. Environ. Microbiol.*, 61, 3105-3112.

[62] Igarashi, K., Hatada, Y., Hagihara, H., Saeki, K., Takaiwa, M., Uemura, T., Ara, K., Ozaki, K., Kawai, S., Kobayashi, T., Ito, S. (1998). Enzymatic properties of a novel liquefying α-amylase from an alkaliphilic *Bacillus* isolate and entire nucleotide and amino acid sequence. *Appl. Environ. Microbiol.*, 64, 3282-3289.

[63] Lin, L.L., Chyau, C.C., Hsu, W.H. (1998). Production and properties of a raw-starch-degrading amylase from the thermophilic and alkaliphilic *Bacillus* sp. TS-23. *Biotechnol. Appl. Biochem.*, 28, 61-68.

[64] Hagihara, H., Hayashi, Y., Endo, K., Igarashi, K., Ozawa, T., Kawai, S., Ozaki, K., Ito, S. (2001). Deduced amino-acid sequence of a calcium-free α-amylase from a strain of *Bacillus*: implications by molecular modeling of high oxidation stability and chelator resistance of the enzyme. *Eur. J. Biochem.*, 268, 3974-3982.

[65] Ballschmiter, M., Armbrecht, M., Ivanova, K., Antranikian, G., Liebl, W. (2005). Amy A, an α-amylase with β-cyclodextrin-forming activity, and amyB from the thermoalkaliphilic organism *Anaerobranca gottschalkii*: two α-amylases adapted to their different cellular localizations. *Appl. Environ. Microbiol.*, 71, 3709-3715.

[66] Henrissat, B. (1991). A classification of glycosyl hydrolases based on amino-acid sequence similarities. *Biochem. J.*, 280, 309-316.

[67] Henrissat, B., Davies, G.J. (1997). Structural and sequence-based classification of glycoside hydrolases. *Curr. Opin. Struct. Biol.*, 7, 637-644.

[68] Kim, C.H., Choi, H.I., Lee, D.S. (1993). Purification and biochemical properties of an alkaline pullulanase from alkalophilic *Bacillus* sp. S-1. *Biosci. Biotechnol. Biochem.*, 57, 1632-1637

[69] Lee, S.P., Morikawa, M., Takagi, M., Imanaka, T. (1994). Cloning of the aapT gene and characterization of its product, α-amylase-pullulanase (AapT), from thermophilic and alkaliphilic *Bacillus* sp. strain XAL601. *Appl. Environ. Microbiol.*, 60, 3764-3773.

[70] Ara, K., Saeki, K., Igarashi, K., Takaiwa, M., Uemura, T., Hagihara, H., Kawai, S., Ito, S. (1995). Purification and characterization of an alkaline amylopullulanase with both α-1,4 and α-1,6 hydrolytic activity from alkalophilic *Bacillus* sp. KSM-1378. *Biochim. Biophys. Acta*, 1243, 315-324.

[71] Hatada, Y., Igarashi, K., Ozaki, K., Ara, K., Hitomi, J., Kobayashi, T., Kawai, S., Watabe, T., Ito, S. (1996). Amino acid sequence and molecular structure of an alkaline amylopullulanase from *Bacillus* that hydrolyzes α-1,4 and α-1,6 linkages in polysaccharides at different active sites. *J. Biol. Chem.*, 271, 24075-24083.

[72] Lee, M.J., Lee, Y.C., Kim, C.H. (1997). Intracellular and extracellular forms of alkaline pullulanase from an alkaliphilic *Bacillus* sp. S-1. *Arch. Biochem. Biophys.*, 337, 308-316.

[73] Hatada, Y., Saito, K., Hagihara, H., Ozaki, K., Ito, S. (2001). Nucleotide and deduced amino acid sequences of an alkaline pullulanase from the alkaliphilic bacterium *Bacillus* sp. KSM-1876. *Biochim. Biophys. Acta*, 1545, 367-371.

[74] Tsukamoto, A., Kimura, K., Ishii, Y., Takano, T., Yamane, K. (1988). Nucleotide sequence of the maltohexaose-producing amylase gene from an alkalophilic *Bacillus* sp. #707 and structural similarity to liquefying type α-amylases. *Biochem. Biophys. Res. Commun.*, 151, 25-31.

[75] Nakamura, N., Horikoshi, K. (1977). Production of Schardinger β-dextrin by soluble and immobilized cyclodextrin glycosyltransferase of an alkalophilic *Bacillus* sp. *Biotechnol. Bioeng.*, 19, 87-99.

[76] Martins, R.F., Delgado, O., Hatti-Kaul, R. (2003). Sequence analysis of cyclodextrin glycosyltransferase from the alkaliphilic *Bacillus agaradhaerens*. *Biotechnol. Lett.*, 25, 1555-1562.

[77] Nakamura, A., Haga, K., Yamane, K. (1993). Three histidine residues in the active center of cyclodextrin glucanotransferase from alkalophilic *Bacillus* sp. 1011: effects of the replacement on pH dependence and transition-state stabilization. *Biochemistry,* 32, 6624-6631.

[78] Borchert, T.V., Lassen, S.F., Svendsen, A., Frantzen, H.B. Oxidation stable amylases for detergents. In Carbohydrate Bioengineering. (Petersen S.B., Svensson B., Pederson S., eds), Amsterdam: Elsevier Science, 1995; 175-179.

[79] Pandey, A., Nigam, P., Soccol, C.R., Soccol, V.T., Singh, D., Mohan, R. (2000). Advances in microbial amylases. *Biotechnol. Appl. Biochem*., 31, 135-152.

[80] Matsuzawa, M., Kawano, M., Nakamura, N., Horikoshi, K. (1975). An improved method for the production of Schardinger β-dextrin on an industrial scale by cyclodextrin glycosyltransferase of an alkalophilic *Bacillus* sp. *Starch*, 27, 410-413.

[81] Ikawa, K., Araki, H., Tsujino, Y., Hayashi, Y., Igarashi, K., Hatada, Y., Hagihara, H., Ozawa, T., Ozaki, K., Kobayashi, T., Ito, S. (1998). Hyperexpression of the gene for a *Bacillus* α-amylase in *Bacillus subtilis* cells: Enzymatic properties and crystallization of the recombinant enzyme. *Biosci. Biotechnol. Biochem*., 62, 1720-1725.

[82] Igarashi, K., Ozawa, T., Ikawa-Kitayama, K., Hayashi, Y., Araki, H., Endo, K., Hagihara, H., Ozaki, K., Kawai, S., Ito, S. (1999). Thermostabilization by proline substitution in an alkaline, liquefying α-amylase from *Bacillus* sp. strain KSM-1378. *Biosci. Biotechnol. Biochem*., 63, 1535-1540.

[83] Hagihara, H., Hatada, Y., Ozawa, T., Igarashi, K., Araki, H., Ozaki, K., Kobayashi, T., Kawai, S., Ito, S. (2003). Oxidative stabilization of an alkaliphilic *Bacillus* α-amylase by replacing a single specific methionine residue by site-directed mutagenesis. *J. Appl. Glycosci*., 50, 367-372.

[84] Igarashi, K., Hatada, Y., Ikawa, K., Araki, H., Ozawa, T., Kobayashi, T., Ozaki, K., Ito, S. (1998). Improved thermostability of a *Bacillus* α-amylase by deletion of an arginine-glycine residue is caused by enhanced calcium binding. *Biochem. Biophys. Res. Commun*., 248, 372-377.

[85] Coughlan, M.P., Ljungdahl, L.G. Comparative biochemistry of fungal and bacterial cellulolytic enzyme systems In Biochemistry and Genetics of Cellulose Degradation. (Aubert, J.-P., Beguin, P., Millet, J. eds.), New York, Academic Press, 1988; 11-30.

[86] Horikoshi, K., Nakao, M., Kurono, Y., Sashihara, N. (1984). Cellulases of an alkalophilic *Bacillus* strain isolated from soil. *Can. J. Microbiol*., 30, 774-779.

[87] Fukumori, F., Kudo, T., Horikoshi, K. (1985). Purification and properties of a cellulase from alkalophilic *Bacillus* sp. no. 1139. *J. Gen. Microbiol*., 131, 3339-3345.

[88] Sashihara, N., Kudo, T., Horikoshi, K. (1984). Molecular cloning and expression of cellulase genes of alkalophilic *Bacillus* sp. strain N-4 in *Escherichia coli*. *J. Bacteriol*., 158, 503-506.

[89] Fukumori, F., Sashihara, N., Kudo, T., Horikoshi, K. (1986). Nucleotide sequences of two cellulase genes from alkalophilic *Bacillus* sp. strain N-4 and their strong homology. *J. Bacteriol*., 168, 479-485.

[90] Nakai, R., Horinouchi, S., Beppu, T. (1988). Cloning and nucleotide sequence of a cellulase gene, casA, from an alkalophilic *Streptomyces* strain. *Gene*, 65, 229-238.

[91] Shikata, S., Saeki, K., Okoshi, H., Yoshimatsu, T., Ozaki, K., Kawai, S., Ito, S. (1990). Alkaline cellulase for laundry detergents: production by alkalophilic strains of *Bacillus* and some properties of the crued enzymes. *Agric. Biol. Chem.*, 54, 91-96.

[92] Sumitomo, N., Ozaki, K., Kawai, S., Ito, S. (1992). Nucleotide sequence of the gene for an alkaline endoglucanase from an alkalophilic *Bacillus* and its expression in *Escherichia coli* and *Bacillus subtilis*. *Biosci. Biotechnol. Biochem.*, 56, 872-877.

[93] Hakamada, Y., Koike, K., Yoshimatsu, T., Mori, H., Kobayashi, T., Ito, S. (1997). Thermostable alkaline cellulase from an alkaliphilic isolate, *Bacillus* sp. KSM-S237. *Extremophiles*, 1, 151-156.

[94] Schulein, M. (1997). Enzymatic properties of cellulases from *Humicola insolens*. *J. Biotechnol.*, 57, 71-81.

[95] Christakopoulos, P., Hatzinikolaou, D.G., Fountoukidis, G., Kekos, D., Claeyssens, M., Macris, B.J. (1999). Purification and mode of action of an alkali-resistant endo-1, 4-β-glucanase from *Bacillus pumilus*. *Arch. Biochem. Biophys.*, 364, 61-66.

[96] Hakamada, Y., Hatada, Y., Koike, K., Yoshimatsu, T., Kawai, S., Kobayashi, T., Ito, S. (2000). Deduced amino acid sequence and possible catalytic residues of a thermostable, alkaline cellulase from an alkaliphilic *Bacillus* strain. *Biosci. Biotechnol. Biochem.*, 64, 2281-2289.

[97] Endo, K., Hakamada, Y., Takizawa, S., Kubota, H., Sumitomo, N., Kobayashi, T., Ito, S. (2001). A novel alkaline endoglucanase from an alkaliphilic *Bacillus* isolate: enzymatic properties, and nucleotide and deduced amino acid sequences. *Appl. Microbiol. Biotechnol.*, 57, 109-116.

[98] Hakamada, Y., Endo, K., Takizawa, S., Kobayashi, T., Shirai, T., Yamane, T., Ito, S. (2002). Enzymatic properties, crystallization, and deduced amino acid sequence of an alkaline endoglucanase from *Bacillus circulans*. *Biochim. Biophys. Acta*, 1570, 174-180.

[99] Singh, J., Batra, N., Sobti, R.C. (2004). Purification and characterisation of alkaline cellulase produced by a novel isolate, *Bacillus sphaericus* JS1. *J. Ind. Microbiol. Biotechnol.*, 31, 51-56.

[100] Kim, J.Y., Hur, S.H., Hong, J.H. (2005). Purification and characterization of an alkaline cellulase from a newly isolated alkalophilic *Bacillus* sp. HSH-810. *Biotechnol. Lett.*, 27, 313-316.

[101] Demain, A.L., Newcomb, M., Wu, J.H. (2005). Cellulase, clostridia, and ethanol. *Microbiol. Mol. Biol. Rev.*, 69, 124-154.

[102] Lynd, L.R., Weimer, P.J., Van Zyl, W.H., Pretorius, I.S. (2002). Microbial cellulose utilization: fundamentals and biotechnology. *Microbiol. Mol. Biol. Rev.*, 66, 506-577.

[103] Ozaki, K., Shikata, S., Kawai, S., Ito, S., Okamoto, K. (1990). Molecular cloning and nucleotide sequence of a gene for alkaline cellulase from *Bacillus* sp. KSM-635. *J. Gen. Microbiol.*, 136, 1327-1334.

[104] Ozaki, K., Hayashi, Y., Sumitomo, N., Kawai, S., Ito, S. (1995). Construction, purification, and properties of a truncated alkaline endoglucanase from *Bacillus* sp. KSM-635. *Biosci. Biotechnol. Biochem.*, 59, 1613-1618.

[105] Park, J.S., Hitomi, J., Horinouchi, S., Beppu, T. (1993). Identification of two amino acids contributing the high enzyme activity in the alkaline pH range of an alkaline endoglucanase from a *Bacillus* sp. *Protein Eng.*, 6, 921-926.

[106] Wong, K.K., Tan, L.U., Saddler, J.N. (1988). Multiplicity of β-1,4-xylanase in microorganisms: functions and applications. *Microbiol. Rev.*, 52, 305-317.

[107] Collins, T., Gerday, C., Feller, G. (2005). Xylanases, xylanase families and extremophilic xylanases. *FEMS Microbiol. Rev.*, 29, 3-23.

[108] Subramaniyan, S., Prema, P. (2002). Biotechnology of microbial xylanases: enzymology, molecular biology, and application. *Crit. Rev. Biotechnol.*, 22, 33-64.

[109] Honda, H., Kudo, T., Ikura, Y., Horikoshi, K. (1985). Two types of xylanases of alkalophilic *Bacillus* sp. no. C-125. *Can. J. Microbiol.*, 31, 538-542.

[110] Nakamura, S., Wakabayashi, K., Nakai, R., Aono, R., Horikoshi, K. (1993). Purification and some properties of an alkaline xylanase from alkaliphilic *Bacillus* sp. strain 41M-1. *Appl. Environ. Microbiol.*, 59, 2311-2316.

[111] Tabernero, C., Sanchez-Torres, J., Perez, P., Santamaria, R.I. (1995). Cloning and DNA sequencing of xyaA, a gene encoding an endo-β-1,4-xylanase from an alkalophilic *Bacillus* strain (N137). *Appl. Environ. Microbiol.*, 61, 2420-2424.

[112] Cho, S., Choi, Y. (1995). Nucleotide sequence analysis of an endo-xylanase gene (xynA) from *Bacillus stearothermophilus*. *J. Miorobiol. Biotechnol.*, 5, 117-124.

[113] Gupta, N., Vohra, R.M., Hoondal, G.S. (1992). A thermostable extracellular xylanase from alkalophilic *Bacillus* sp. NG-27. *Biotechnol. Lett.*, 14, 1045-1046.

[114] Akiba, T., Horikoshi, K. (1988). Xylanases of alkalophilic thermophilic *Bacillus*. *Methods in Enzymol.*, 160, 655-659.

[115] Bansod, S.M., Dutta-Choudhary, M., Srinivasan, M.C., Rele, M.V. (1993). Xylanase active at high pH from an alkalotolerant *Cephalosporium* species. *Biotechnol. Lett.*, 15, 965-970.

[116] Kang, M.K., Maeng, P.J., Rhee, Y.H. (1996). Purification and characterization of two xylanases from alkalophilic *Cephalosporium* sp. strain RYM-202. *Appl. Environ. Microbiol.*, 62, 3480-3482.

[117] Gessesse, A. (1998). Purification and properties of two thermostable alkaline xylanases from alkaliphilic *Bacillus* sp. *Appl. Environ. Microbiol.*, 64, 3533-3535.

[118] Gessesse, A. Mamo, G. (1998). Purification and characterization of an alkaline xylanase from alkaliphilic *Micrococcus* sp AR-135. *J. Ind. Microbiol. Biotechnol.*, 20, 210-214.

[119] Beg, Q.K., Bhushan, B., Kapoor, M., Hoondal, G.S. (2000). Production and characterization of thermostable xylanase and pectinase from *Streptomyces* sp. QG-11-3. *J. Ind. Microbiol. Biotechnol.*, 24, 396-402.

[120] Balakrishnan, H., Kamal Kumar, B., Dutta-Choudhury, M., Rele, M.V. (2002). Characterization of alkaline thermoactive cellulase-free xylanases from alkalophilic *Bacillus* (NCL 87-6-10). *J. Biochem. Mol. Biol. Biophys.*, 6, 325-334.

[121] Taneja, K., Gupta, S., Kuhad, R.C. (2002). Properties and application of a partially purified alkaline xylanase from an alkalophilic fungus *Aspergillus nidulans* KK-99. *Bioresour. Technol.*, 85, 39-42.

[122] Gupta, S., Bhushan, B., Hoondel, G.S. (2000). Isolation, purification and characterization of xylanase from *Staphylococcus* sp. SG-13 and its application in biobleaching of Kraft pulp. *J. Appl. Microbiol.* 88, 325-334.

[123] Sabini, E., Sulzenbacher, G., Dauter, M., Dauter, Z., Jorgensen, P.L., Schulein, M., Dupont, C., Davies, G.J., Wilson, K.S. (1999). Catalysis and specificity in enzymatic glycoside hydrolysis: a 2,5B conformation for the glycosyl-enzyme intermediate

revealed by the structure of the *Bacillus agaradherans* family 11 xylanase. *Chem. Biol.*, 6, 483-492.

[124] Teplitsky, A., Feinberg, H., Gilboa, R., Lapidot, A., Mechaly, A., Stojanoff, V., Capel, M., Shoham, Y., Shoham, G. (1997). Crystallization and preliminary X-ray analysis of the thermostable alkaline-tolerant xylanase from *Bacillus stearothermophilus* T-6. *Acta Crystallogr. D*, 53, 608-611.

[125] Zolotnitsky, G., Cogan, U., Adir, N., Solomon, V., Shoham, G., Shoham, Y. (2004). Mapping glycoside hydrolase substrate subsites by isothermal titration calorimetry. *Proc. Natl. Acad. Sci. USA*, 101, 11275-11280.

[126] Torronen, A., Rouvinen, J. (1995). Structural comparison of two major endo-1,4-xylanases from *Trichoderma reesei*. *Biochemistry*, 34, 847-856.

[127] Torronen, A., Rouvinen, J. (1997). Structural and functional properties of low molecular weight endo-1,4-β-xylanases. *J. Biotechnol.*, 57, 137-149.

[128] Fushinobu, S., Ito, K., Konno, M., Wakagi, T., Matsuzawa, H. (1998). Crystallographic and mutational analyses of an extremely acidophilic and acid-stable xylanase: biased distribution of acidic residues and importance of Asp37 for catalysis at low pH. *Protein Eng.*, 11, 1121-1128.

[129] Liu, L., Li, X., Li, X., Shao, W. (2004). Computational analysis of responsible dipeptides for optimum pH in G/11 xylanase. *Biochem. Biophys. Res. Commun.*, 321, 391-396.

[130] de Vries, R.P., Visser, J. (2001). *Aspergillus* enzymes involved in degradation of plant cell wall polysaccharides. *Microbiol. Mol. Biol. Rev.*, 65, 497-522.

[131] Herron, S.R., Benen, J.A., Scavetta, R.D., Visser, J., Jurnak, F. (2000). Structure and function of pectic enzymes: virulence factors of plant pathogens. *Proc. Natl. Acad. Sci. USA.,* 97, 8762-8769.

[132] Kelly, C.T., Fogarty, W.M. (1978). Production and properties of poly-galacturonate lyase by an alkalophilic microorganism *Bacillus* sp. RK9. *Can. J. Microbiol.*, 24, 1164-1172.

[133] Yoshihara, K., Kobayashi, Y. (1982). Retting of mitsumata bast by alkalophilic *Bacillus* in paper making. *Agric. Biol. Chem.*, 46, 109-117.

[134] Cao, J., Zheng, L., Chen, S. (1992). Screening of pectinase producer from alkalophilic bacteria and study on its potential application in degumming of ramie. *Enzyme Microb. Technol.*, 14, 1013-1016.

[135] Kobayashi, T., Koike, K., Yoshimatsu, T., Higaki, N., Suzumatsu, A., Ozawa, T., Hatada, Y., Ito, S. (1999). Purification and properties of a low-molecular-weight, high-alkaline pectate lyase from an alkaliphilic strain of *Bacillus*. *Biosci. Biotechnol. Biochem.*, 63, 65-72.

[136] Kobayashi, T., Hatada, Y., Higaki, N., Lusterio, D.D., Ozawa, T., Koike, K., Kawai, S., Ito, S. (1999). Enzymatic properties and deduced amino acid sequence of a high-alkaline pectate lyase from an alkaliphilic *Bacillus* isolate. *Biochim. Biophys. Acta.*, 1427, 145-154.

[137] Hatada, Y., Higaki, N., Saito, K., Ogawa, A., Sawada, K., Ozawa, T., Hakamada, Y., Kobayashi, T., Ito, S. (1999). Cloning and sequencing of a high-alkaline pectate lyase gene from an alkaliphilic *Bacillus* isolate. *Biosci. Biotechnol. Biochem.*, 63, 998-1005.

[138] Hatada, Y., Saito, K., Koike, K., Yoshimatsu, T., Ozawa, T., Kobayashi, T., Ito, S. (2000). Deduced amino-acid sequence and possible catalytic residues of a novel pectate lyase from an alkaliphilic strain of *Bacillus*. *Eur. J. Biochem*., 267, 2268-2275.

[139] Ogawa, A., Sawada, K., Saito, K., Hakamada, Y., Sumitomo, N., Hatada, Y., Kobayashi, T., Ito, S. (2000). A new high-alkaline and high-molecular-weight pectate lyase from a *Bacillus* isolate: enzymatic properties and cloning of the gene for the enzyme. *Biosci. Biotechnol. Biochem*., 64, 1133-1341.

[140] Kobayashi, T., Hatada, Y., Suzumatsu, A., Saeki, K., Hakamada, Y., Ito, S. (2000). Highly alkaline pectate lyase Pel-4A from alkaliphilic *Bacillus* sp. strain P-4-N: its catalytic properties and deduced amino acid sequence. *Extremophiles*, 4, 377-383.

[141] Hatada, Y., Kobayashi, T., Ito, S. (2001). Enzymatic properties of the highly thermophilic and alkaline pectate lyase Pel-4B from alkaliphilic *Bacillus* sp. strain P-4-N and the entire nucleotide and amino acid sequences. *Extremophiles,* 5, 127-133.

[142] Tanabe, H., Kobayashi, Y., Akamatsu, I. (1988). Pretreatment of pectic wastewater with pectate lyase from an alkalophilic *Bacillus* sp. *Agric. Biol. Chem*., 52, 1855-1856.

[143] Greenberg, D.M., Ichihara, A. Further studies on the pathway of serine formation from carbohydrate. *J. Biol. Chem*., 224, 331-340.

[144] Manwell, C. (1957). Alkaline denaturation of hemoglobin of postlarval and adult *Scorpaenichthys marmoratus*. *Science*, 126, 1175-1176.

[145] Kuwajima K, Ogawa Y, Sugai S. (1981). Role of the interaction between ionizable groups in the folding of bovine α-lactalbumin. *J. Biochem*., 89, 759-770.

[146] Chen, H.M., You, J.L., Markin, V.S., Tsong, T.Y. (1991). Kinetic analysis of the acid and the alkaline unfolded states of staphylococcal nuclease. *J. Mol. Biol*., 220, 771-778.

[147] Bergamini, C.M., Dean, M., Matteucci, G., Hanau, S., Tanfani, F., Ferrari, C., Boggian, M., Scatturin, A. (1999). Conformational stability of human erythrocyte transglutaminase. Patterns of thermal unfolding at acid and alkaline pH. *Eur. J. Biochem*., 266, 575-582.

[148] Federici, G., Dupre. S., Barboni. E., Fiori, A., Costa, M. (1973). Insulin dissociation at alkaline pH. *FEBS Lett.* 32, 27-29.

[149] Homchaudhuri, L., Kumar, S., Swaminathan, R. (2006). Slow aggregation of lysozyme in alkaline pH monitored in real time employing the fluorescence anisotropy of covalently labelled dansyl probe. *FEBS Lett*. 580, 2097-2101.

[150] Geiger, T., Clarke, S. (1987). Deamidation, isomerization, and racemization at asparaginyl and aspartyl residues in peptides. *J. Biol. Chem*., 262, 786-794.

[151] Lura, R., Schirch, V. (1988). Role of peptide conformation in the rate and mechanism of deamidation of asparaginyl residues. *Biochemistry*, 27, 7671-7677.

[152] Gulich, S., Linhult, M., Stahl, S., Hober, S. (2002). Engineering streptococcal protein G for increased alkaline stability. *Protein Eng*., 15, 835-842.

[153] Russell, A.J., Fersht. A.R. (1987). Prediction of electrostatic effects of engineering of protein charges. *Nature*, 328, 496-500.

[154] Joshi, M.D., Sidhu, G., Pot, I., Brayer, G.D., Withers, S.G., Mclntosh, L.P. (2000). Hydrogen bonding and catalysis: a novel explanation for how a single amino acid substitution can change the pH optimum of a glycosidase. *J. Mol. Biol*., 299, 255-279.

[155] Liu, X., Qu, Y., You, F., Liu, Y. (2002). Studies on the key amino acid residues responsible for the alkali-tolerance of the xylanase by site-directed or random mutagenesis. *J. Mol. Catal. B*, 18, 307-313.

[156] Fang, T.Y., Ford, C. (1998). Protein engineering of *Aspergillus awamori* glucoamylase to increase its pH optimum. *Protein Eng.*, 11, 383-388.

[157] Joshi, M.D., Sidhu, G., Nielsen, J.E., Brayer, G.D., Withers, S.G., McIntosh, L.P. (2001). Dissecting the electrostatic interactions and pH-dependent activity of a family 11 glycosidase. *Biochemistry*, 40, 10115-10139.

[158] Becker, D., Braet, C., Brumer, H. III, Claeyssens, M., Divne, C., Fagerstrom, B.R., Harris, M., Jones, T.A., Kleywegt, G.J., Koivula, A., Mahdi, S., Piens, K., Sinnott, M.L., Stahlberg, J., Teeri, T.T., Underwood, M., Wohlfahrt, G. (2001). Engineering of a glycosidase Family 7 cellobiohydrolase to more alkaline pH optimum: the pH behaviour of *Trichoderma reesei* Cel7A and its E223S/A224H/L225V/T226A/D262G mutant. *Biochem. J.,* 356, 19-30.

[159] Shiraki, K., Norioka, S., Li,.S., Yokota, K., Sakiyama, F. (2002). Electrostatic role of aromatic ring stacking in the pH-sensitive modulation of a chymotrypsin-type serine protease, *Achromobacter* protease I. *Eur. J. Biochem.*, 269, 4152-4158.

[160] Chen, Y.L., Tang, T.Y., Cheng, K.J. (2001). Directed evolution to produce an alkalophilic variant from a *Neocallimastix pariciarum* xylanase. *Can. J. Microbiol.*, 47, 1088-1094.

[161] Lu, Y., Zen, K.C., Muthukrishnan, S., Kramer, K.J. (2002). Site-directed mutagenesis and functional analysis of active site acidic amino acid residues D142, D144 and E146 in *Manduca sexta* (tobacco hornworm) chitinase. *Insect Biochem. Mol. Biol.*, 32, 1369-1382.

[162] Nielsen, J.E., Borchert, T.V., Vriend, G. (2001). The determinants of α-amylase pH-activity profiles. *Protein Eng.,* 14, 505-152.

[163] Wind, R.D., Uitdehaag, J.C., Buitelaar, R.M., Dijkstra, B.W., Dijkhuizen, L. (1998). Engineering of cyclodextrin product specificity and pH optima of the thermostable cyclodextrin glycosyltransferase from *Thermoanaerobacterium thermosulfurigenes* EM1. *J. Biol. Chem.*, 273, 5771-5779.

[164] Nielsen, J.E., Beier, L., Otzen, D., Borchert, T.V., Frantzen, H.B., Andersen, K.V., Svendsen, A. (1999). Electrostatics in the active site of an α-amylase. *Eur. J. Biochem.*, 264, 816-824.

[165] Hicks, D.B., Krulwich, T.A. (1995). The respiratory chain of alkaliphilic bacteria. *Biochim. Biophys. Acta.*, 1229, 303-314.

[166] Kang, S.K., Kudo, T., Horikoshi, K. (1992). Molecular cloning and characterization of an alkalophilic *Bacillus* sp. C125 gene homologous to *Bacillus subtilis* sec Y. *J. Gen. Microbiol.*, 138, 1365-1370.

[167] Bohm, G., Jaenicke, R. (1994). A structural-based model for the halophilic adaptation of dihydrofolate reductase from *Halobacterium volcanii*. *Protein Eng.*, 7, 213-220.

[168] Richard, S.B., Madern, D., Garcin, E., Zaccai, G. (2000). Halophilic adaptation: novel solvent protein interactions observed in the 2.9 and 2.6 Å resolution structures of the wild type and a mutant of malate dehydrogenase from *Haloarcula marismortui*. *Biochemistry*, 39, 992-1000.

[169] Bessler, C., Schmitt, J., Maurer, K.H., Schmid, R.D. (2003). Directed evolution of a bacterial α-amylase: toward enhanced pH-performance and higher specific activity. *Protein Sci.,* 12, 2141-2149.

[170] Takase, K. (1993). Effect of mutation of an amino acid residue near the catalytic site on the activity of *Bacillus stearothermopjilus* α-amylase. *Eur. J. Biochem.*, 211, 899-902.

[171] Nakamura, A., Haga, K., Yamane, K. (1994). Four aromatic residues in the active center of cyclodextrin glucanotransferase from alkalophilic *Bacillus* sp. 1011: effects of replacements on substrate binding and cyclization characteristics. *Biochemistry*, 33, 9929-9936.

[172] Wang, T., Liu, X., Yu, Q., Zhang, X., Qu, Y., Gao, P., Wang, T. (2005). Directed evolution for engineering pH profile of endoglucanase III from T*richoderma reesei. Biomol. Eng.*, 22, 89-94.

[173] Wohlfahrt, G., Pellikka, T., Boer, H., Teeri, T.T., Koivula, A. (2003). Probing pH-dependent functional elements in proteins: modification of carboxylic acid pairs in *Trichoderma reesei* cellobiohydrolase Cel6A. *Biochemistry*, 42, 10095-10103.

[174] Shibuya, H., Kaneko, S., Hayashi, K. (2005). A single amino acid substitution enhances the catalytic activity of family 11 xylanase at alkaline pH. *Biosci. Biotechnol. Biochem.*, 69, 1492-1497.

[175] Turunen, O., Vnorio, M., Fenel, F., Leisola, M. (2002). Engineering of multiple arginines into the Ser/Thr surface of *Trichoderma reesei* endo-1,4-β-xylanase II increases the thermotolerance and shifts the pH optimum towards alkaline pH. *Protein Eng.*, 15, 141-145.

[176] Fenel, F., Zitting, A.J., Kantelinen, A. (2006). Increased alkali stability in *Trichoderma reesei* endo-1, 4-β-xylanase II by site directed mutagenesis. *J. Biotechnol.*, 121, 102-107.

[177] Sakoda, H., Imanaka, T. (1992). Cloning and sequencing of the gene coding for alcohol dehydrogenase of *Bacillus stearothermophilus* and rational shift of the optimum pH. *J. Bacteriol.,* 174, 1397-1402.

[178] Sugino, M., Kajita, S., Banno, K., Shirai, T., Yamane, T., Kato, M., Kobayashi, T., Tsukagoshi, N. (2002). Upward shift of the pH optimum of *Acremonium* ascorbate oxidase. *Biochim. Biophys. Acta.,* 1596, 36-46.

[179] Nakamura, A., Fukumori, F., Horinouchi, S., Masaki, H., Kudo, T., Uozumi, T., Horikoshi, K., Beppu, T. (1991). Construction and characterization of the chimeric enzymes between the *Bacillus subtilis* cellulase and an alkalophilic *Bacillus* cellulase. *J. Biol. Chem.*, 266, 1579-1583.

[180] Nishimoto, M., Kitaoka, M., Hayashi, K. (2002). Employing chimeric xylanases to identify regions of an alkaline xylanase participating in enzyme activity at basic pH. *J. Biosci. Bioeng.*, 94, 395-400.

[181] Krulwich, T.A., Ito, M., Gilmour, R., Guffanti, A.A. (1997). Mechanisms of cytoplasmic pH regulation in alkaliphilic strains of *Bacillus*. *Extremophiles*, 1, 163-169.

[182] Liu, J., Xue, Y., Wang, Q., Wei, Y., Swartz, T.H., Hicks, D.B., Ito, M., Ma, Y., Krulwich, T.A. (2005). The activity profile of the NhaD-type $Na^+(Li^+)/H^+$ antiporter from the soda lake haloalkaliphile *Alkalimonas amylolytica* is adaptive for the extreme environment. *J. Bacteriol.,* 187, 7589-7595.

[183] Blanco-Rivero, A., Leganes, F., Fernandez-Valiente, E., Calle, P., Fernandez-Pinas, F. (2005). *mrpA*, a gene with roles in resistance to Na^+ and adaptation to alkaline pH in the cyanobacterium *Anabaena* sp. PCC7120. *Microbiology*, 151, 1671-1682.

[184] Muntyan, M.S., Popova, I.V., Bloch, D.A., Skripnikova, E.V., Ustiyan, V.S. (2005). Energetics of alkalophilic representatives of the genus *Bacillus*. *Biochemistry (Mosc)*, 70, 137-142.

[185] Goto, T., Matsuno, T., Hishinuma-Narisawa, M., Yamazaki, K., Matsuyama, H., Inoue, N., Yumoto, I. (2005). Cytochrome c and bioenergetic hypothetical model for alkaliphilic *Bacillus* spp. *J. Biosci. Bioeng*., 100, 365-379.

[186] Takami, H., Nakasone, K., Takaki, Y., Maeno, G., Sasaki, R., Masui, N., Fuji, F., Hirama, C., Nakamura, Y., Ogasawara, N., Kuhara, S., Horikoshi, K. (2000). Complete genome sequence of the alkaliphilic bacterium *Bacillus halodurans* and genomic sequence comparison with *Bacillus subtilis*. *Nucleic Acids Res.,* 28, 4317-4331.

[187] Takami H, Takaki Y, Uchiyama I. (2002). Genome sequence of *Oceanobacillus iheyensis* isolated from the Iheya Ridge and its unexpected adaptive capabilities to extreme environments. *Nucleic Acids Res*., 30, 3927-3935.

INDEX

A

B

C

D

E

F

G

H

I

K

L

M

Q

R

S

T

U

V

W

X

Y